A CULTURAL HISTORY OF THE HUMAN BODY

VOLUME 4

A Cultural History of the Human Body
General Editors: Linda Kalof and William Bynum

Volume 1
A Cultural History of the Human Body in Antiquity
Edited by Daniel H. Garrison

Volume 2
A Cultural History of the Human Body in the Medieval Age
Edited by Linda Kalof

Volume 3
A Cultural History of the Human Body in the Renaissance
Edited by Linda Kalof and William Bynum

Volume 4
A Cultural History of the Human Body in the Age of Enlightenment
Edited by Carole Reeves

Volume 5
A Cultural History of the Human Body in the Age of Empire
Edited by Michael Sappol and Stephen P. Rice

Volume 6
A Cultural History of the Human Body in the Modern Age
Edited by Ivan Crozier

A CULTURAL HISTORY OF THE HUMAN BODY

IN THE AGE OF ENLIGHTENMENT

Edited by Carole Reeves

BERG
Oxford • New York

English edition
First published in 2010 by
Berg

Editorial offices:
First Floor, Angel Court, 81 St Clements Street, Oxford OX4 1AW, UK
175 Fifth Avenue, New York, NY 10010, USA

Berg is the imprint of Oxford International Publishers Ltd.

Library of Congress Cataloging-in-Publication Data

A catalogue record for this book is available from the Library of Congress.

British Library Cataloguing-in-Publication Data

A catalogue record for this book is available from the British Library.

ISBN 978 1 84788 791 7 (volume 4)

978 1 84520 495 2 (set)

Typeset by Apex CoVantage, LLC, Madison, WI, USA.

Printed in the UK by the MPG Books Group

www.bergpublishers.com

CONTENTS

ILLUSTRATIONS

CHAPTER 1

CHAPTER 2

CHAPTER 3

CHAPTER 4

CHAPTER 5

CHAPTER 6

CHAPTER 7

CHAPTER 8

CHAPTER 9

CHAPTER 10

ACKNOWLEDGMENTS

I would like to pay tribute to the late Roy Porter who influenced my interest in the eighteenth century and whose archive and bibliography I assembled after his untimely death in 2002. Bill Bynum encouraged me to assume editorship of this volume, and I have greatly enjoyed working with the exceptional scholars who have contributed chapters. I am grateful to the Wellcome Library, London, for outstanding research facilities and for providing all the images used in the book. My gratitude also goes to the libraries of University College London, and the British Library, and to my colleagues at the Wellcome Trust Centre for the History of Medicine, University College London.

Carole Reeves
July 2009

SERIES PREFACE

A Cultural History of the Human Body is a six-volume series reviewing the changing cultural construction of the human body throughout history. Each volume follows the same basic structure and begins with an outline account of the human body in the period under consideration. Next, specialists examine major aspects of the human body under seven key headings: birth/death, health/disease, sex, medical knowledge/technology, popular beliefs, beauty/concepts of the ideal, marked bodies of gender/race/class, marked bodies of the bestial/divine, cultural representations and self and society. Thus, readers can choose a synchronic or a diachronic approach to the material—a single volume can be read to obtain a thorough knowledge of the body in a given period, or one of the seven themes can be followed through time by reading the relevant chapters of all six volumes, thus providing a thematic understanding of changes and developments over the long term. The six volumes divide the history of the body as follows:

Volume 1: A Cultural History of the Human Body in Antiquity (750 B.C.E.–1000 C.E.)
Volume 2: A Cultural History of the Human Body in the Medieval Age (500–1500)
Volume 3: A Cultural History of the Human Body in the Renaissance (1400–1650)
Volume 4: A Cultural History of the Human Body in the Age of Enlightenment (1650–1800)
Volume 5: A Cultural History of the Human Body in the Age of Empire (1800–1920)
Volume 6: A Cultural History of the Human Body in the Modern Age (1920–21st Century)

General Editors, Linda Kalof and William Bynum

Introduction

Enlightenment Bodies

CAROLE REEVES

The Enlightenment was a time when people began to take stock of their intrinsic worth as individuals. This appraisal was different of course from ownership of one's body. Slaves were still property, servants and apprentices were indentured, daughters belonged to fathers and brothers (and wives to husbands), and paupers were tethered to their parish. But there was change in the air and a new optimism born of a freedom to think and the right to challenge. Against a worldwide demographic explosion between the years 1650 and 1800, with its associated urbanization and industrialization, which increased the migratory potential of individuals, families, and groups, persuasive notions of national and personal identity began to develop. These ideas filtered down to even the poorest street beggars, as portrayed so elegantly in Tim Hitchcock's study of beggars and their bodies in eighteenth-century London. The chapters in this book, written by historians of culture, literature, science, and medicine, explore the human body in all its guises from conception and birth to death and beyond, both as a physical and symbolic entity. They reveal its amorphous nature in an age of turbulence and transition as the highways to modernity were forged inexorably through the Western world. The epithets *pliable, fragile, susceptible, manipulative, entrepreneurial, imaginative, jocular, deceptive, vain, proud, passionate, mutilated, stereotyped, alert, pox-ridden, rickety, wrong-looking, incestuous,* and *dangerously reproductive* have been used by the authors to describe Enlightenment bodies. The Enlightenment celebrated nonconformity as it erased the traces of old mentalities and encouraged individuals to remold themselves as originals.

Fiction emerged as the medium for rethinking the self and trying out new identities,[1] though a great cluster of these narratives explored the lives and deaths of ordinary people.[2] Popular tales from abroad, particularly the Orient, introduced readers to unfamiliar bodies including the enclosed seraglio body, the foot-bound body, the reincarnated body, and the castrated (eunuchoid) body. The *Arabian Nights Entertainments, Turkish Tales, Chinese Tales,* and *Mogul Tales* were hybrid East-meets-West fables, but their popularity spawned successful European versions such as Montesquieu's *Persian Letters* (1722), discussed by Lisa Forman Cody in her chapter on "The Body in Birth and Death," and Oliver Goldsmith's *Citizen of the World* (1760–1761).[3] The cult of the Gothic, which spawned Frankenstein's mismatched but misunderstood monster, was heralded by Horace Walpole's *The Castle of Otranto* (1765). Goethe's *Die Braut von Korinth* (1797) capitalized on the epidemic waves of vampirism that swept mid-century Central Europe, although his blood-sucking creature bore little resemblance to the humbler embodied revenants purported to inhabit rural villages.[4] Nevertheless, the whole issue of resurrection within one's own earthly body and the possibility of resuscitation of the so-called undead were hotly debated, not least among theologians such as the French Dominican Augustin Calmet, who refused to dismiss vampires in principle, and Benito Jerónimo Feijóo, professor of theology at Oviedo, who considered them a mere effect of the imagination.[5] Thomas Paine's Deist stance that the resurrection of Jesus was a fraud if considered as a miracle but might be explained as a case of apparent death followed by resuscitation,[6] could theoretically be extended to embodied revenants, especially if they had been victims of violent death or drowning. Being buried alive might also explain their reported forceful resistance to being staked through the heart and decapitated when unearthed. A Society for the Recovery of the Apparently Drowned, founded in Amsterdam in 1767, and its London equivalent, The Royal Humane Society (1774), were the medical profession's response to the resuscitation debate. They aimed to instruct people in life-restoring techniques such as inflating the lungs with bellows, applying electrical stimulation to various parts of the body, and fumigating the rectum with tobacco enemas.

The macabre mystery of death and the possibility of resuscitation drew crowds to executions. From the mid-eighteenth century, enlightened penal reformers in England campaigned to bring order, if not dignity, to the business of capital punishment and the ensuing scramble for a touch of the hanged corpse, proclaimed by folk wisdom to have miraculous healing properties.[7] Public hangings at Tyburn, London's spectacle of retribution since circa 1300, were abolished in 1783, and public executions in England ceased totally under Queen Victoria in 1888. Conversely, the frequently staged Tyburn riot by ordinary citizens against the surgeons emphasizes the disapprobation and fierce resistance to the practice of anatomizing hanged murderers at Surgeons' Hall.[8] As Ruth Richardson discusses in her chapter on the dead body, there was little difference in the public mind between the legal acquisition of bodies from the gibbet and the sordid (and illegal) involvement of anatomists with resurrectionists or body snatchers. Although the rich were seldom the victims of grave robbers, they were not inviolable, as proclaimed by the London surgeon Sir Astley Cooper (1768–1841), who arrogantly maintained that he could procure any body he liked provided the price was right.[9] Enlightened opinion also abandoned punitiveness for pity in the case of those who chose to take their own lives, whose bodies had, during medieval times (in Christian theology), been denied resurrection by being buried face down in a cemetery and, after the Reformation, at a crossroads with a stake through the heart or mutilated in some way.[10] Suicide was secularized and medicalized as sympathetic coroners returned verdicts of non compos mentis, which sanctioned churchyard burials and, almost as importantly, prevented the victims' assets being forfeited to the Crown.[11] The Enlightenment (preceded by the Scientific Revolution of the seventeenth century) is associated with a general shift from divine dramatics to more secular practices and naturalistic meanings. As suggested by the anthropologist Mary Douglas, Western medicine gradually separated itself from spiritual matters as part of this process, and intellectuals distanced themselves from old magical medicine.[12] After Queen Anne (1702–1714), for example, British monarchs stopped the custom of touch-healing sufferers of scrofula, known as the King's Evil, although in France the Bourbons continued the practice until 1830.[13] Robert Burton, in his *Anatomy of Melancholy* (1621), had entertained no doubt that Satan personally and directly visited sinful man with sickness,[14] and the pious trope of vile bodies—dubbed "boxes of poison" by John Donne (1572–1631), poet and dean of St. Paul's Cathedral, London[15]—could be authenticated by the revelations of the microscope, that seventeenth-century dilettante toy, which exposed the teeming mass of repulsive bugs and grubs feeding off them.[16] Plague, consumption, syphilis, and other deadly and disfiguring afflictions proved the infirmities of the flesh and confirmed the association between lust, sin, and suffering.[17] By the end of the eighteenth century, however, such explanations and associations were becoming something of a joke among the elite. When the physician Erasmus Darwin, a beacon of the Enlightenment, was forced to cancel his attendance at a meeting of the Lunar Society, he blamed the devil for

> sending the measles with peripneumony amongst nine beautiful children of Lord Pagets's. For I suppose it is a work of the devil! Surely the Lord could never think of amusing himself by setting nine innocent little animals to cough their hearts up?[18]

Witchcraft was similarly debunked. The witch was identifiable by the *stigmata diaboli* on her body—moles, warts, birthmarks, blemishes, scars—to be flushed out by the expert gaze of priests, witch finders, and the courts. In accusations of witchcraft, the presence or absence of these distinguishing marks could be a matter of life or death.[19] Like Jane Wenham of Hertfordshire, the last English witch to be condemned (in 1712), most accused of *maleficium* were poor, eccentric, elderly women living in rural communities. As well as a resentment among rationalists against the clergy responsible for whipping up fears about witchcraft for their own power game, there was, as suggested by Daniel Schäffer, a revaluation of old age in the early Enlightenment, which may have contributed to it being recognized as a state of illness and neediness and not only as a natural phase of decay.[20] Indeed, as Lisa Forman Cody discusses in the opening chapter of this volume, the medicalization of human existence from birth to death is rooted in the eighteenth century, as are the scientific underpinnings of sexual deviance, racial diversity, criminality, and madness. The early modern insane body was portrayed in theatrical, artistic, and literary conventions as bestial, naked, or clad only in rags, its hair disheveled and matted with straw. More often than not, a stone bulged from its forehead. The stage fool and the court jester were traditionally clad in motley, caps and bells, sporting bladder and pinwheel, the carnivalesque accoutrements of folly. In humoral pathology, a victim of excess black bile (the melancholic) could be identified by swarthiness of skin, dark hair, and eyes or "black looks"—the demonizing daub of blackness.[21] The maniac's high color might be due to an excess of yellow bile in a choleric personality. With the possible exception of syphilis, physical appearance in madness advertised what was believed to be going on inside the body more stereotypically than did any other condition. As part of the secularization of psychology, there was a renaissance in the Greek art of physiognomy, the reading of character from facial features. Artists such as Charles Le Brun and artist/anatomists such as Sir Charles Bell made physiognomical studies of the emotions—fear, joy, rage, jealousy, anger, anguish, and so forth—that could be used as bona fide diagnostic tools by enlightened doctors wishing to distance themselves from supernatural causes. The centuries-old public spectacle of lunacy at London's Bethlam Hospital, better known as Bedlam, the byword for chaos and confusion, was ended around 1770 when its doors were closed to paying sightseers.[22] Like Tyburn, it ceased to be on the tourist itinerary. At the same time, many private and charitable lunatic asylums were established, some of which, like the Quaker-run York Retreat, employed the new enlightened moral therapy, which substituted reason, kindness, and good example for restraint, neglect, and cruelty. Not all the private establishments, of course, followed this modus operandi.[23]

While the mad body disappeared from public view, human marvels, monstrosities, and oddities were on display at fairs and freak shows to a greater extent than ever before as enlightened curiosity replaced the medieval Christian endorsement of wonder, with its fear of forbidden knowledge.[24] Conjoined (Siamese) twins, individuals bearing the remnants of a parasitic twin, limbless torsos, giants, dwarfs, wild men and bearded women, exotic erotic specimens such as Saartjie (Sarah) Baartman the Hottentot Venus, pickled malformed fetuses, and people with serious (yet undiscovered and unnamed) diseases and disabilities such as elephantiasis, neurofibromatosis, and vitiligo, were exhibited to a paying public increasingly accustomed to window shopping in the new retail consumer market.[25] Exploitation was undoubtedly the name of the game in many cases of exhibitionism but not in all as the new culture of body identity and personal ownership attests, and as is reinforced by Tim Hitchcock's analysis of London beggars who used their bodies in a self-conscious and deliberate way to weave a story as they waved for attention. Nevertheless, in an age that disputed ownership of a dead body, so-called sports of nature risked being acquired (and therefore denied burial) by the new and increasing breed of collector for anatomical and pathological display. The woeful tale of Charles Byrne, the Irish giant, stalked and purchased after death for the reputedly enormous sum of £500 ($60,000 today) by the surgeon-anatomist John Hunter (1728–1793) for his London museum, is well known, but Hunter was also not averse to securing the rights of the bodies of his patients postmortem in exchange for repairing them antemortem.[26] Jeremy Bentham (1748–1832), the founder of Utilitarianism and an affirmed atheist, bequeathed his own body to posterity. He was the first person known to have volunteered to be dissected, and this took place during the passage of the Anatomy Bill (1832) through Parliament, a bill he helped draft to secure bodies for dissection by medical students. Following dissection—for the edification of surgery or physic and not for determining cause of death—Bentham's body was to be embalmed and displayed fully clothed as a so-called auto-icon, which he considered to be a cheaper and more realistic alternative to statuary. Bentham commended others to do likewise.[27] His skeleton, supporting a waxwork effigy (the embalming attempt failed), is displayed in a glass-fronted wooden cabinet in the South Cloisters of University College London, the enlightened nonconformist university that he helped to found in 1826. As Andrew Cunningham has pointed out, it is "the body of the complete exhibitionist."[28] It probably comes as no surprise that Bentham's auto-icon is unique because nobody before or since has taken up the challenge to be three-dimensionalized in this way. Indeed, its crudity would undoubtedly have offended many who aspired to the new enlightened modes of gentility, which raised the threshold of shame and prized politeness and the closed body that went with it.

There had been close links between the early modern body and the environment. Pain and disease were held to be mobile entities that could be brought to the body's surface by bleeding, cupping, purging, and vomiting. Evacuations would be viewed and discussed by family and friends as well as the attending practitioner. Eventually, all of the body's products, except tears, became unmentionable in decent society, and the new emphasis on propriety in turn prefigured and pointed in the direction of the stricter codes of bodily control associated with evangelicalism and Victorianism.[29]

As always, however, there are cultural dichotomies, and the eighteenth century witnessed the birth of the body's subversive potential, irresistible to a rebellious media with scant regard for doctors, be they learned or otherwise. Bodies and body language featured extensively in political cartoons and caricatures after the lapse of the Licensing Act in 1695 put an end to censorship and facilitated the rise of the Grub Street press.[30] Artists and engravers such as Thomas Rowlandson, James Gillray, and George Cruikshank vied with each other to produce the most salacious and scatological visual punning. Society doctors representing politicians purged, anatomized, amputated, condemned, and certified opponents, royalty, and even entire countries. Asses and enemas, phalluses and farts, breasts and bloodletting, and bowels and butchery were used to maximum effect in the double entendre. The body politic could be reduced to a repulsive victim of overzealous doctoring or it could, alternatively, become transformed into a grotesque mob-monster, devouring everything in its path. The resultant imagery was generally both lewd and ludicrous. In addition, new lifestyle magazines such as the *Spectator* (founded in 1711) aimed to rewrite the eighteenth-century body, using a mix of satire, silliness, and reflective prose. The body was a trope for all the usual social mishmash of the day, but at the same time, the *Spectator*, as Roy Porter has shown in his own posthumously published work, was an early champion of Locke's idea that consciousness, not substance, was the location of the person, and that man did not seem born to enjoy life but merely to deliver it down to others.[31]

The fusion of fine art and anatomy was also formalized during the Enlightenment. London's Royal Academy of Arts, founded in 1768, appointed a professor of anatomy, William Hunter (1718–83), the older and more socially adept brother of John, to instruct the students of its life classes. Hunter's job was to demonstrate both surface anatomy (on living models) and dissected anatomy. Flayed (skinned) bodies, for example, revealed musculature and tendon attachment to joints, while deeper dissection explored the relationship between the organs and the underlying skeleton. As explored by Ruth Richardson, Hunter's own great oeuvre, *The Anatomy of the Human Gravid Uterus* (1774), exemplifies the synthesis of master dissector and anatomical illustrator, in this case the Dutch artist, Jan Van Rymsdyk. Needless to say, corpses obtained for artistic instruction and for medical books were invariably commissioned from grave robbers. At precisely the same period, a shift was taking place from the concept of holistic disease, with its roots in humoral pathology, to disease localization, based on postmortem findings associated with signs and symptoms during illness, a technique pioneered by the Italian pathologist Giovanni Battista Morgagni. This foreshadowed a much more hands-on approach to diagnosis, with the patient being removed from the central role of informant to that of passive recipient of the doctor's physical examination.[32] It also removed the personal narrative of bodily experience as expressed by patients in consultation letters to eminent physicians such as Morgagni, Francesco Torti,[33] and Samuel Auguste Tissot of Lausanne. Tissot's correspondents were aware of their bodies' uniqueness even when they talked of their bodies as machines (recalling iatromechanistic theories of the body examined by Jessica Riskin in her chapter, "Medical Knowledge: The Adventures of Mr. Machine, with Morals"), but they also paid much attention to the six nonnaturals—air, diet, exercise, sleep, bodily excretions, and passions of the soul.[34] Purely physical symptoms were very often associated with the passions of the soul and a good economy of the passions was viewed as necessary for maintaining bodily health. Although Thomas Dixon warns against interchanging the word "passion" with "emotion" at this period,[35] the bodily nature of emotion in its widest sense (encompassing all mental states including passions) was an important aspect of the mind/body/soul relation. Fay Bound Alberti has suggested that a major shift in medico-scientific emotion theory at this time was the erosion of the traditional model of emotions, with the heart and vascular system at its center, and the subsequent domination of the brain and central nervous system. Such a transition was the necessary context for the scientific redefinition of emotions as measurable, quantifiable experiences manifested in and on the body, and the rise of professions uniquely equipped to deal with emotions as mental phenomena.[36] As the nervous system was mapped onto the body,[37] the brain incorporated the mind but not the soul. Sensitivity (feeling) and irritability (movement) gradually became key words in the expression of bodily sensation, the former being closely linked with the presumption of a delicate and refined body.[38] A number of Tissot's correspondents, nevertheless, preferred to use their own highly personal language to describe their symptoms and bodily sensations. Monsieur Thomassin, writing in 1775, explained why:

> My disease is interior, only I feel it; I also thought that only I could describe it; this is why I have not taken a doctor from the faculty as interpreter, who in using the terms of the art, would explain things to me perhaps less well than I can speak in my weak patois.[39]

There is a palpable self-obsession in many patient letters of this period, an inevitable by-product perhaps of the cultural narcissism encouraged in polite society, which fed into the commercialization of medication and all that passed for it, and included much that was habit-forming. Hypochondria, hysteria, and the addictive personality were created in this "age of feeling," which arguably lowered the pain threshold.[40] Although doctors realized that the desire for drugs was often itself a disease, many profited very handsomely from it. Others encouraged the idea that certain diseases of the self such as hypochondria, tuberculosis, and the exquisitely painful gout (often itself the end-product of gluttonous overindulgence) bespoke identity, pedigree, and status. The agony of throbbing inflamed toes could thus be borne with pride and their shock-absorbing bandages worn as dressings of distinction.[41]

While physicians medicated the body, surgeons penetrated its fabric, often performing painful, mutilating, daring, and dangerous procedures. Nevertheless, as Emma Spary has shown for French surgery of the Enlightenment, the performance was as important as the end result. In simple bloodletting for example, even elite surgeons played to the sick person and spectators by making the blood flow in an arc, knowing that the procedure was just as good when the blood ran along the arm.[42] Since the ideal surgical body was perfectly passive and immobile, however, surgical narratives gradually erased the sick person from the scene, with descriptions containing only accounts of body parts and instruments moving in relation to one another. Operations described were idealized accounts of surgical encounters and have remained so to this day.

For most of the eighteenth century, patients were still individual bodies and not yet statistics. Doctors used both their private and charitable patients as potential stories of individual disorders,[43] and W. F. Bynum has dated "multiple case reporting" from the mid-eighteenth century, and "systematic record keeping" from the late eighteenth century. Prior to 1750, doctors frequently described "only single instances of disease and therapies, and only 'cures.' "[44] This began to change in the later eighteenth century as admissions registers were established and this information was condensed into numerical tables for printed annual reports. Relaying the utility of medical charity to lay supporters was an important function of annual reports, but it also meant that the ongoing collection and publication of clinical data became an integral part of institutional medicine before the close of the eighteenth century.[45] The transformation to localizing pathology also meant that the diseased body could be read in a different, less personal way. A severe deforming birthmark such as a hemangioma, for example, might not be a sign of an adulterous conception but could be the visual evidence of an underlying vascular pathology.

However, doctors who probed now represented a threat to the body—diagnostically, therapeutically, and sexually—and none more so than the man-midwife, that "newly discovered animal" of the Enlightenment.[46] In Isaac Cruikshank's well-known frontispiece to Samuel Fores's *Man-Midwifery Dissected . . . also of Their Cunning, Indecent and Cruel Practices . . .* (1793)—reproduced on the cover of this book—the male accoucheur is depicted with the instruments of his trade including forceps, giant cutters, and a two-ended hook for destroying and removing fetal parts in obstructed labor. Medicaments and love potions (for his own use) complete the picture of a monstrous sexual predator. In contrast, the female midwife, standing in front of a cozy fire—on which she would presumably boil her wash water—holds in her hand a simple feeding cup.[47] In another (anonymous) illustration titled "The Man-Midwife, or Female Delicacy after Marriage," published by S. Hooper, London (1773), a man-midwife kneels in front of a heavily pregnant woman with his hand up her skirt while the woman's husband is led protesting out of the bed chamber by a servant. In many eighteenth-century scenarios, clinical consultations are featured as nothing more than erotic skirmishes, obviously a gross exaggeration but made plausible by the fact that sex was publicly flaunted in a historically atypical manner.[48] The Enlightenment celebrated the body as it exalted the passions.[49] John Cleland's *Memoirs of a Woman of Pleasure* (1749) had his heroine, Fanny Hill, the welcoming recipient of "yard" upon "yard" and making a handsome living in the process.[50] There were many witnesses, of course, to the ghastly fate of those who all too frequently fell victim to sexually transmitted diseases, so graphically portrayed by Susan Staves in her chapter on the pox-marked body. And Laura Gowing, in the chapter "Marked Bodies and Social Meanings," reveals how tolerant even the law was on cross-dressers and the burgeoning gay subculture. Toleration of homosexuality was advocated by Jeremy Bentham, who, according to Andrew Cunningham, is not known to have had sex at all during his lifetime,[51] but who nevertheless considered sexual nonconformity a victimless crime, "free from worry to third persons," and believed that it should be penalized no more than religious dissent.[52] An explosion in the production of erotica was augmented by sexual advice literature, which maintained that sex was a pleasure in its own right and a contribution to conjugal bliss.[53] James Graham (1745–1794), London's leading sexologist, whose Temple of Health and Hymen boasted a monstrous vibrating blue and purple satin "celestial bed," claimed that his inventions would exquisitely delight the organs of sensation.[54] But in an Enlightenment volte-face, the possibility of reproducing asexually through such contraptions as a wind machine designed to trap "floating embryos" was offered to the scientific community as a technology of the future, for reasons including the denial of sexual pleasure altogether. As observed by George Rousseau in his chapter titled "Sexual Knowledge: Panspermist Jokes, Reproductive Technologies, and Virgin Births," such jests have turned out to be anything but preposterous.

A beauty industry, described in detail by David M. Turner in his chapter, "The Body Beautiful," arose to meet the needs of women (and to a lesser extent men) who wished to decorate and display their bodies or to hide the ravages of time or disease. The revered body was straight, symmetrical, light-skinned, graceful, elegant, and for the most part female. In a man's world, which the Enlightenment decidedly was, despite some rumblings toward a greater intimacy between the sexes,[55] the accent inevitably fell on the display of female bodily assets as an erotic commodity. In her disturbing analysis of incest, sexual violence, and fratricide, Ruth Perry reveals that the expectation that sisters would be protected and cherished by their brothers, and that brothers would be responsible for their sisters, diminished during the course of the eighteenth century. As Roy Porter has commented, the mutilated Horatio Nelson was a hero; his lover, Emma Hamilton, whom he famously bequeathed to the nation, was just a body.[56]

The eighteenth century witnessed an increasing emphasis on personal cleanliness as the route to health and beauty. In Europe, public bathing had declined from the Middle Ages as it came to be associated with prostitution, while private bathing was considered dangerous because it opened the pores to pestilential air. Clothing was brushed or sponged rather than washed, infestation with lice and parasites was common, and there were varying standards of waste disposal.[57] During the eighteenth century, French nobility in particular adopted cleansing rituals as part of a move toward civilized manners, later copied by the bourgeoisie. The new bourgeois body was to be created through practices of hygiene and healthful behavior, instituted by science and policed by the individual.[58] Advice literature, written by doctors, was a new and popular genre, largely aimed at women as the focus of the household economy, although both sexes took enormous interest in matters of health and felt themselves able to participate in their own diagnoses and cures. William Buchan's *Domestic Medicine* (1769), a best-seller for more than thirty years, was one of the first to associate the poor with dirt, disease, and danger to others. He suggested that a cause of putrid fevers was uncleanliness among the inhabitants of overcrowded, dirty houses who breathed bad air and wore filthy clothes. It was not sufficient, he claimed, to be clean oneself, if dirty neighbors spread infections far and wide.[59] This so-called moral biology, whereby poverty, class, and race determined constitution, health, and the ability to generate disease, is explored at length by Kevin Siena in his chapter on pliable bodies. Homemade soaps prepared from lye or potash and fat, or natural soap from plants such as soapwort (*saponaria officinalis*), were replaced by commercial products. Pears's soap, created in the 1790s by Andrew Pears, a London barber, was fiercely marketed in the United States after 1865 by Thomas Barratt. In the north of England, William Lever built a hygienic town (Port Sunlight) for his employees on the proceeds of selling Sunlight soap. Soap was also recommended by one writer as a cure for corpulence—the bête noir of the overconsuming eighteenth-century body—not by scrubbing fat from the body's exterior but by dissolving it from the inside through daily ingestion of suds in water.[60] Narratives of corpulence, as discussed by David M. Turner, and the resultant calls for temperance and self-control played a part in the prescriptive domestic self-help literature. They warned not only about the consequences of overembodiment but tackled widely shared anxieties about the unsettling and potentially disruptive forces of consumer society.[61]

The modern progressive self nevertheless rested on a healthy and disciplined body, although later, for William Godwin, what mattered for progress was mind control. The triumph of reason would render physical bodies redundant. Not so for Robert Owen, who also championed Locke and reason in the form of mass education but relied on programmed bodies to run the new mechanized society that he helped create in the spirit of the Enlightenment. For Jeremy Bentham, too, the virtuous—and liberated—body was the laboring body, although it could (and should) be controlled by supposedly naturally ruling bodies, in Bentham's case the emerging industrial middle class. In his Panopticon, a theoretical penitentiary conceived during a visit to Russia in 1786, the poor and the deviant would work alone fourteen hours a day under direct surveillance of an inspector who could see into every cell from a single vantage point but could not himself be seen.[62] Bentham described the Panopticon as a body with the inspection lodge as its heart and the passageways lined with cells as its nerves and arteries. The inspector's gaze, rather than physical punishment, would control the functions of the cell contents, that is, the prisoners, who would in turn be adequately fed and clothed and would enjoy a greater happiness than that afforded by a life of crime and undisciplined indigence. Bentham's concept of personal liberty relied on a fundamental security even if the roof over one's head was a prison.

The Enlightenment preached and encouraged the quests for self-identity, self-discovery, self-motivation, and even self-obsession. There has been no reversal of these fundamental rights of personal freedom.

CHAPTER ONE

The Body in Birth and Death

LISA FORMAN CODY

While the twenty-first-century Western demographic profile of low infant mortality rates and long life expectancies has attenuated the distance between the body in birth and the body in death, this was not yet so during the Enlightenment. Well into the nineteenth century, birth and death were intertwined, as they had been for millennia in the West. Not only had there been spiritual resonances between the beginning and the end—"We have a winding sheet in our mother's womb," preached John Donne—early moderns also feared that newborns and their mothers might perish in birth or soon after.[1] Charles Dickens captured the entanglement of birth and death in his 1844 satire of Mrs. Sairey Gamp, combination midwife and funeral assistant, whom he described as: "Setting aside her natural predilections as a woman, she went to a lying-in or a laying-out with equal zest and relish." The gin-sodden, goggle-eyed, and corpulent Gamp herself noted, when called to help prepare Anthony Chuzzlewit's corpse: "And so the gentleman's dead, sir! Ah! The more's the pity. . . . But it's what we must all come to. It's as certain as being born, except that we can't make our calculations as exact."[2]

Despite death's hovering over birth for the long durée, new sensibilities about reproductive and dying bodies had emerged by the time Dickens published *Martin Chuzzlewit*. His depiction of Gamp as fleshy and grotesque, inappropriately sexed when he qualified her "equal" enthusiasm for managing birth and death as something that a woman's "natural predilections" would

FIGURE 1.1: Sairey Gamp and Betsy Prig prepare Anthony Chuzzlewit's corpse. Half-tone reproduction, 1906. Wellcome Library, London.

prohibit, betrayed new nineteenth-century assumptions about gender's relationship to birthing and dying bodies. Although they had long handled the corporeal aspects of lying-in and laying-out, by 1800, women were considered too delicate for the mess and smells of bodies, especially those in labor. Women who continued to manage these indelicate matters were themselves considered indelicate. Yet this was an ironic development. By 1800, Europeans described the female body as fundamentally reproductive, and the female soul as wrenchingly funereal, so it seemed then that their essential roles as mothers and mourners would well qualify them to deliver babies and prepare the dead. But eighteenth-century medical men and others argued otherwise. Emphasizing the perils of birth and the mysteries of death, Enlightenment authorities argued that educated, professionalized men of science could best manage humanity's mortal entrances and exits, while simultaneously explaining the truths of gender, sex, and the body.

WOMEN'S CHANGING ROLES

Enlightenment claims about gender, perhaps more than any other development, dramatically altered Western beliefs about the body in all its aspects, including its status during birth and death. The rise of science and medicine and the Enlightenment's emphasis on maternity and romanticizing of childhood all paradoxically combined forces to sever women from their age-old role as the natural attendants to birthing bodies. Male authorities' claims to knowledge about existential matters and undertakers' eighteenth-century commercialization of funerals did not fully remove women from laying out corpses but instead contributed to bifurcated beliefs about the sexes' proper spheres. While men were seen as being able to acquire reasonable authority over the body in life and death, women were increasingly viewed as so overcome with reproductive desire and melancholy attachment to the living and the dead that they could be harmed—and do harm—by taking part in the corporeal aspects of birth and death. It was only among the poor that women continued to control the entrance and exit of life, mostly among their own class.

The contrasts between 1650 and 1800 were dramatic. In 1650, most babies across classes were delivered by women and their mothers and were tended by female relations afterward—in cities, midwives tended to be literate, well respected, and from economically solid households.[3] By 1800 this had changed among upper-class British, American, and northern European mothers, who had begun using men-midwives to deliver their children because they believed masculine science offered a safer alternative to traditional midwives, who were typically depicted as coarse and lower class. Not only did men take over delivering elite babies, they also began focusing on reproduction in natural philosophy and population theory, claiming to lay bare the inner truths of the sexed body and its relation to society.

Likewise, in 1650, most corpses were washed and tended by women. Though the male clergy commemorated the souls of the departed, women's role in handling dead bodies was no less important. Few Europeans could afford to be embalmed, and so the females who prepared corpses cleaned and covered the exterior, but eighteenth-century male authorities increasingly looked inside bodies as coroners, anatomists, embalmers, and philosophers. Though women did not stop attending the dying and dead by 1800, medical men and the new eighteenth-century professional undertakers played an ever greater role in the physical management of death. Theologians and philosophers developed new perspectives on the body's inner truths. After all, dying was more than corporeal decomposition; it was also a matter of the soul and spirit. Added to Protestants' and Catholics' disagreements about spiritual matters, philosophers from René Descartes onward offered new (and often conflicting) models of the soul, mind, and body that altered, at least for some elite Europeans, how they viewed death as well as life.

THE DEMOGRAPHIC CONTEXT

Enlightenment bodies did not realize it, but they took part in a dramatic demographic revolution. Since the fourteenth century when bubonic plague had decimated up to a third of the population between the Levant and London, Europe had maintained an overall stable relation between birth and death rates. In the eighteenth century, population growth began to accelerate. Overall, it is estimated that the European population grew from 95 million in 1700 to 111 million in 1750 and then to 146 million people in 1800. The English population grew from roughly 5 million to 9 million between 1700 and 1800. The North American colonies grew from 360,000 Europeans in 1713 to 5.3 million in 1800, thanks in large part to immigration. Expansion was not as rapid in France and other parts of Europe as it was in Britain, but it was notable nonetheless. France grew from roughly 20 to 26 million people between 1700 and 1800, while Sweden grew from 1.5 to 2.3 million people between 1700 and 1790. The eighteenth-century population surge occurred also in Asia. There numbers expanded from 433 to 631 million people across the century. However, Africa and Oceania saw decreases from 107 to 102 million and 3 to 2 million, respectively, between 1700 and 1800.[4]

Beginning in the 1670s, the political arithmeticians William Petty and John Graunt began counting heads and analyzing population's effect on the economy. Their conclusions were essentially mercantilist: like the exchange of land, specie, and other goods, population growth and decline were generally zero-sum games. Reproduction seemed to be static and therefore secondary to variations in mortality rate. If death rates could be improved, then the population might grow, but these authors said little of increasing the birth rate, no doubt because the food supplies to support expansion appeared limited. Seventeenth-century theorists, including Petty and Graunt, emphasized the importance of male laborers, soldiers, and sailors, and so unsurprisingly the first state-sponsored censuses in Norway and Denmark in the 1660s and 1701 only tallied men. Eighteenth-century authorities shifted toward calculating both sexes and all ages, beginning in 1719 in Prussia, 1749 in Sweden, 1769 in Norway, and 1790 in the United States. France had first surveyed the European population in Quebec, Canada, in 1665–1666 but did not undertake a systematic census of its own numbers until 1801, the same year Parliament produced the first census for England and Wales.

Some Enlightenment observers believed the population was shrinking. While Petty and Graunt noted that London seemed to be growing—like a monstrous, inflating head—it seemed to expand at the expense of the countryside, which was depicted as an atrophying body that sent waves of immigrants to the insalubrious capital.[5] Montesquieu argued in *The Persian Letters* (1721) that the contemporary world was only one-tenth its ancient size because "nature" had lost "the prodigious fertility that she had originally" and was "already in old age."[6] Tapping the common trope that the entire population could be metonymically represented by a single figure, Montesquieu depicted nature as a menopausal body, but others depicted population as fecund. In a polemical

1755 pamphlet designed to scare British policymakers, Benjamin Franklin argued that the North American colonies were doubling every twenty-five years; he used the metaphor of fast-breeding fennel and of the amazing polypus, a sea hydra that continental biologist Abraham Trembley had recently discovered reproduced itself infinitely when chopped up. Either way, whether arguing that population had shrunk or shot up, nearly all Enlightenment commentators agreed that the greater the numbers, the stronger the nation.

Thomas Robert Malthus, Anglican clergyman and Britain's first political economy professor, revolutionized these Enlightenment views. He overturned both the fear of depopulation and the faith in fertility as an unqualified good in his blockbuster work, *An Essay on the Principle of Population* (1798). Redolent with references to dangerously reproductive and sickly, dying bodies,[7] the *Essay* argued that although population can increase geometrically, food supplies can only grow arithmetically. When population expands dangerously, famines and disease check growth. By 1801, the results of the first English census showed that the nation had nearly doubled its numbers from the seventeenth century, from 5 to 9 million. The so-called Malthusian trap had appeared to snap, but these teeming numbers seemed a lower-class menace to policy makers, who, in 1834, ended welfare provisions for unmarried mothers[8] and, later in nineteenth-century Britain and the United States, even proposed eugenic policies to discourage the poor from reproducing at all.[9]

Eighteenth-century demographic growth had many causes and was supported by the intensification of agricultural production.[10] Partly, populations grew because life expectancy increased, especially in such countries as Sweden. Some diseases, such as syphilis, diminished in strength while others retreated from parts of the world. The last European plague epidemic occurred in 1720 Marseilles, but from a global perspective, eighteenth-century colonial ambitions brought smallpox and other epidemics, decimating native populations in the Americas, Oceania, and central Asia.[11] In northern Europe some improvements in nutrition, including the widespread adoption of the potato from the 1770s forward; the spread of soap and cotton cloth, which allowed for superior hygiene; and the encouragement of mothers to breast-feed their own infants rather than send them out to wet-nurses, all of which had positive effects on the health of the population. Smallpox, which may have accounted typically for 10 to 15 percent of European nations' deaths in the early eighteenth century was somewhat controlled by inoculation, introduced into England by Lady Mary Wortley Montague from Turkey in 1721. By the 1790s, Edward Jenner greatly advanced the technique by vaccinating with the milder strain of cowpox, which conferred immunity against virulent smallpox as well. Despite many of the century's demographic improvements, there were weaknesses. England's fast-growing industrial cities lacked the infrastructure to provide a healthy environment for the poor, and some regions of the world from

France to Bengal suffered significantly from various famines, especially in the early 1770s. Health may have improved generally in the period, but there were demographic pockets of great misery.[12]

Improving life expectancy was not the only cause of the Enlightenment's demographic boom. Benjamin Franklin attributed the North American colonies' growth to immigration and early marriage, made possible by the seemingly endless resources of that continent. Malthus criticized the dropping age of marriage among the English poor, which he argued was fueled by welfare provisions that encouraged the lower classes to couple before they were financially self-sufficient. Both Franklin and Malthus correctly identified the causes of expansion in their worlds. The cause of steep population growth in eighteenth-century England thus resulted not so much from increased life expectancy, but rather from a rapid drop in the age of marriage: the mean age of marriage dropped from 26.2 to 24.0 between the periods 1650–1699 and 1750–1799,[13] plus illegitimacy and prenuptial pregnancy rates at least doubled in the same period.[14]

BIRTH AND THE EARLY MODERN REPRODUCTIVE BODY

In the late seventeenth century, political arithmeticians John Graunt and William Petty were not alone among men for ignoring fertility. Generation generally held little interest for doctors and natural philosophers in this period, except in emergencies or rare cases of monstrous births. By the mid-eighteenth century, though, male authorities were investigating all aspects of reproduction with great zeal. They proclaimed an objective point of view over the reproductive body, which they claimed women could not acquire. Male science thus fundamentally changed how the body in birth was represented and understood. Whereas seventeenth-century popular and medical models of the reproductive body overlapped, the two models diverged as men entered the field of midwifery and explored reproductive biology more broadly.

In 1650, most people lived close to the land and to food production; work and the passage of time followed the circular passage of the seasons. Jewish, Catholic, and even Protestant worship was centered on annual cycles; seventeenth-century Europeans' religious customs interwove with agricultural rhythms and the vacillation between feasting and fasting. Europeans understood human lives and the bodily experiences of birth and death as tied to the seasons and nature. The ages of man seemed, for example, to pass through the summer of youth and the autumn of adulthood, and as plants and the elderly withered in the winter, babies were born in the spring. In addition to these analogies describing the body's stages, Europeans relied on several other models to explain the relationships between all things and creatures. These included the hierarchical Great Chain of Being, which linked all from angels to oysters and heavenly, orbital maps superimposed on diagrams of the body, with the head at the center and the limbs reaching out into further orbits. These chains, orbits, and bodies modeled, for example, the patriarchal household, a microcosm for society and the universe. The father was the center, and he was also the head to the body of the family, just as people orbited around and were commanded by their monarchs.

Early moderns had several ways of modeling the sexes within these and other schemes. To the degree that the two were bifurcated in the early modern world, women were considered less controllable and more carnal, humorally moist and cool than men, who were seen as more reasonable and contained, warm and dry. According to centuries-old theories, which had survived into the Enlightenment, women's minds were liable to their bodies' influence—wombs could wander and cause hysteria; women's orifices exuded fluids; their bodies and brains were highly permeable to outside forces, including the occult and wicked. Men appeared to have carefully delineated bodies and minds, which made them more resistant than women to external forces and internal rumblings. Yet the sexes were not opposite in all ways. Galen's second-century claim that women's reproductive anatomy was an inside-out version of men's outward genitalia remained a dominant anatomical explanation into the first decades of the eighteenth century. Here, women and men were not visualized as opposites but as versions of each other, and, indeed, some even believed that a person's sex could change. The cultural historian Thomas Laqueur has explained these beliefs by arguing that early moderns maintained a one-sex model that placed gender on a continuum rather than as polarized opposites as they came to be seen after 1800.[15]

Many sources confirm Laqueur's early modern model of gender on a continuum, but other historians have shown that early moderns viewed the sexes as opposites, particularly in terms of birth.[16] Cosmological diagrams, humoral bifurcation, and comparisons to seasons and the natural world all offer competing and sometimes contradictory models of sex, gender, birth, and the body. Late seventeenth-century authors ricocheted between classical and new models of conception, between Galenic traditions and the microscopic discoveries of Antony van Leeuwenhoek, between gender on a continuum and women and men as opposites. These multidimensional depictions of the body survived into the eighteenth century but became considerably less common by 1800, as contemporary authorities more often insisted that one model was correct and others were false. In 1760, it still could be plausible to maintain, as midwife Elizabeth Nihell did, that "one truth cannot hurt or exclude another truth, and that all truths may very well coexist."[17] But the response to her work in the press marked a turning point as her critics dismissed such an argument as indefensibly illogical.

When early modern and popular authors described the birthing body, they evoked vivid imagery. In the spirit of hierarchical, naturalistic schemes, women were described as subordinate to their rulers, imagined, for instance, as the ground on which a father raised his stock. Poets and plebeians described female bodies as the fertile land that husbands tilled, or on which they planted their seeds or watered. Mothers were trees with budding fruit; their big bellies seemed like hills or mountains. As the unseen child grew in utero, it was said to "flower," or alternately to "ferment" or "bake." The early modern metaphorical language describing reproduction referenced the garden, landscape, and kitchen. Mary Fissell has noted how a new trope appeared in the 1650s that emphasized men's power to make and master the world. Texts compared reproduction to glass-blowing, printing, metal-casting, and other masculine crafts, including especially alchemy. Fissell argues that such metaphors correlated with men's emerging interest in reproduction. But whether texts used the new reference of craftwork or such traditional clichés as a fetus as the fruit of the womb, these comparisons helped portray reproduction's hidden processes in an age when few ever saw an image of or an actual dissection revealing the internal workings of the birthing body.[18]

Late seventeenth-century midwifery, medical, and erotic texts used pastoral, domestic, and mechanical metaphors to describe the reproductive body, and these comparisons survived well past 1800 in *Aristotle's Masterpiece* and other reprinted popular texts.[19] But it is difficult to say whether this is how women (and men) experienced their own bodily experiences in relation to reproduction. Certain metaphors—menstruation as flowers, the uterus as an animal or an oven[20]—were commonplace and crossed language groups, but did pregnant women feel that their unborn children were forming inside them like cheese, as Jane Sharp described the process of conception and embryonic development?[21]

Perhaps women envisioned congealing curds or rising bread as their bellies grew, but it seems that the experience of reproduction was not apprehended primarily through analogical comparison, but through claims to the qualities of apprehension itself. Barbara Duden's analysis of over 1,800 medical cases recorded by the German physician Johannes Storch between 1720 and 1741 suggest that the women who consulted him did not emphasize the visual and concrete metaphors of midwifery and medical texts. Instead, they described experience as deeply felt sensation, space, and movement.[22] This value, placed on sensibility and self-apprehension, also informed the widespread belief that pregnancy could only be determined by the mother herself when she felt the baby quicken at about four months. This was when it was popularly believed the fetus acquired life. Thus, despite the disapproval of theologians and late eighteenth-century doctors on this point, few Europeans before 1800 considered terminating a pregnancy before quickening an act of abortion because the fetus was believed to lack a soul.[23]

When English midwives asserted their authority over the reproductive body, they likewise emphasized feeling as a form of knowledge. As English midwife Jane Sharp explained in her treatise: "It is not hard words that perform the work"; rather, it was "the natural propriety of women to be much seeing into that Art."[24] Midwifery's epistemological position thus arose from subjective feeling, sympathy for women, and hands-on experience.[25] Elizabeth Nihell distinguished women's knowledge of reproduction as sympathetic compared to the supposed scientific detachment of male doctors: '"Women . . . have more bowels for women: they feel for those of their own sex so much, that that feeling operates in them like an irresistible instinct, both in favor of the pregnant mother and of the child."[26] Because maternal feeling and experience, not words or images, created knowledge, both Sharp and Nihell described labor and delivery cursorily. Aside from texts referencing the pain of childbirth, often through the shorthand of God's curse on Eve in Genesis, little was said in early modern texts about the body's struggle during birth.

Seventeenth- and early eighteenth-century Europeans idealized midwifery and birth as belonging to an entirely female arena, separated from men and even routine family life. In northern Europe, the ideal birthing room was warm and dark because cold air and bright light were harmful, plus such a space seemed itself womblike and protective. In theory, men and the intrusions of the household were banned except in emergencies. European travelers were fascinated with other cultures' birth rituals, highlighting the powerful role of midwives and the separation of sexes during childbirth around the world. Many Enlightenment authors also mobilized culturally distinct reproductive rituals, especially the seemingly irrational, to order cultures and races hierarchically. As female midwives lost authority in Europe in the eighteenth century, these armchair travelers considered midwives' authority elsewhere to be a sign of unsophisticated nations.[27]

MEN'S ENLIGHTENMENT VIEW OF THE BODY IN BIRTH

Ninety-two percent of labors were considered normal, even by eighteenth-century men-midwives.[28] Although doctors often carped that midwives bungled difficult or unusually slow labors, many of them, especially in urban areas, did manage complications quite successfully, from turning breech babies in utero to even rescuing infants by Cesarean section.[29] Despite most midwives' skill at delivering newborns, they nonetheless confronted intractable cases that required surgeons to take drastic measures, including the dismantling of unborn, full-term babies. In the seventeenth century, medical men mostly appeared in cases of grave danger, and their presence usually signified that mother, child, or both would die.

Here, birth intersected with life, a sadly common conjunction in the early modern world. Although 99 mothers survived for every 100 deliveries, given the number of children most women had, the average northern European mother before 1800 had about a 5 percent cumulative chance of dying in childbirth or from other reproductive complications (such as expiring from postpartum fever). Adrian Wilson has argued that the fear of death in childbirth ultimately led to the rise of the male midwife.[30] While surgeons were associated with reproductive tragedy in the seventeenth century, news spread in Britain and elsewhere by the early eighteenth century that some male attendants had special techniques and tools—primarily the obstetrical forceps—that did not always result in death. Women chose obstetricians because they trusted men's ability to intervene more successfully than traditional midwives could in dangerous deliveries. Although some historians have rejected Wilson's argument,[31] his work offers important insights into the eighteenth-century birthing body: the arrival of the man-midwife, whatever the causes were, resulted in a new way of representing and establishing authority over women's bodies.

Midwives, for example, may well have noticed that as a neonate emerged in the birth canal its head typically turned sideways. Yet no texts mentioned this before the 1740s, when Irish man-midwife Fielding Ould and Scotsman William Smellie contemporaneously discovered so-called fetal rotation. Seemingly a small point, this recognition, based on careful measurements and calculations about the proportions of the maternal pelvis and fetal head, allowed practitioners to manage difficult births as problems in physics and engineering. Discovering the baby's head did not squeeze straight out of a symmetrical chute, but rather compressed itself as it rotated out of a more complicated passageway allowed Smellie and others to develop mechanical interventions when the baby was positioned oddly and could not begin its natural twist outward.[32] These men-midwives consequently established entirely new depictions of the body in birth. Their descriptions and illustrations of childbirth were visual, mechanical, and mathematical, with few analogies and metaphors. Even some Enlightenment-era female midwives, including Madame du Coudray, who taught thousands of French provincial women the art of midwifery, relied on using models of fabric-covered pelvises to demonstrate visually and mechanically the otherwise invisible processes of both normal and abnormal deliveries.[33]

Whereas Jane Sharp and Elizabeth Nihell emphasized language's inability to communicate subjective knowledge, Enlightenment doctors highlighted objective, visual, and mechanical pedagogy. By the time the obstetrical atlas *The Anatomy of the Human Gravid Uterus* (1774), by William Hunter and artist Jan Van Rymsdyk, appeared, Hunter described pictures as "an universal language" that could "conve[y] clearer ideas of most natural objects, than words

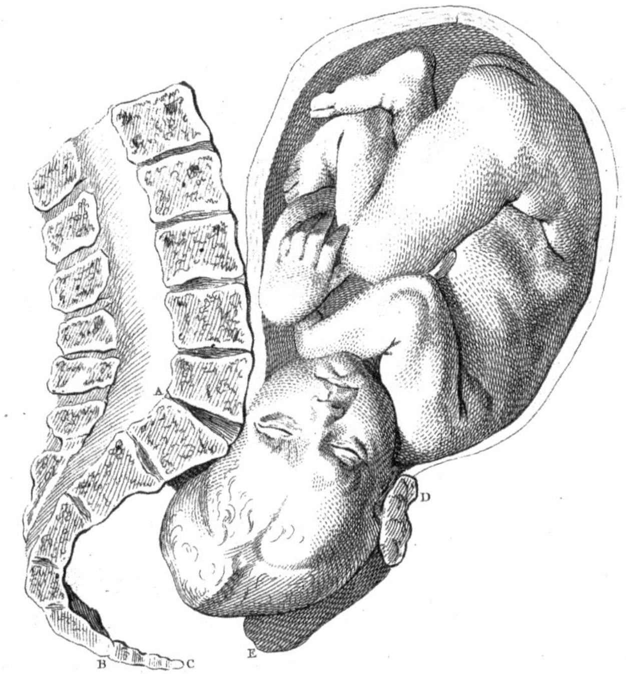

FIGURE 1.2: Fetal rotation. The baby's head turns sideways as it enters the birth canal. In William Smellie, *Anatomical Tables . . .* (Edinburgh: 1787). Rare Books, Wellcome Library, London.

can express." Knowledge of the birthing body derived from observing and representing birth as an anatomist, not from being a mother. This is where women's own experiences worked against them. Though many obstetricians worked with and learned from female practitioners, by the 1770s, obstetricians typically dismissed midwives as feeling too much and projecting their own reproductive experiences onto other women. Sympathy was thus dangerous, according to several men-midwives, as it prevented women from properly separating themselves from others.[34]

Enlightenment authors also asserted authority over the birthing body by peering deep inside to make reproduction reveal its secrets. In 1650, most theories of conception and embryology derived from Greek, Roman, and Arabic theories, including those of Aristotle and Galen. For instance, nearly all authors portrayed males and females as contributing liquid seed that coagulated like cheese in the uterus, with the major organs soon forming. These and other schemes influenced early modern European beliefs about the nature of a fetus's development, but after 1650, natural philosophers' investigations into the nature of plant and animal reproduction led to new theories, some of which were based on microscopic and anatomical observation and experimentation.

Anatomists and natural philosophers explored more than only living women in labor. From the late seventeenth-century, Dutch microscopists Nicholas Hartsoeker and Antony van Leeuwenhoek who looked at male ejaculate, to continental biologists Abraham Trembley, Lazzaro Spallanzani, and others who experimented with sea hydras, frogs, dogs, and rabbits, to the Scottish brothers William and John Hunter who worked on pregnant and infant cadavers, dozens of scientists examined what the sexes contributed to conception, how creatures developed in utero and in eggs, what the relation between mother and unborn was, and why reproductive accidents occurred.[35]

Leeuwenhoek and Hartsoeker promoted the theory of preformation, the idea that a fully formed future human being lay encased in either an ovum or sperm cell. Leeuwenhoek argued that men's testicles contained future human beings fully encased in their sperm, but generally spermatic preformation was not as popular as ovism, partly on theological grounds: what sort of God would allow so many millions of future creatures to be squandered in each act of sex? For this reason and others, eighteenth-century Europeans preferred the theory of ovist preformation, which did not seem to waste so many potential lives. Europeans of the seventeenth and early eighteenth centuries also maintained that if conception only occurred when men climaxed and ejaculated, so too must women orgasm to activate an egg and conceive.[36] While ovism might seem to justify on scientific grounds why women needed to climax and release their eggs to procreate, by the 1750s, ovists described women's eggs and bodies as passive, ultimately helping to undermine the rationale for female orgasm. John Hunter's successfully artificial insemination of a married woman in 1790 further demonstrated that female pleasure had little role in conception.[37] By the 1820s, when the Estonian anatomist Karl Ernst von Baer charted the process of ovulation in female dogs, and the medical profession emphasized menstruation as the mark of femininity, fertility became completely tied to the fluctuations of the female cycle. Women no longer needed to enjoy sex to reproduce.

In the early modern world, women were porous, easily influenced by outside forces, and so were their in-utero offspring. Europeans entertained many explanations for anomalies and tragedies, including the supernatural and cosmological. Magical explanations waned among European elites after 1700, but many of all classes continued to maintain the view that women's imaginations could physically mold the unborn because their minds and wombs were somehow deeply interconnected—through humors, nerves, blood, or something else. Contemporaries frequently reported cases of so-called maternal imagination, in which mothers marked their babies or even made monsters when they craved a food too strongly or were overly frightened by a terrible sight, such as an execution. The period's most famous case occurred in 1726 when Mary Toft, a poor English peasant, claimed to give birth to seventeen rabbits because

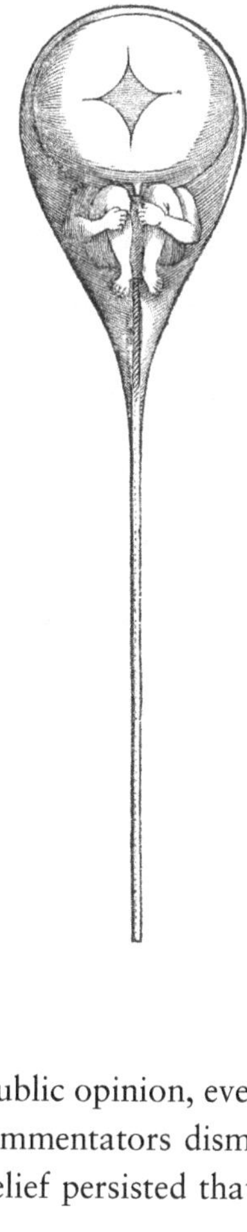

FIGURE 1.3: Preformed human in spermatozoon. In Nicolas Hartsoeker, *Essay de Dioptrique* (Paris: 1694). Rare Books, Wellcome Library, London.

of her insatiable hunger for hare. Her claim divided public opinion, even after it was exposed as a hoax. Only by the 1760s did commentators dismiss the case as an absurdity, yet far beyond 1800 popular belief persisted that monstrosity had a root cause in maternal imagination.[38]

THE BODY IN DEATH

In the period 1650 to 1800, dead bodies came in all shapes and sizes. Though many adults succumbed to the same causes as in the modern Western world—warfare, suicide, accidents, cancer, heart disease, old age—death spread itself across all age groups, with the very young hit hard by disease, mishaps, and inadequate nutrition. For example, in eighteenth-century France, only 55 percent of babies lived to age ten,[39] and while the late twentieth-century European infant mortality rate within the first year was 9 per 1,000, before 1750 it was 135, 200, and 296 per 1,000 in England, France, and Geneva, respectively.[40] Life expectancy for newborns increased after 1750 in London, and life expectancy in general edged up a few years in some, but not all, European regions during the Enlightenment. Yet declining mortality rates were not steady or

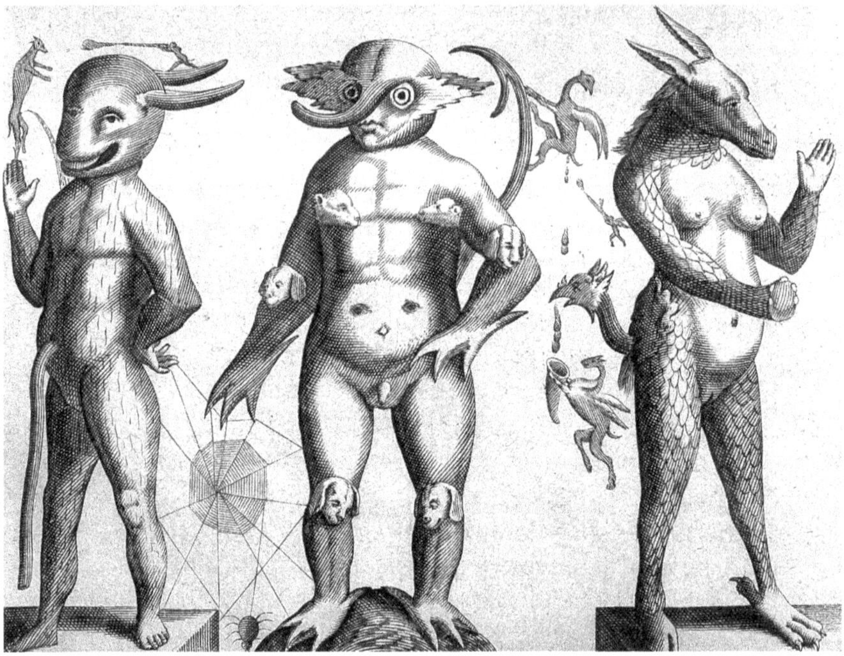

FIGURE 1.4: "Monsters" produced through maternal imagination. In Fortunius Licetus, *De Monstris . . .* (Padua: 1668). Rare Books, Wellcome Library, London.

progressive in all places. In some parts of Germany, the chance of dying within 6 weeks after childbirth almost tripled—from 4 women in 1,000 between 1690 and 1724, to 11 per 1,000 between 1760 and 1794. Here, the increased mortality rate arose from new postpartum practices that greatly shortened mothers' customary lying-in period.[41]

Death was interwoven with daily life. Because most died in their own or their relations' homes, everybody saw corpses. People suddenly took ill or had an accident—coroners' records and newspapers reported suffocated babies left to die in privies, children and workers run over in the street, not to mention countless random drownings and freak accidents. In times of mass fatalities, for instance in the 1665 London bubonic plague outbreak in which nearly 80,000 died, or the 1755 Lisbon earthquake that killed 90,000, observers described piles of corpses rotting faster than they could be buried. Although such horrors did not occur every year, they happened frequently enough that many people in the period of 1650 to 1800 could personally recall such scenes of despair. Capital punishment existed in all Western cultures, and the numbers of executed were in the hundreds and thousands annually in most European countries. The corpses of highwaymen rotted in cages along turnpikes; the heads of traitors were seen on pikes at London's Temple Bar until the 1770s and were paraded through Parisian streets from

FIGURE 1.5: Hanged bodies suspended from a gibbet while others are mutilated. The man tied to a wooden cross is being disemboweled alive. Engraving with etching, circa 1660. Iconographic Collections, Wellcome Library, London.

the summer of 1789 onward. Although torture was illegal in England, it was not in other places and was considered a fit spectacle for the public, as in the 1757 case of Damien the would-be regicide, whose limbs were tied to and pulled off his torso by four horses, while he was still alive. Even without torture, thousands turned out for public hangings at such sites as Tyburn Tree in London.

These spectacles of death alarmed Enlightenment philosophers. The Italian theorist Cesare Beccaria condemned capital punishment and torture in a 1764 treatise, inspiring the Habsburg Grand Duke Leopold II to abolish the death penalty in Tuscany in 1786. Many French philosophers also opposed the monarch's use of corporeal force, but ultimately when the monarch was overthrown and himself executed in January of 1793, the radical French revolutionaries had fully embraced state-sanctioned violence and capital punishment. The guillotine, promoted by the Enlightenment physician J.-I. Guillotin to execute the condemned painlessly and swiftly, ironically became a symbol of the Revolution's worst brutality.

Throughout the Enlightenment, dead bodies imparted their lessons. As in centuries past, clerics used ordinary peoples' deaths and the righteous execution of criminals to admonish their flocks. Catholics and Protestants both promised potential life after death, but they relied on different routes to grace.

For Enlightenment Catholics, a temporary stay in purgatory before going to heaven was facilitated by sacramental participation while on earth, good works, and the prayers of the living. The body in birth and death served as a central vessel for salvation. From the sacramental rites of baptism to last unction, priests' touching of Catholics' bodies linked them to the Church. During the mass, Christ's dead body mystically appeared with every act of transubstantiation, and for the faithful, the Eucharist connected their living bodies with Christ's. Protestants denounced most of the sacraments, and though theologians debated the fine points, most refuted the notion that the communion bread miraculously became Christ's flesh, believing instead that it symbolized his sacrifice. Though most Protestants deemphasized corporeal rituals, including, for instance, the renunciation of flesh during Lent, they did sometimes find evidence of salvation through the body. For evangelicals, the corporeal ecstasy of being born again signified the possibility of true salvation, and sometimes this bodily reassurance occurred on the deathbeds of the faithful.[42] Without the feeling, one might not be saved.

Though perhaps Catholics placed greater weight on visceral experience and the real presence of Christ's body than Protestants did, both branches treated dead bodies with reverence, ordering that corpses be cared for until burial. Although cremation had been practiced for millennia in other parts of the world, Europeans recoiled at the suggestion. Few, other than physician Sir Thomas Browne in his *Hydriotaphia: Urn Burial* (1669), argued for any disposal of human remains other than conventional burial until well after 1800.[43] Christian theologians debated whether the bodies or only the souls of the dead would be resurrected someday, but reluctance to cremate must also be attributed to the popular belief that the body continued to have life to it, even after death. Not only did a corpse's hair and nails continue to grow, so too might the body of a murdered person bleed when near the murderer. And as nearly everyone believed, body parts of dead humans and animals had curative and magical properties. These ideas survived well into the eighteenth century in both Protestant and Catholic regions, not to mention other parts of the world.[44]

When the dying approached the end, their bodies offered signs: eyes glazed, throats made their death rattle, hands clutched, or the sphincters relaxed with bodily waste spewing. Nursing the dying and cleaning the mess was left to women, although the eyes were often closed by family members, both female and male. Fearing premature burial in the period, many families waited twenty-four hours or more. But because the body decomposed rapidly, letting off putrefying smells and swelling sometimes to the point of bursting, relatives were eager to bury quickly. The vast majority of early modern corpses were interred within four days, and the intervening time of watching (and smelling) the body around the clock was left to servants and family members, with men often taking on the night watch in particular.

The corpses of poor, ordinary Europeans were cleaned up and wrapped in shrouds by female attendants. Dead adults of the middling and elite ranks were placed in bran- or straw-filled coffins, but most could not afford even this, so they were simply wound in cloth and buried with others recently deceased, several bodies deep. By the 1730s in London, the professional male undertaker had begun to orchestrate funerals for families.[45] Although women, including Dickens's Mrs. Gamp, helped lay out corpses, they increasingly were employed by undertakers rather than independently by the families of the dead. During the period, London's authorities also hired destitute, old women of the parish as so-called searchers to tabulate the causes of death for the *Bills of Mortality*, which were published between 1665 and the 1830s. Searchers were not forensic authorities of course, yet medical officers did not fully take over determining and enumerating death's details in London and other cities until the 1840s.

Few of the dead before 1800 could afford embalming. Arresting decomposition and preserving outward appearance was practiced on royal and elite corpses throughout the early modern period for public funerals. Embalming was generally not popular before the eighteenth century thanks to the expense and the disfiguring effects on the corpse. In the eighteenth century, William Hunter developed elaborate wax injections to preserve anatomical specimens (as well as the wife of a colleague),[46] while embalmers prepared corpses in wine and spirits. These new techniques allowed dead bodies to appear more lifelike than what the old methods of using resins and linen could achieve.[47]

Postmortem dissections were far more controversial than embalming. Governments had long decreed that surgeons and medical students had the privilege of dissecting a small number of executed criminals' bodies throughout Europe. Yet these few corpses failed to satisfy the demand of medical professionals, who were often accused during the Enlightenment of supporting grave-robbing, a lucrative occupation for such late eighteenth-century characters such as Jerry Cruncher in Dickens's *Tale of Two Cities* (1859), and real-life traffickers such as the Edinburgh resurrectionists William Burke and William Hare of the 1820s. Grave-robbing terrified the living, who did support, on the other hand, autopsies in cases of apparent murder, suicide, and accidents, or when requested by the family. Sometimes, according to Philippe Ariès, the dying asked for a postmortem examination to guarantee that the end had truly come—after all, scalpel incisions had been known to wake the only putatively dead.[48] From the late seventeenth century onward, the scientifically inclined offered their bodies to medical men, proposing that their cadavers might be illuminating. Coroners' and surgeons' autopsies in suspicious deaths caused little dismay during the period, but the cadavers they opened were often those of the poor, the marginal, and the unclaimed.

As a practical matter, there was inadequate space for all the dead to find separate graves, and until the nineteenth century few thought that they should. Into the eighteenth century, as graveyards became crowded, old bones were dug up, piled together in charnel houses, or even intermingled with the church's paving stones. Places like Paris had central cemeteries designed to accommodate all but the most elite dead bodies. Those who could afford to, or who lived in small communities, might be buried with their relatives in the parish churchyard; often mothers and babies who perished in childbirth together were buried together. As graveyards and the ground beneath churches overflowed with dead bodies, municipal officials began placing cemeteries wherever they could find land cheaply. In 1763, authorities in Paris, influenced both by Enlightenment discussions of public sanitation and the stench of the city's abundant corpses, began to transfer cemeteries outside the old city walls. During winter nights between 1785 and 1787, they even had all the bodies from the city's central cemetery removed because the putrefaction had become too noxious. Other European cities would also establish large public cemeteries in the years after 1800.

For most Europeans between 1650 and 1800, there was more to the body in life and in death than mere flesh. There was also the matter of the soul. Theologians and philosophers during the period debated what and where it was. For the most learned, it seemed to be an immaterial substance, perhaps even separate from the body and mind, as in Descartes's arguments. The reproductive research of Abraham Trembley in the 1740s on the infinitely divisible polypus raised fresh theories of the nature of the soul. Julien Offray de La Mettrie was among those materialists who ultimately proposed that there was no soul.

This was not how most people between 1650 and 1800 viewed the issue, however. Most Europeans in the period were religious and hoped for eternal life. Though some must have agreed with the philosophers that the soul was immaterial, people often spoke of it having a personal, even visceral nature. Yet anatomists could not exactly locate where it was. Descartes proposed it might be in the pineal gland at the base of the brain, a spot that nicely served as the nexus between body and mind, and most eighteenth-century texts suggest that this ethereal, transcendent thing was in the brain or nerves or a part of the mind's function. A handful of atheists and agnostics of the age such as Denis Diderot suggested that the soul was nothing more than memory, and according to John McManners, most learned eighteenth-century Frenchmen took a stoical view that whatever the soul might be, humans lived on in their loved ones' memories or, if really prominent, in posterity.[49] Ordinary people seem to have had a much more visceral sense of these matters, for instance believing that the soul came into being in a baby when its mother felt it kick in utero, or sensing that it still lingered on in graveyards. What happened after death was not clear, but the persistent beliefs in ghosts and postmortem miracles, even among Enlightenment-era Protestants, suggest the powerful hope that the dead, perhaps without a body, nonetheless carried on in some vaguely corporeal way.[50]

As fleeting as the experiences of both birth and death ultimately were, they were made understandable through concrete metaphors and the intense corporeal experiences of the body itself. These aspects of birthing and dying were consistent across the period 1650 to 1800. But the management of birthing and dead bodies underwent major changes during the period. Women had a vital role to play in the corporeal entrance and exit of life in the seventeenth century. Sairey Gamp was a nineteenth-century caricature, but countless women before her combined midwifery and attending the dead. Though the majority of Europeans, mostly poor and struggling, continued to rely on women's labor as midwives, searchers, and dressers of the dead far into the nineteenth century, elites came to place their birthing and dying bodies under the purview of science by 1800. Science proclaimed its authority over all bodies throughout the nineteenth century, and by the twentieth century, the funeral and birth industries grew enormously, removing birth and death from peoples' homes and neighborhoods. As birth and death became both scientized and commercialized in the modern world, they became separated from each other and the natural order of the home and society.

CHAPTER TWO

Pliable Bodies

The Moral Biology of Health and Disease

KEVIN SIENA

When it came to matters medical, Enlightenment bodies were beset on all sides. Never before was there so much medical advice, yet never before did there appear to be so many health risks. For all the confidence of the age, the body was considered fragile and susceptible. Science advanced exponentially, yet little of that knowledge actually made people healthier, and much of the mounting corpus of information only highlighted how precarious human health was.

Enlightenment discussions of disease represent a mix of new and old ideas. Novel constructions of pathology and anatomy undoubtedly changed the medical landscape by the eighteenth century. New theories about sickness were part and parcel of the period's lauded epistemic revolution. However, often overlooked is how very old ideas still counted in the eighteenth century. Enlightenment medicine, therefore, presents many contradictions. Medical thought was no more linear than was the era's political thought. Like other Enlightenment pursuits, medicine was characterized by multiple viewpoints and debates moving in many directions at once.

Enlightenment science thus provided a vast palette of ideas from which medical commentators could draw. The body produced by this wide-ranging discourse was extremely plastic. Doctors writing on what made a body prone to sickness inevitably inscribed values onto that body. Perhaps nowhere were the physical and the moral so intimately bound; doctors describing people, places, or lifestyles prone to disease wrote profound moral claims onto a body that was very much a blank slate awaiting their handiwork.

Many themes bear this out. One of the more intriguing is that of the environment. While Enlightenment thinkers adored abstraction, medical commentators rarely considered bodies divorced from their settings. Bodies did not float freely in the ether but were situated in contexts. To many doctors, one's place determined one's health. However, these settings were as much moral environments as physical ones. The intimate connection between bodies and environments was founded on the principle that the body was porous; what happened outside of it affected what happened inside of it and vice versa. The body was also profoundly mutable, for physiology changed according to factors like climate, lifestyle, and emotions. Theories of disease fused the physical and the moral, allowing Enlightenment pathologists to produce what I want to call a moral biology, in which physiology itself was recast by moral action. Pathology, therefore, contributed greatly to the construction of the Enlightened individual, capable of rational choice but very much responsible for those actions. Finally, one of the Enlightenment's many contradictions demonstrates that while the individual still played a vital role in debates on disease, the body's position as a site for the inscription of values meant that social markers like race, gender, and class—indiscriminate of individuality—informed disease theories profoundly. In a variety of ways, then, pathology contributed to much deeper Enlightenment debates.

FLUIDS, FIBERS, AND NERVES: EVOLVING VISIONS OF THE BODY

Before charting new ideas it is imperative to stress the continued importance of an old vision of health that still dominated Enlightenment medicine, the so-called six nonnaturals. Dating back to the ancients, these were six factors thought to influence health: (a) air, (b) sleeping and waking, (c) food and drink, (d) rest and exercise, (e) excretion and retention, and (f) the emotions or passions. Health in the Galenic system was a matter of balance. Each body had a constitution, characterized by its particular mix of the four humors, which had to be kept in balance to prevent slipping into a diseased state. The nonnaturals—the quality of the air one breathed, the food one ate, and so forth—affected this balance, hence managing them formed the basis for traditional prophylaxis.[1]

The humoral system did not survive into the Enlightenment. By 1650 Vesalius's depictions of dissections and Harvey's discovery of circulation had destroyed Galen's anatomy, and the movement was on to forge new models. Yet the nonnaturals continued to thrive. Enlightenment doctors continued to stress that moderation of the nonnaturals was crucial to health. The healthiest regimens were those in which one neither overindulged nor totally abstained from the nonnaturals. Too much rest was as bad as too little, and overly strenuous exercise was as dangerous as lethargy.[2] Moreover, despite the forward-looking mood of the age, Enlightenment physicians lauded classical medical authorities.[3] Such conservatism has been ascribed in part to the pressures of the market, where doctors catered to patients still yearning for traditional treatments.[4]

Conservatism notwithstanding, medical theory was changing. Iatrochemistry and iatromechanism presented significant challenges to Galenism in the later seventeenth century. Heirs to earlier theories of Paracelsus and van Helmont, iatrochemists looked away from Galen's humors and instead strove to understand how different kinds of matter, especially acids, alkalis, and salts, affected the body.[5] In England, the chemical physicians reached their zenith by the 1660s, but their failure to establish an institutional base, and the aforementioned conservatism of a public wary of curious therapeutics, limited their success.[6] However, one of their most important contributions was to advance a corpuscular vision in medicine, presenting chemical substances as small particles that penetrated the body, enhancing the intimate relationship between body and environment.

Medical mechanics or iatromechanism was more influential by the close of the century.[7] Medical contemporaries of Newton quickly applied their intense respect for mathematics to the body and like good Newtonians rejected the chemists' stance that different kinds of matter held different qualities. Physicians like Archibald Pitcairne, Richard Mead, and George Cheyne presented the body as a machine, functioning according to predictable laws.[8] It took decades to take form, but various contributions were molded into a system by one of the leading medical figures of the age, Leiden's Herman Boerhaave.[9] His influential textbook, *Institutiones medicae* (1708), emphasized the body's solid structures, moving once and for all beyond the Galenic body's amorphous slush of fluids. The Boerhaavian body had fluids, most importantly blood and nervous fluid, but they moved about within a concrete plumbing of tubing. In this hydraulic model, the movement of fluids was governed by the fibers of the solids that contained them. Health demanded that one's fibers remain sound. If fibers were too loose or too rigid, circulation was sluggish or overly intense. Disease became increasingly described in these terms. For example, fever, the era's most common disease, was said to arise from flawed circulation. Physician Jeremiah Wainewright proclaimed: "a Fever is but the increased Circulation of the Blood."[10]

Boerhaave emphasized arterial circulation, but later theorists like Albrecht von Haller and especially Edinburgh's William Cullen, whose teachings surpassed Boerhaave's as the most influential system in the later eighteenth century, instead stressed nervous fluid.[11] By the mid-eighteenth century, the nerves were central in medicine. For doctors, the concept of irritability loomed large as a pathological counterpoint to the increasingly important concept of sensibility. Stimulation, whether too much or too little, caused the nervous fibers to irritate the body's tissues, disrupting circulation and causing disease.[12] The nerves' importance enhanced the role of the sixth nonnatural, the passions, allowing doctors to contribute to larger Enlightenment debates about the mind/body question, sensibility, and gender.[13] Through the nerve/fiber matrix, emotions like fear or dread could, by themselves, predispose the body to sickness or even throw it into disease altogether.[14]

So, for both Boerhaave and Cullen, disease resulted from disrupted circulation, but what, aside from emotional breakdown, might cause such disruption? To answer this question, doctors of all theoretical stripes emphasized foreign particles entering the body. Alternatively calling such substances poisons, viruses, or effluvia, doctors after 1650 intensified the movement already underway in the Renaissance (thanks to plague and syphilis) to point to external substances as the main cause of disease.[15] Such particles were frequently thought to affect bodily tissues by transferring their own malignant qualities.

FIGURE 2.1: "Sensibility." Engraving by Richard Earlom after George Romney, 1789. Iconographic Collections, Wellcome Library, London.

By corrupting the body's fibers bit by tiny bit, minute substances could wreak havoc on a healthy body. Even Thomas Sydenham, renowned for emphasizing practical observation over fanciful theory, stressed the impact of foreign particles on the body:

> Therefore I think all that Malignity which is in Epidemicks . . . consists of very hot and spirituous Particles . . . because only such Particles can so soon alter the Humours, as we see they are in malignant Diseases. I also think . . . every active Particle is busied in creating its like, and in turning and accommodating whatever is contrary to its own Genius. So Fire generates Fire.[16]

Heralded studies in microscopy by Robert Hooke, Marcello Malpighi, and Antony van Leeuwenhoek undoubtedly advanced this focus on the miniscule. Boerhaave himself warned of toxic substances, the smallest speck of which could render "The whole Mass of Blood . . . no other than downright Pus of a malignant Nature." If this happened: "the whole body goes to Wreck, every Particle having acquired such a Virus as to be capable of infecting a healthy Person with the like Contagion." Such poisons could enter the lungs through breath, the stomach through ingestion, or could even seep through the skin.[17] The porous human hide provided poor armor in a world teeming with invisible toxins.

The focus on invading particles advanced what many scholars correctly see as a modern conception of disease. In the Galenic model disease was not an entity, but a bodily state. One was diseased when one's humors fell into imbalance. Diseases did not exist independently of bodies. Moreover, diseases lacked uniformity. Because each person's constitution was unique, each humoral imbalance—hence each occurrence of disease—was also unique. As humoralism faded, movement began toward what many term an ontological notion of disease. In the modern schema, diseases become independent entities with uniform characteristics. The emphasis on the invasion of the body by poisons represented a major development toward this view, as did the increasing portrayal of diseases as commodities, exchanged between nations ever more linked by global trade.[18]

DISEASE TAXONOMY AND THE INDIVIDUAL CONSTITUTION

Advances in medical statistics and taxonomy also influenced the development of an ontological view of disease. Especially important was a focus on health within the nascent science of political economy.[19] John Graunt's *Natural and Political Observations upon the Bills of Mortality* (1662) is landmark in this regard. The *Bills of Mortality* registered deaths by various causes. Scores of diseases were listed, and the *Bills* became regular publications, inviting frequent analysis.[20] Presenting diseases as items to be counted advanced the sense that they were tangible things. Moreover, the act of quantifying hundreds of deaths under a single heading implied that all of these occurrences of illness were the same. The Galenic view, that each occurrence of illness represented a unique phenomenon, could not allow for this kind of quantification, an approach to knowledge that gathered steam as the Enlightenment advanced.

Quantification accompanied taxonomy. The taxonomic impulse was one of the Enlightenment's most profound features, providing powerful new schemas to organize the world.[21] It was botanists who first developed taxonomies to arrange newly discovered flora. In medicine, this development, usually associated with the eighteenth century,[22] was well underway quite early. Long before Linnaeus, the seventeenth-century medical icon Sydenham was already imploring doctors to organize diseases: "[according] to certain and definite Species, with the diligence we see it is done by Botanick Writers in their Herbals."[23] Doctors pursued this project throughout the next century, most notably Cullen, whose nosology formed the basis for medical teaching at Edinburgh, the epicenter of instruction in the later eighteenth century.

Accounting for disease due to invading agents, conceptualizing them as commodities, categorizing them, and evaluating them statistically all represented steps toward a modern view that diseases were independent entities that affected bodies uniformly. Yet for all these developments, medicine resisted this view of disease in its entirety; the shift to an ontological conception of disease was incomplete.[24] Here again vestiges of the classical body continued their influence long into the eighteenth century. Mitigating the impact of these developments was the continued centrality of the individual constitution. The constitution was now described in terms of the fitness of one's fibers rather than their humoral balance, but that made little difference. Through to the nineteenth century doctors agreed that each body had a unique constitution. Diseases could not yet be uniform because bodies were not yet uniform.

Moreover, disease remained a bodily state rather than a freely existing entity. Boerhaave famously defined disease thus: "Whatever State of the Human Body doth disorder the Vital, Natural, or even the Animal Functions of the Same, is called a Disease."[25] As late as 1800, physicians such as Thomas Garnett still described disease as a corporeal state: "Health is the easy and pleasant functions of the body and mind; and disease consists in the uneasy and disproportioned exercise of all, or some of the functions."[26] Bodies were invaded by foreign particles, but they were not yet invaded by *diseases*. Concepts like the poison or the effluvia were vital conceptual tools to distinguish between the two.[27] Toxic substances pervaded the environment; no one could hope to avoid them. What mattered in the fight against disease was the strength of the constitution, for when people came into contact with a disease-causing agent, the health of their constitutions, sometimes described as their temperament or habit of body, determined whether their bodies succumbed to its effects and fell into disease.

Professional aspirations contributed to the constitution's continuing importance. Physicians had long argued that only they were trained to read the baffling phenomena of disease, which varied from patient to patient based on their humoral make-up. As competition in the marketplace heated up, thanks to the explosion of cheap print,[28] doctors clung to this argument. Writing on smallpox in 1743, James Kilpatrick emphasized that quality care required "a considerable sagacity in discerning the particular Habits of different bodies." Only a "competent physician"—to be distinguished from unlicensed empirics and quacks—"may judge of the different Preparation appropriate to different Habits and Temperaments."[29]

Kilpatrick's point concerned treatment, but the issue of the individual constitution also informed theories on who would fall sick in the first place. As a result, the *predilection* for infection dominated discussions of health. Making just this point, John Huxham explained how some constitutions were armored against even the worst diseases:

> effluvia affects different persons in a very different manner . . . probably not only from the different Crasis of the animal Spirits, but also from the different Tension, Strength, &c. of the nervous Fibrillae . . . It may perhaps be in part owing to the peculiar Disposition of the Nerves and animal Spirits, that some are very readily infected by the Plague, Small-Pox &c. and some never, although multitudes suffer around them.[30]

While some bodies resisted the effects of contagious matter, others provided fertile soil for invading particles to do their worst. Disease could only take hold in bodies *prepared* to receive infection. Henry Burdon argued that airborne particles "can only infect such Bodies as are fit to Receive that Kind of Impureness,"[31] while Swiss physician Johann Georg Zimmermann put it succinctly: "a certain concourse of causes, before-hand internally existing in the human body, is requisite to produce disease." This stood to reason, for why else in epidemics did only certain people fall ill even though infectious miasma pervaded everywhere? For Zimmermann, only constitutional differences present before an epidemic hit could explain the phenomenon.[32] Finally, Samuel Brown's 1797 dissertation on yellow fever displays the continued currency of these beliefs at the end of the century: "There are many directly and indirectly debilitating powers, which prepare the constitution for the reception of disease, these are called Remote and Predisposing Causes." Hippocrates himself could have predicted these "predisposing causes," which included food and drink, sleep, evacuations, and the passions.[33] The nonnaturals strike again.

Some factors were beyond one's control. Because age was said to affect one's fibers, children and the elderly had weaker constitutions than did hearty adults. Sex played a similar role; women had more delicate fibers than men. The sensitivity of women's systems, which were more easily affected by emotion, informed debates on gender politics and the need to cloister women from the rough and tumble public sphere, for their own good.[34] Sex also informed pathology in other ways. The enshrinement of sexual dimorphism in Enlightenment anatomy, which presented sexual difference as firmly fixed in Laqueur's famous two-sex model, had two principle influences on discourses of disease.[35] On the one hand, anatomical difference now focused on the structures of reproduction rather than on the fluid humoral tendencies of the Galenic body. Thus the womb, long pathologized,[36] played an even greater role in disease-causation, exemplified by the distinctly feminine disease hysteria[37] but also in the presentation of the womb as the site for the generation of syphilis.[38] On the other hand, the two-sex model was also part of the larger cultural movement to define women according to bourgeois domesticity by emphasizing their bodies' capacity for reproduction and lactation.[39] Treatises on women's diseases focus almost exclusively on reproductive health. For example, despite its general title, Jean Astruc's two-volume *Traité des maladies des femmes*, spent the entirety of volume one exploring "The Diseases of Women Caused by the Menses" and all of volume two on "The Diseases of Women Depending on the State of the Uterus."[40] For the female body, the only diseases that seemed to matter were those affecting its capacity to mother. Treatises on women's diseases thus performed the same cultural work as anatomical texts. When it came to gender, Enlightenment pathology formed one strand in an emerging science of sex that depicted women's bodies in the narrow terms of reproductivity.

While factors influencing health like age or sex were beyond an individual's control, other variables were not. Constitutions were hardly set in stone. Quite the opposite: the nonnaturals suggested, they were malleable. Issues of environment, emotions, and lifestyle could all mold the constitution, meaning that not only was the Enlightenment body pliable, but more importantly its health was a matter of moral responsibility.

Because bodies had to be predisposed for disease to take hold, pathology was wide open for moral claims about behaviors that made bodies vulnerable. This dynamic raises an important Enlightenment contradiction. Advancing statistical and rational schemas promoted universalist claims, yet the age also witnessed an enormous emphasis on individualism, famously championed by the liberal political economies of Locke and Smith. This tension was felt in pathology, as it was throughout Enlightenment thought, but the continued importance of the individual constitution—restraining universalist visions of diseases that affect all people uniformly—represents the triumph of bourgeois individualism in medicine. For just as the individual citizen had to balance his rights with responsibilities in the political sphere, so too did the medical citizen bear responsibility for his or her own health. Rational individuals made lifestyle choices that literally changed their physiology. Right living promised good health; poor living predisposed the body to disease. Invoking a great metaphor of the age, the clock, physician Robert Robertson emphasized the individual's responsibility for health. Disease was

> the natural consequence of health injured or ruined by the folly and indiscretion of man. As I impute a complete time-piece's being irregular, to the person's mismanagement who had the care of it; in like manner I consider bad health (generally) as the consequence of the sick's misconduct or negligence in preserving good health.[41]

Health was thus linked to righteousness. Doctors constantly invoked temperance, a concept understood as essentially the moderation of the six non-naturals. Temperance was crucial, because its opposite, intemperance (or, frequently, irregularity), weakened the fibers and rendered the body susceptible to disease. For example, among the "remote causes" of fever, Robertson included "intemperance" in eating and drinking, as well as "slothful indolence" and "immoderate venery."[42] John Huxham's list of behaviors predisposing bodies to disease demonstrates how a range of factors mixed in the construction of the Enlightenment's moral biology. Huxham blended moral actions with physical, environmental, and even emotional factors in a motley stew of disease causation, punctuated by an etcetera that allowed readers to fill in other moral failings that struck their imaginations:

> [Fever] most commonly attacks Persons of weak Nerves, a lax Habit of Body, and poor thin Blood; those who have suffered great Evacuations, a long Dejection of Spirits, immoderate Watchings, Studies, Fatigue and the like; and also those who have used much crude unwholesome Food, vapid impure Drinks, or have been confined long in damp, foul air; that have broken the Vigor of their Constitutions by Salivations, too frequent Purging, immoderate venery &c.[43]

In a word, vice crippled the constitution rendering it prone to disease.

Huxham's attention to "damp, foul air" returns us to the issue of the porous body and its environment. Enlightenment pathology attests to a profound focus on medical environmentalism,[44] with much attention given to the quality of air. "Many men's bodies are like a weather glass, subject to the least alterations of the air," wrote physician John Tylston in 1701, and few doctors disagreed.[45] Because it might become rife with infectious particles, vapors, or effluvia, air could become pathogenic. Indeed, treatises like John Arbuthnot's authoritative

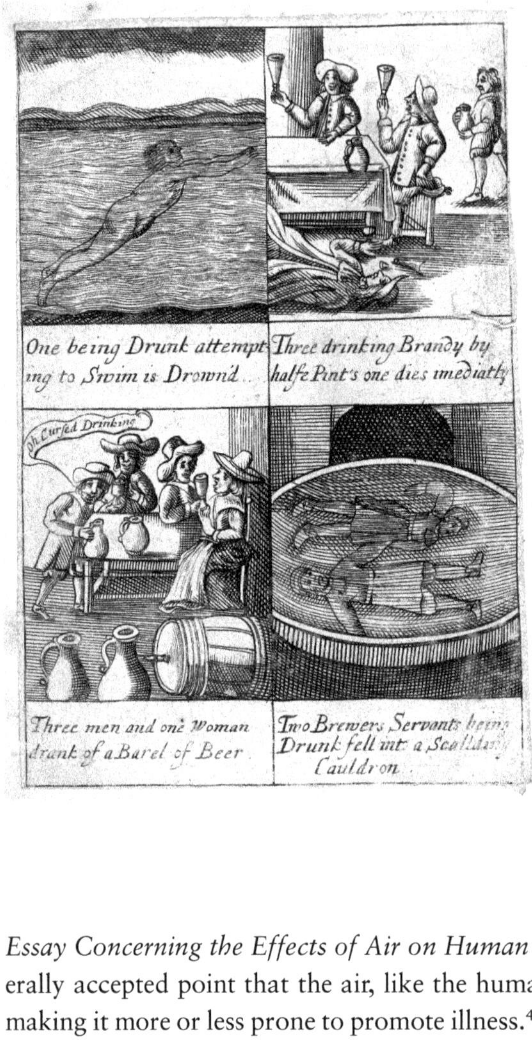

FIGURE 2.2: Health warnings to drunkards. In Samuel Ward et al., *A Warning-Piece to All Drunkards and Health-Drinkers* . . . (London: 1682). Rare Books, Wellcome Library, London.

Essay Concerning the Effects of Air on Human Bodies (1733) related the generally accepted point that the air, like the human body, had a "constitution," making it more or less prone to promote illness.[46] The air, especially if stagnant, heated, or contained in a small area, could become rife with disease-causing effluvia from the decay of organic matter, the breath of diseased people, or merely the respiration of too many people in a confined space.

RACE, COLONIALISM, AND THE ADAPTABLE BODY

Arbuthnot focused on European examples, but Europe's growing connection to colonies enhanced the importance of tropical medicine, a genre typified by James Lind's *An Essay on Diseases Incidental to Europeans in Hot Climates* (1768). While depictions of healthy paradises can be found, most medical presentations of tropical settings were cautionary tales of pathogenic lands where excessive heat and moisture filled the air with toxins.[47] Marshes, overly damp and prone to putrefaction, had long been considered sources of poisonous miasma.[48] Now, based on the same principles, much of the newly colonized

FIGURE 2.3: Daniel Massiah suffering from the glandular disease of Barbados (elephantiasis). In James Hendy, *A Treatise on the Glandular Disease of Barbadoes . . .* (London: 1784). Rare Books, Wellcome Library, London.

globe was pathologized. In a sweeping accusation, surgeon Dale Ingram blamed the continent of Africa for outbreaks of plague everywhere else. Winds blowing across the Dark Continent picked up "all the unwholesome effluvia of stagnating lakes, as well as those cadaverous animals and vegetables, which die and rot upon that tract of land" and transported this toxic stuff around the globe.[49] Writers warned that newcomers to the "torrid zone" were particularly susceptible to disease. Europeans' constitutions were poorly equipped for such hostile climates.[50]

Yet the colonial project could not stop. Thankfully, the pliable Enlightenment body could change. Not every European settler perished, and the fact that newcomers died most frequently signaled that the body might adapt. Because changes in the nonnaturals could alter the constitution, bodies could become acclimatized to the air, water, and food of new homes. Physicians believed that heat and moisture could slowly transform the consistency of the body's fibers, enabling it to stay healthy in the tropics. Newly introduced bodies could thrive if, like transplanted flora, special care was taken to acclimatize them.[51] Called "seasoning," acclimatization was core health advice for travelers.[52] Seasoning is notorious for its role in the slave trade, which saw a proto-industrial method develop to process the commodified bodies of Africans by introducing them to American climates gradually. The bodies of thus-seasoned Africans were thought to be more resistant to New World diseases and fetched higher prices at auction.[53]

Settlers were instructed to seek out favorable spots to fix their camps, avoiding low-lying lands and areas with stagnant waters. High grounds and coastal plains with air-cleansing winds offered Europeans the best chance of surviving the shock of transplantation, giving their constitutions time to transform. However, Europeans were not just instructed to sit on hilltops and wait for their bodies to change. If the Enlightenment worldview counted for anything it meant that human action could effect change. Besides instructing settlers where to live, doctors told settlers to govern their behavior to help gird themselves against disease. In such advice, the line between the physical and the moral was obscured entirely.

Indeed, because the tropics were so dangerous, personal behavior was all the more vital. As a result, doctors emphasized immorality as a primary cause of tropical illness.[54] John Bell, writing about soldiers in the West Indies, stressed that "irregular" soldiers, by whom he meant drinkers and the undisciplined, were the most susceptible to illness. Bell went so far as to exonerate the West Indian climate, underscoring the importance of morality. "The climate of Jamaica is not of so destructive a nature as has been represented," he argued, assuring readers that "irregularity" was "the principal predisposing cause of disease" in the Caribbean.[55] Bell targeted drinking, but all vices weakened the constitution and were potentially deadly in hot climates. Thomas Trotter railed against idleness, arguing that laziness relaxed the fibers and predisposed seamen's bodies to scurvy, noting the sailors' proverb that "the first scorbutics are skulkers."[56]

The success of the imperial project, then, rested on discipline. Enlightened individuals had to govern themselves. Only disciplined bodies could withstand the ravages of the harsh climates that Europe hoped to conquer. Naval surgeon John Atkins emphasized the importance of self-control when he lamented the effects of sailors' indiscretions in West Africa. Discussing the troops' sickness he complained that the men were "more at their own Wills and Disposals then . . . they should ever be trusted with; being ungovernable in their Actions and Appetites."[57] Health advice in tropical medicine treatises began to take on the tone of the morality play, as disciplined bodies escaped disease while immoral ones became corrupted. So it was that surgeon Colin Chisholm called temperance "the great prophylaxis" for sailors in the West Indies.[58]

Because Europeans reacted to New World diseases differently than did natives, tropical medical writers were also forced to confront the complex question of race, and here again we witness a powerful moral biology at work. There were two main camps in Enlightenment discussions of race.[59] For polygenists, multiple races descended from separate ancestors, likened to different species. They were a minority before 1800, outnumbered by monogenists like Samuel Stanhope Smith, who, often from a Christian perspective, believed that humans descended from the same ancestors. Monogenists explained bodily differences like skin color as the influences of the environmental forces that we have already encountered. What stands out about discussions of health within these early discourses of race is the extent to which both camps advanced similar arguments about the body's moral biology.

Aforementioned naval surgeon John Atkins was a polygenist. His descriptions of Africans' propensity for certain diseases contributed to the early genesis of hardened views on racial biology that would triumph in the nineteenth century.[60] Atkins analyzed the "sleepy distemper," to which African bodies seemed uniquely prone. Atkins deemphasized environmental factors and instead blamed Africans' "indolence," which, combined with their unhealthy diet, predisposed their constitutions to the disease. Moreover, Atkins went further and suggested that Africans had naturally weak brains arising from limited intellectual stimulation. From this, he likened Africans to children, and like children their bodies had naturally weaker fibers than adult Europeans.[61] Rather than focusing on the individual constitution, Atkins posited a larger group constitution arising from moral assumptions about Africans' culture. "Where Ignorance and Stupidity reign, therefore, and neither Sciences nor Mechanics are planted for exercising the Faculties, the Brain must grow weak, and such a State of Thoughtlessness and Inactivity dispose it for the reception of Serosities," Atkins argued, concluding that "the natural weakness of the Brain . . . [is] the principal Cause of this Distemper."[62]

Yet, monogenists like Smith also used health to construct non-Europeans as inferior and also sustained their positions with claims about culture. Smith attributed racial difference to the effects of two forces, climate and "the state of society." In fact, Smith considered the role of society so important as to trump the effects of climate. For him, civilization protected the body from the effects of the environment.[63] Uncivilized "savages" living in Africa or the Americas were exposed more directly to the effects of hot and cold, effects which were "degenerative," a term on which Smith and his contemporaries relied heavily.[64] When compared to the stance of polygenists that races constituted separate species, monogenism held possibilities for egalitarianism. Yet Smith clearly arranged races hierarchically, and health was central in his taxonomy. As people degenerated, their constitutions became weaker and more prone to disease. For example, he attributed the darkening of skin to the effects of bile augmented by heat. Because darker people had bile "shed through the whole mass of the body" they became "indolent" and subject to diseases.[65] Moreover, like Atkins, Smith believed that changes to the constitution—including disease—became permanent over time: for Smith a race's "figure, stature, complexion, features, diseases and even powers of the mind, become hereditary."[66] Transplanted Europeans would no doubt be affected by the same climate as their indigenous neighbors. However, like Buffon before him, Smith assured readers that Europeans' "manners" and "civilization" protected their bodies from ever degenerating to the state of natives.[67]

Smith's emphasis on the influence of social practices on health and disease also points toward a most remarkable example of Enlightenment moral biology. It is not surprising to find discussions of race or gender emphasizing corporeal difference with corresponding points for health. Of course, research has demonstrated that those categories were hardly separate and that, especially in science, Enlightenment race and gender were intimately linked.[68] But the fusion of the moral and the physical in medicine is perhaps even more demonstrable through an exploration of class, which, in pathology, was every bit as physiological as race or gender.

CLASS CONSTITUTIONS AND THE TRANSFORMATION OF PHYSIOLOGY

Smith himself demonstrates how the line between race and class was entirely unclear in the eighteenth century. Smith repeatedly used the example of poor and laboring people to make comparative points about so-called savages. Like natives, the poor were also more exposed to the effects of climate, more dependent on a poor diet, and less influenced by the effects of manners, art, and civilization than were their social betters. As a result, their bodies resembled those of natives. They, too, were more "swarthy," "course," and less "robust" than were well-off whites. They, too, were distinguished by their "thin habit" and their "uncouth features," and they were "ill-formed in their limbs."[69] Many of Smith's claims focused on outward appearances, but beauty and health implied one another in the eighteenth century (see Chapter 6 of this volume). It was standard belief that beauty reflected inner health and that sickness revealed itself on the body's surface.[70] Beauty was no less a symptom of health in the eighteenth century than was long life.[71] Indeed Smith's claims about the physical appearance of the poor assumed and often sat side-by-side with claims about their poor health and weak constitutions. Most tellingly, Smith commented on these factors in natives and workers interchangeably.

> The hardships of their [i.e., natives *and* workers] condition, that weaken and exhaust the principle of life—their scanty and meager fare, which wants the succulence and nourishment which give freshness and vigor

> to the constitution—the uncertainty of their provision, which sometimes leaves them to languish in want, and sometimes enables them to overstrain themselves by a surfeit—and their entire inattention to personal and domestic cleanliness, all have a prodigious effect to darken the complexion, to relax and emaciate the constitution.[72]

The physiological differences between rich and poor were as profound as the differences between whites and savages. So fundamental were the constitutional distinctions between classes that Smith conceded repeatedly that, while his monogenist position could not accept it, he understood why "some writers would have ranged them [i.e., the classes] in different species."[73]

Smith's assumptions about the working-class body were standard. In northern temperate climates, environmental forces were less intense; ironically this meant that disease hinged even more on morality. Samuel Brown assured Bostonians that diseases in their temperate city "must be more in consequence of idleness, neglect of cleanliness, internal and external, or some moral default, than of situation."[74] Whenever Enlightenment physicians emphasized morality in disease causation, class was paramount. Diseases like plague had long been linked to the poor,[75] and it was still commonly assumed that their weak constitutions predisposed them more than others to the effects of contagion. Commenting on the 1720 epidemic in Marseilles, Dale Ingram proclaimed that it was "well know[n] that the poor in general are the first to receive the original attacks of any contagious distemper, for a depravity of the blood, coarse diet, uncleanliness, &c. contribute not a little to such diseases."[76]

FIGURE 2.4: The poor, depicted as constitutionally and physically weak, with "uncouth features." Engraving by T. Sanders after John Collier, 1775. Iconographic Collections, Wellcome Library, London.

Environments populated by the poor were thus worrisome. It was widely held that concentrations of their unhealthy bodies fouled the air with effluvia and rendered it pathogenic. So profound was this belief that the poor had their very own disease, so-called jail fever. Eighteenth-century institutions were crowded, poorly ventilated, unhygienic by modern standards, and suffered badly when infectious diseases took hold. Witnessing the dreadful instances of institutional epidemics, doctors invented jail fever. This oft-discussed disease was said to be generated by the breath of many people in a confined (and therefore usually hot) space, filling the air with noxious effluvia that became more toxic as it became concentrated.

A number of issues stand out in discussions of jail fever. For starters, a moral biology was again at work in the explanation of why prisoners' bodies were particularly prone to generate the disease. Describing the kinds of people who might inhabit a jail, physician Daniel Peter Layard wrote:

> by the depravity of their minds; and vice rooted in their hearts; keeping them bound in the chains of wickedness, they are totally changed in constitution, as much as in principles; and both filth and disease are become as natural to them as cleanliness and health are to the virtuous.[77]

Thus the bodies that came to inhabit spaces like a prison were already toxic from their lifestyle, their physiology transformed by vice and filth before they set foot in a prison. The effluvia they exhaled were therefore more poisonous than those of a respectable citizen. John Mason Good's winning entry in a 1795 contest for the best essay on the question: "What are the Diseases most frequent in Workhouses, Poor-houses, and similar Institutions?" emphasized that the kinds of people in poorhouses were corrupted before they ever arrived. Good argued that institutions witnessed two classes of disease: those arising within institutions themselves and those "solely introduced by those who enter, in consequence of prior vice, misfortune, or uncleanliness."[78] The poor were renowned, he claimed, for such loathsome maladies as syphilis and "the itch," stressing that this was especially true of industrial workers in urban manufacturing districts.[79] He went on to specify the class-specific factors that predisposed bodies to disease:

> The poor are, in general, but little habituated to cleanliness; they are liable to a thousand accidents, and a thousand temptations, which every superior rank of life is free from; and they feel not, from want of education, the same happy exertion of delicacy, honor, and moral sentiment, which every where else is to be met with. It is not surprising, therefore, that such diseases as the above should be frequent in almost every prison, and every poor-house, in the kingdom.[80]

Moreover, the pathogenic effect of so many of poor bodies together was exponential. "Such a motly set of beings crouded together, breathing the same infectious air, which daily grows more pestilential . . . produce such an acrid, powerful and destructive change in it, that no wonder the very walls are infected with noxious vapours."[81]

It is vital to note, however, that this moral biology was not restricted to criminals. It was extended to the general plebeian class, for jail fever was not confined to jails. The same principles that generated disease in prisons were said to be at work in other environments populated by the working poor. The terms "jail fever," "hospital fever," and "workhouse fever" were interchangeable.[82] Beyond these, still other plebeian environments generated the disease. John Pringle described outbreaks of jail fever in army barracks; Lind warned of jail fever in naval ships; John Alderson discussed jail fever in cotton mills; and John Heysham described it in working-class neighborhoods.[83] The workhouse. The factory. The hospital. The neighborhood. Anywhere that poor bodies congregated—bodies fed on bad diets, languishing in filth, and steeped in debauchery—could produce this disease. The relationship between the poor and their environments was moreover reciprocal, resulting in a pathogenic vicious cycle: poor bodies fouled the air, the breathing of which made them sicker, further intensifying the toxicity of their respiration, rendering the air more poisonous, and so on. Not even the diligent laborer who shunned the "intemperence in drinking" and "excess venery" included among the predisposing causes of the disease was safe.[84] By merely inhabiting the same space as his filthy and immoral neighbors, his physiology took on the same pathogenic qualities; he, too, became toxic. Here the tension between individual constitutions and group constitutions is profound. Enlightenment pathology held out agency to the individual; good health was his as long as he heeded the advice of rational science and lived morally. Yet that individuality seems a luxury not afforded the poor, the nonwhite, or women, whose proclivities for health were beyond their control. The constitution was a central unifying concept in the Enlightenment medicine, but greater forces molded plebeians' constitutions for them, inscribing culture on their very fibers, one and all.

Furthermore, when they emerged from their environments, the poor posed a dire risk to elite bodies. Discussions of fevers that ravaged American cities in the 1790s consistently told of disease arising and intensifying in immigrant neighborhoods before striking middle-class bodies that, despite their superior constitutions, could not fend it off. Demonstrating nativist and class prejudices, which would intensify in coming decades, American physician E. H. Smith recounted the mixture of filth and debauchery in Irish slums that produced a fever that went on to grip Sheffield, Massachusetts.[85] Proper living among city elites could strengthen their constitutions, but only so much. Smith warned:

FIGURE 2.5: Rowdy overcrowded poverty. Eighteenth-century etching. Iconographic Collections, Wellcome Library, London.

> Expose the most temperate and cleanly native [i.e., native-born Anglo-American], to the full influence of all the causes predisposing to this fever, besides those which his birth, cleanliness, and temperance, would disarm, and his chance for continued health must have been small.[86]

Medical writers offered numerous strategies to combat poverty-driven disease, especially calls for hygiene and ventilation to combat filth and cleanse the air.[87] However, the sense of risk embodied in the Enlightenment moral biology of class also drove strategies to simply keep the infectious bodies of the poor at bay. Environments where the rich and poor came together, such as courtrooms, were rife with danger. The bodies of respectable men like judges had to be protected from the bodies of the poor facing trial. Thus doctors like Lind and Layard prescribed not just schemes to cleanse and ventilate courtrooms but even suggested architectures of separation. In addition to clogging their nostrils with lint dipped in vinegar to prevent inhaling prisoners' exhalations,[88] court officers were instructed to build special chairs for judges to guard them from infection, and to use windows for interrogation, keeping patrician and plebeian bodies safely on opposite sides.[89] By the end of the century, such screening was also used in workhouses where fear of contagion was answered by keeping the bodies of poor relief applicants and parish officers safely apart.[90] Moreover, doctors advised that the bodies of accused paupers be shut up tight when brought into courtrooms, encased in garments that closed tight at the neck, wrists, and ankles, trapping contagion within and mitigating the threat they posed to those who sat in judgment of them.[91]

Class-specific conceptions of disease sat aside class-specific conceptions of health. For starters, labor was constructed as healthy because it combated the vice of idleness, already seen to weaken the constitution. But as the eighteenth century wore on, health itself was increasingly defined according to the body's ability to work. If health, like disease, was merely a state of the body, it was difficult to prove its presence. When it came to nonelite bodies, more and more the ability to work was the key litmus test. Two kinds of discourse are instructive. First, social welfare debates around poor relief established ablebodiedness as a primary category in discussions of entitlement. Initially, the term "disabled" was constructed to describe soldiers whose injuries rendered them eligible for pensions.[92] In poor relief schemes the concept of disability extended beyond soldiers to the poor at large, whose eligibility for relief was similarly determined by an evaluation of their bodies' ability to labor. Poor relief agents had long complained about so-called sturdy beggars, whose lack of work stemmed from a failure of will rather than physical incapacity (see Chapter 9 in this volume). But the so-called workhouse test, which evaluated applicants' capacity to labor, became institutionalized in the eighteenth century, establishing the ability to work as an important definition of health in the earliest stages of industrialization.[93]

Medical treatises demonstrate this emerging definition. In something of a narrative trope, doctors' discussions of epidemics among working people held up the point at which people ceased to be able to work as a key marker that disease had gripped them.[94] Not only did the inability to work testify to disease, but the ability to return to work after illness signified health. This was especially common in case studies. Thomas Prior closed the treatment narrative of a tape weaver, reminding readers that although consumption had "so weakened the Poor man that he was not able to work; [he] now looks brisk and gets his bread comfortably." Emphasizing the economic contingencies implied by this working man's health, he added: "And to use his own words, he makes his Paws maintain his Jaws."[95] This dynamic is particularly apparent in discussions of soldiers' health. It was typical for physicians like John Clark to end treatment narratives of soldiers with the announcement that they returned to work, holding up labor as the ultimate proof of health. Clark punctuates one narrative after another with the same point—that the patient was finally again "fit for duty"—often leaving these as the very last words of the case.[96] Clark is revealing because, while he described the health of gunner mates, carpenters, and common seamen in these terms, he described officers' health differently, noting that "gentlemen" were simply "restored to usual health" or "soon recovered."[97]

A mixture of assumptions thus produced a distinctly classed body which, likened to its own species, had a unique physiological capacity for disease and its own litmus test for health. Perhaps even more than race or gender—both

FIGURE 2.6: Robust native inhabitants of the American Antilles. In Jean Baptiste du Terre, *Histoire Generale des Antilles Habitées par les François,* vol. 2 (Paris: 1667). Rare Books, Wellcome Library, London.

of which are still often conceptualized as physiologically fixed—the moral biology of something like class demonstrates the profound pliability of the Enlightenment body. Able to mobilize a range of factors, whether moral, physical, emotional, or environmental, disease theorists made of the body what they wanted. Certainly, one can find Enlightenment presentations of health distinct from the examples adduced here: noble savages in pristine environments said to be healthier than Europeans[98] or workers whose labor made them stronger than the luxurious rich who fell prey to their own diseases.[99] But, like exceptions that prove the rule, such cases only demonstrate further how commentators on health and disease inscribed shifting values onto the Enlightenment body, making it their primary vehicle for contributions to pivotal debates of the age. By applying the discourses of seemingly objective hard science, disease theorists welded cultural assumptions and categories of difference onto the body more thoroughly than ever before. In that way, theories on health and disease were central to a much larger Enlightenment project.

CHAPTER THREE

Sexual Knowledge

Panspermist Jokes, Reproductive Technologies, and Virgin Births

GEORGE ROUSSEAU

Freud, in one of his memorable pronouncements, decreed that all jokes have a subconscious purpose.[1] This usually reveals the speaker's deep-layer intention or ambivalence about the ostensible subject of the joke. Not all jokes, of course, deal with persons. Some extend to subjects: topics apart from persons, including historical figures, abstract topics, and now even virtual topics in our cyberspace reality. The joke about historical epochs when I was a graduate student in America in the 1960s was that earnest types became Victorianists, dripping sentimentalists became Romanticists, and rollicking, fun-loving types—like me—were drawn to the eighteenth century. It wasn't true—the Enlightenment had many faces to those who experienced it—but the joke persisted and students often selected their fields of specialty according to these edicts, lending further credence to the myth.

More recently, psychoanalyst Jacques Lacan has located the birth of human culture in the knowledge that heterosexual intercourse produces babies. Myths begin with this fundamental awareness, he claims, that is not exactly a joke but that functions as one.[2] Lacan's knowledge invites further enquiry; for if heterosexual intercourse leads to reproduction and necessarily involves the union of a sperm and egg, the ensuing joke is that culture itself can be undone without them, and this, presumably, is Lacan's point. But it isn't perfectly clear what the undoing of culture means. Further analysis, however, reveals that the issue is not preeminently biological: it is less a matter of one sperm and one egg—one customer to each—but rather that even when limited to *one* client per sperm and egg, the offspring is not created in, or reducible to, their exact image.[3] The joke then is the maddening paradox that one and one do *not* equal one: the *one and only* created wholly in the likeness of its maker. Instead, *one and one* seems to equal many *others,* when indeed, a benevolent nature, such as biology, ought to have decreed offspring in the likeness of its two makers.

John Hill's (1714–1775) panspermist jokes—the subject of this chapter—lie close to these linkages.[4] Hill's knowledge (if one can repeat Lacan's introductory trope about reproduction and the birth of culture) was that even the hierarchization of causes was incapable of separating out the body's role in reproduction from its (i.e., the body's) memorable part in politics, ideology, and technology. It is a complex sequence in need of unpacking; all the more so in that Hill succeeded a generation of projectors—groups such as the Scriblerians—who had also proposed extravagant schemes for generation.[5] In Hill's version, the technology equaled the body in importance: the joke began and ended with the technological—with his mad "wind-machines." His further joke, as we shall see, was the notion that Georgian reproductive life could continue *apart from* biological reproduction. Viewed three centuries later it seems a plausible fantasy: the stuff of posthumanism.[6] Today, more empirically inclined mindsets will inquire how life could have continued without the prerequisite heterosexual intercourse that produced babies. The slightest flicker reminds us that we have returned to Lacan's originating knowledge with this additional caveat: Lacan died in 1980, just before the arrival of posthumanism (as distinct from postmodernism). Had he lived on to turn posthumanist himself, he may have revised his *first classical principle* (bred in Freudian spirit about the birth of human culture in heterosexual reproduction) to *asexual reproduction*: that is, reproduction by the *solitary* self. Odd as it appears, this asexual principle is a predominant view of recent posthumanists from Harroway and Halberstam.[7] They do not mean, of course, *literal asexual* reproduction. Half jesters at heart, like Hill, their posthumanist quip is that it is capitalism, culture, professors, and institutions that shape *reproduction*, not babies.[8] Ridley Scott's landmark posthumanist horror film *Alien* (1979) actually claims, capturing the spirit of Lacan's birth of cultures, that "the film works to take us outside of the logic of 'the human', to imagine other (alien) systems of reproduction, other (alien) logics of identity," among which the asexual body is prime.[9]

CAPTURING THE REPRODUCTIVE LANDSCAPE

Hill's joke about asexual reproduction was perpetrated just as biology took a turn in dissociating female orgasm from generation. Thomas Laqueur has recounted this development so well that it requires no repetition here,[10] but it is worth citing his summary:

> Put briefly, the model of hierarchical difference, based on a set of differences between male and female reproductive systems . . . gave way to a model of complementary difference, which stressed the binary oppositions between the two physiologies. The hierarchical model that held sway from ancient times until the eighteenth century . . . interpreted the female body as merely an inferior and inverted version of the male body, all of the woman's reproductive orgasms simply underdeveloped homologues of male organs. . . . Such a view assumed, moreover, that female orgasm, just like male orgasm, was necessary for generation and that orgasm derived from pleasurable stimulation.

Laqueur traces the new

> reproductive biology stressing the opposition of male and female bodies, the woman's automatic reproductive cycle, and her lack of sexual feeling. . . . The new opposition of male and female turns into an opposition of desire and nondesire. Whereas it was thought normal for women to be ruled in all of their mental states by activities of their reproductive organs, it was also thought abnormal for them to have pleasurable sexual sensations.[11]

Sperm and egg remained necessary, but the female's role became so minimal that orgasm was not necessary for conception. This reduction did not *equal* asexual conception but was proximate enough to appear so in popular imagination. Women now copulated under false pretenses: thinking they were at least half the cause of reproduction. Before this biological revolution, they putatively derived pleasure, but in the new biology the effect of such forced copulation was reproduction *against* their wills. Only the male's volition counted: his sequence of arousal, ejaculation, and attachment of sperm to egg.[12] Set the chronological dials to 1750—when Hill published *Lucina*—and female orgasm plays no part in generation. This tide would turn in the mid-nineteenth century. For now, in the 1750s, Hill's joke about asexual reproduction was enabled because biological confusion was starting to flourish.[13] His panspermist joke captured the reproductive landscape, not dissimilarly from our posthumanist jokes: that is, that *cultural* birthing far exceeds the importance of biological reproduction in an overpopulated world, and that the institutions of cultural transformation are predominantly *asexual*. *Non*-posthumanists among us wonder how this can be. The joke is that agents of transformation aren't *quite* asexual: Western capitalism, culture, the professions, the makers of knowledge—all *commingle* to reproduce culture.

There was further slippage in Hill's joke: *Lucina sine concubitu*—"Lucina conceives without a man." The old Italian goddess who brings children into the world and whose Roman cult fed into Juno's larger one elicited thoroughly polarized responses: anger and offense versus uproarious laughter and hilarity. The collective furious response came from Fellows of the Royal Society who were outraged at Hill's pot shots and dismissed *Lucina* as a Dunciadic lampoon.[14] The laughing reaction interpreted Hill's satire as a stroke of wit about the reproductive body, assuming his joke to be that the product of heterosexual intercourse was misunderstood. Traditional wisdom held that the heterosexual body alone must be privileged for its role in human reproduction.[15] Now, in 1750, the upstart Hill came along, exploiting the new ambiguity diminishing the female body, and purporting that neither the heterosexual nor sodomitical male body was paramount,[16] but that the *asexual* body was. Two decades later John Hunter's artificial insemination was proof of Hill's visionary insight. Yet even without Hunter's technology, Hill intuited what was coming, saw what the Fellows of the Royal Society did not: that asexual reproduction lay outside the female womb, as it does today. In 1750 it was a wickedly naughty idea.

SIBLY, HUNTER, STERNE, AND THE MECHANICS OF HUMAN REPRODUCTION

Half a century later, just four years before the old century was over, when millenarian predictions ran high in 1796, that most curious of Enlightenment medical astrologers, Ebenezer Sibly, published a book as disturbing to many readers as Hill's had been in 1750. Sibly's *Medical Mirror* was weird: in form and content, and not least for its explicit illustrations.[17] Polite readers must have judged Sibly's treatise on "the impregnation of the human female"—its subtitle—more pernicious than many French revolutionary tracts inciting readers to riot. Sibly's 198-page book discussed the reproductive organs, especially the anatomical penis and uterus, engraved them, and explained the "evolution of the human genitals,"[18] following semen's pathway from its rise in the phallus to its arrival, through coition, in the vagina; discussed the moment of conception when sperm and egg met for the first time; included illustrations of a cherubic nude male showing his genitals; and waxed lyrical on the anatomical female vagina. Sibly also traced the role of fantastical influences: the moon, sun, and planets. He claimed that the sun ("solar tincture") increased seminal discharge, while the moon enlarged in women the "fertility of menstrual discharge,"[19] and he explained how important the genitals had been in human history.[20] Not least, he discussed Hunter's "anatomy of the uterus in impregnation,"[21] referring to Hunter's theories of reproduction but not his 1776 experiments in artificial insemination. Sibly appears not to have known about them.[22]

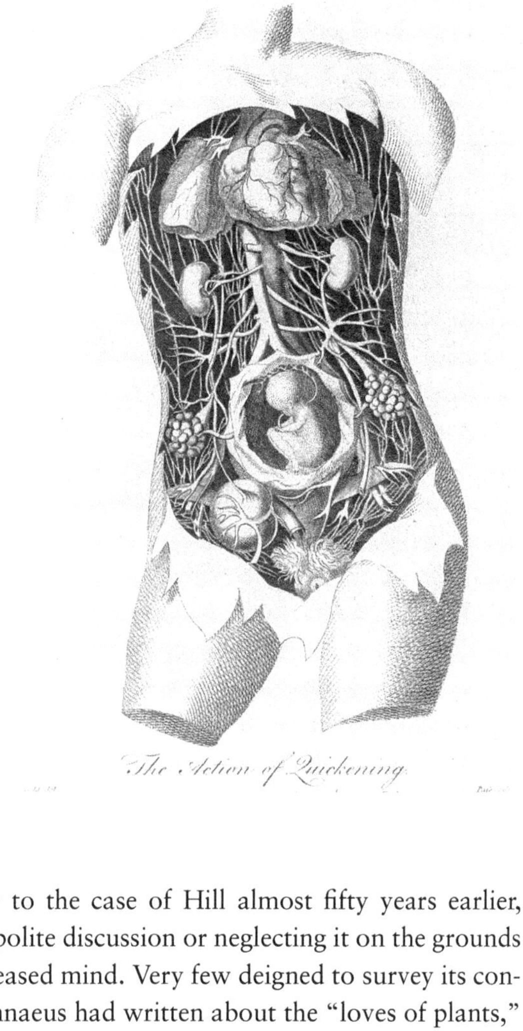

FIGURE 3.1: Pregnancy at about the fourth month when the mother first feels the baby's movements (quickening). In Ebenezer Sibley, *The Medical Mirror. Or Treatise on the Impregnation of the Human Female: Shewing the Origin of Diseases, and the Principles of Life and Death* (London: 1796). Rare Books, Wellcome Library, London.

Reviewers reacted similarly to the case of Hill almost fifty years earlier, either dismissing it as unfit for polite discussion or neglecting it on the grounds that it was the product of a diseased mind. Very few deigned to survey its contents. Earlier in the century, Linnaeus had written about the "loves of plants," and Erasmus Darwin, the polymath doctor of Lichfield, had versified them in *Zoonomia* (1786). But these were inoffensive sallies into merry-May garden blooms, not explicit analyses of labia, wombs, and vaginas. If readers of Sibly in 1796 (assuming there *were* readers) cried out in disapprobation, their attitude hardened Sibly in his view that esoteric astrological influences on fertility were justified. Disapproval and neglect legitimated his book's title, "the medical mirror": he merely reflected through a prism what he had seen inside the vaginal cavity. He flaunted his reproductive erudition, and although the *Medical Mirror* contained no notes or index, it enlisted earlier naturalists on these matters: Leeuwenhoek, Buffon, Maupertuis, Drake, Friend, Stahl, and Cullen. Sibly presents himself as the natural historian of the organs of sexual reproduction, tracing the "diseases Adam transmitted to posterity."[23]

The linkage of Hill and Sibly lies in their concealed jests: slippage occurs in both because their makers can touch a deep nerve in contemporary society without sacrificing their true intentions. They can be, by turns, ironic and sarcastic, but neither author ever surrenders belief in his theory. Yet neither was aware of the moment in eighteenth-century reproductive technology when real science confronted jest and slippage. This was John Hunter's (1728–1793) remarkable, even path-breaking, attempt at artificial insemination, and it was left to his bother-in-law, the controversial surgeon Sir Everard Home (1763–1832), to describe it for posterity. Home remained closely associated with Hunter after leaving Cambridge.[24] Home himself had cultivated reproductive oddities, and would soon turn his attention to "hermaphroditical dogs," "sodomitical cats," and "deformed kangaroos."[25] His exposition of Hunter's attempts at artificial insemination dismayed members of the medical profession. Even Hunter was recalcitrant to write about them in the climate of advancing reticence about so-called private parts and reproductive technologies. His biographers have adjudged his caution as justified, and his reputation remained untarnished.[26] Dr. Hunter, according to Home, apparently had been sought out by a husband whose wife was unable to conceive. Eager to help them, Hunter devised an infusion from the semen of a dog and injected her. *Mirabile dictu*, she conceived, but Hunter proceeded no further with his experiment in, his discovery of, artificial insemination. Home waited twenty years, until 1799, to write it up.[27] One can imagine what different treatment it would have received under Sibly's—or for that matter Hill's—ironic pen. Home, in contrast, limited himself to scientific experiments: he described the event and experimenter, explained the physics of interaction between (dog) sperm and female ovary in neutral terms, and limited himself to crumbs of speculation about the diseases that could arise from putrid male seed.[28]

Even so, and apart from the treatment the more colorful Sibly would have given it, the topic disturbed British readers in 1799. The public had other items on their mind, ranging from reports of revolution across the English Channel to instabilities at home. And even if they had not yet been introduced to Frankenstein—that other awesome marvel of reproductive technology—they had already been indoctrinated to other indelicacies of reproduction. Three decades earlier, in the 1760s, no less memorable a satirist than Laurence Sterne had opened his best-selling fiction, *Tristram Shandy*, on an identical theme: a description of the precise meeting of the sperm and egg at the moment of conception, having been conveyed through the genitals by the almighty animal spirits. However ironic Sterne's treatment of the animal spirits' passage through the penis and their attachment to the egg in the ovary, he made it plain from the first sentence of the book that this subject was one of great importance: "I wish either my father or my mother, or indeed both of them, as they were in duty both equally bound to it, had minded what they were about when they begot me."[29] It is unknown whether Hunter read Sterne's ribald fiction in which the novelist takes a stand on these reproductive controversies, but it is ultimately beside the point: all three accounts (Sibly, Hunter, Sterne) make plain that the mechanics of human reproduction—the moment of conception and physiology of genitals—were being widely discussed in the second half of the eighteenth century, even if sotto voce.[30]

The notion that human life could begin *asexually*—by a solitary person—or with the aid of machines and canine seminal injections, as in Hunter's experiments, violated every providential principle, as well as offended common sense. Yet miracle and progress permeated the air. Anything was possible in that era of perfection. Still, it was one thing to be put off by the prurient and pornographic—the idea that male and female private parts were now as available for public discourse as the weather—and another to uproot the fabric of Christian history with the nonsense that heterosexual intercourse was *unnecessary* for the continuation of the species. Hunter's experiments proved that the forlorn baker's wife could have children if she would just *forego* heterosexual intercourse. This undoing of the oldest act in history seemed to be put on trial as if it were just another uncertain hypothesis. Perhaps this is why Hunter knew better than to write up these experiments. The villain was technology, far more than *phalli* and *pudendi*, its slippage the weird notion that asexual conception was somehow more advanced, and progressive, than its heterosexual competitor. Immaculate conception was by then centuries old, laden with layers of theological significance. What changed now, in the second half of the eighteenth century, especially in Hill's naughty satire intended as revenge, was its unexpected reception.

ASEXUAL CONCEPTION: A DOCTRINE OF WINDS AND EMBRYOS

Hill's joke appeared in the form of a shilling book, a prose satire of fifty pages titled *Lucina sine concubitu*, subtitled "A Letter Humbly address'd to the Royal Society," signed "by Abraham Johnson," a nom de plume. Johnson and his alleged adversary Richard Roe, a name conjured for its alliterative ring, are invented for the purpose of puffery, self-advertisement in an era when it was becoming necessary for success. By 1753 Hill changed these to female names as he entered a cross-gendered phase of his writing career.[32] He paid the London printer Mary Cooper to publish *Lucina* in January 1750: ingenious Cooper had her own agenda contra authority and, even if she herself did not read Hill's lampoon, she basked in the promise of its bombshells that would make her famous. Lesser jokes complemented greater ones in *Lucina*'s appearance, and Hill as Johnson cheekily inserted two jibes in his postscript.[33] First, he announced that the Fellows of the Royal Society were diverted from replying to his "modest proposal" because London was threatened by three earthquakes on the eighth day of the first three months of 1750,[34] and, secondly, he invited "All Ladies willing to breed [i.e., asexually] to favour me with their Companies before the 20th of *April* next."[35] He described himself as a "midwife" and country practitioner of "physic," who had "stumbled" upon his discovery when attending the "sick daughter of a Gentleman," whom he delivered of child without her "having had commerce with a man."

The father confirmed his daughter's story. Startled, Hill claimed to have rummaged through natural philosophy and medical history looking for similar cases. While reading William Wollaston's (1659–1724) *Religion of Nature* (the edition of 1726), and specifically while immersed in the passages about traducianism (the entry of souls into bodies), Hill discovered the account of another young woman who had bred by inhaling "seeds" (*animalculae*) from the air that "served as a certain hot Bed for them to dilate and expand themselves."[36] Hill paused, he writes,[37] at the notion of this type of asexual breeding, wondering whether the seeds circulated through the air, or whether they were "lodged in the Loins of Males." Lucretius had explained how these animalculae were dispersed during the atomic creation of the world, but Wollaston omitted the Lucretian account. If in male loins, then how, Hill wondered, did the seeds proceed from loins into the wind that blew them about? A fantasist might allow his imagination to run riot at the myriad possibilities (onanism among them), but Hill looked elsewhere for the explanation. He remembered that Virgil had written in the *Georgics* that the west wind impregnated "certain Mares of his Acquaintance."[38] Persuaded that these "floating Embryos"[39] permeated the air, he set about in good Newtonian fashion—as a dedicated projector—to invent a machine capable of catching them: "a cylindrical, catoptrical, rotundo-concavo-convex Machine."[40]

A dedicated naturalist, the narrator further recounts how he reached for his microscope to discover the "little Men and Women inside the animalcules."[41] He spent a further year—he claims—repeating the experiments on the seeds, and eventually decided that the likeliest explanation was "a Doctrine of Winds and Embryos"[42]; wind, which in good Stoic fashion circulated all elements in the air over seven seas and continents, also scattered these "floating embryos." Still unsatisfied that he had found the solution, he located a sick chambermaid on whom to try his experiment: in solid Hunterian fashion, yet without injecting her, Hill made her inhale millions of "floating embryos" collected in his machine.[43] The chambermaid became pregnant, case proved, conception without a man.

Hill/Johnson ought to have desisted here but instead continued for another twenty pages[44] to the detriment of his sustained satiric joke, turning next to his hobby horses—Bishop Warburton, plagiarism, and venereal disease—without augmenting his case for the virtues of asexual, nonpleasurable, reproductive technology. He also worried that his readers would think he had "stolen the discovery" (the embryo-catching machines) from others. So he ransacked

FIGURE 3.2: Rural scene from the *Georgics*. A mare (top right) impregnated by the west wind. In Vergil, *Opera Vergiliana* (Paris: 1515). Wellcome Library, London.

ancient literature to determine whether immaculate conceptions had been known apart from Mary's birth. He found discussions in Galen and Hippocrates and Ovid's *Fasti*, when Juno eats cabbage as the impregnated Flora sits close by "ravaged by Zephyrus"[45]—the "fecund wind" that will impregnate her. It was this Ovidian passage that Milton had in mind when versifying the same idea in *L'Allegro*: "The frolic Wind that breathes the Spring,/Zephyr with Aurora playing," only to meet her once and "fill'd her with thee a Daughter fair,"/ So buxom, blithe, and debonair."[46]

Or was Hill thinking of another technological device, such as Antoine Réaumur's (1683–1757) chicken-breeding ovens that reproduced these avians without roosters? Much was made of these ovens in Hill's generation: a technology no less problematic than Hill's embryo-catching machines.[47] For the third penchant—venereal disease—Hill waxed lyrical and pedantic by turns. First he offers a concise history of this medical condition,[48] paradoxically followed by a descent into bathos in the name of sublimity: the claim that he has elevated his style to be able to describe the malady's ravage on the human body. And he concludes this section of digressions with a dig at the ever-increasing tribe of cross-dressers in his day, noting that not all those clothed as women actually *are*: "If all in female Shape (for I dare not call them all Women)."[49] This irrelevance has no place in his yarn about virgin birth. Yet within three years, by 1753, Hill altered his all-male repertory of nom de plumes to include females, adopting the name "Susannah Juliana Seymour" as his favorite. It was another stunt in the art of self-puffery.[50] And it may have fed further into the imaginative thrust of his joke about asexual parturition.

Only at the end does Hill clinch his case for the benefits of immaculate conception by querying whether "single-distill'd Infants" will be as healthy as those "who pass thro' the seminal Vessels of both Sexes in the old way of Generation."[51] This structural trick he may have learned from Swift in the *Modest Proposal* (1729), who also saved it for the end of his sarcastic satire. Always anatomically alert, Hill may have also been thinking, à la Laqueur, of the new theories about the diminished role of orgasm in conception. More likely, it is asexual generation he wants to quip about: the old-fashioned immaculate conception that has sustained him throughout *Lucina*. His conclusion is that asexually produced and technologically induced children would be vital: his new machine would produce "Offspring robust and healthy" as any generated in the old heterosexual way.[52] For this reason he "will apply for a Patent" for his machine to guarantee the trinity of "maidenly virtue," the abolition of matrimony, and the demise of sexually contagious diseases.[53]

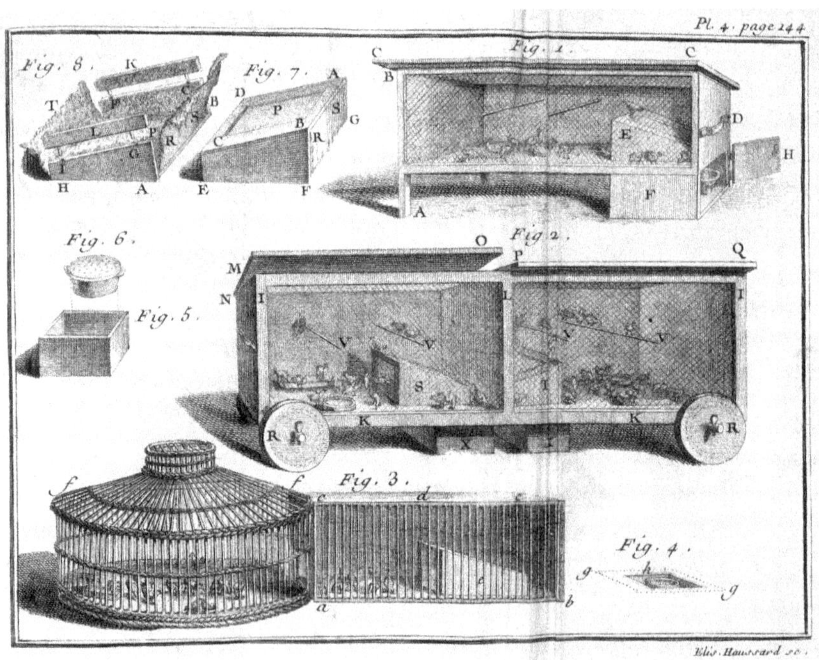

FIGURE 3.3: Chicken-breeding ovens. In René-Antoine Réaumur, *Pratique de l'art de Faire Eclorre et d'élever en toute Saison des Oiseaux Domestiques* . . . (Paris: 1751). Rare Books, Wellcome Library, London.

ANIMACULISTS, OVISTS, AND THE MYTHS OF PROCREATION

To call *Lucina*'s 1751 reception frosty is an understatement—it was icy. That powerful engine of sponsorship, the Royal Society, whose patronage Hill needed if he was to advance, turned up the temperature of its fury against its fair-haired boy. Dr. James Parsons (1705–1770), Fellow of the Royal Society and organizer of the famous Wednesday Circle, had himself been the author of a noted little treatise on "seeds."[54] A decade older than Hill, he had earlier mentored him, now replacing cordiality with venom. *Lucina*'s obsession with "seeds in the wind" could be interpreted as directed at Parsons, a personal affront. Parsons was imaginative, subtle, and never the run-of-the-mill Royal Society member: by 1750 he had composed an insightful treatise on the anatomy of hermaphrodites that interlinked anatomical and moral considerations[55]—no wonder Hill had latched on to him. But now the gossip increased, the rumors about Hill—his ambition, aversion to plain truth, outrageous pride—becoming rank. He was written out, never again to be elected to their ranks. Within months a paper war began as one Royal Society fellow after another attacked him, most often in letters dripping with scandal about this maniac who had been set loose against the cause of scientific advancement. The circumstances leading up to this warfare are too intricate to narrate here.[56] Suffice it to say that *Lucina* was the game board on which the dice of Hill's destiny were rolled. Even so, another context than *Lucina*'s reception needs to be developed: one that demonstrates how closely attuned to the lineage of Hunter and Sibly, Hill's vision of Lucina—the mysteriously fertile maiden—was.

The pivotal concept was male sperm and its place in the Enlightenment wars over reproduction. Without sperm, Hill's wind machines were dysfunctional and Hunter's artificial insemination was inexplicable. But first we need to understand how these battles over biological reproduction originated—Laqueur's landscape of model change—and what was at stake in the different positions; for without these heated debates *Lucina* could not have touched the nerve it did, leading to its literary cachet. Male sperm, or spermatozoa, was at issue.[57] Ficino, Vesalius, and the Renaissance anatomists had thought sperm resided in the blood, but Leeuwenhoek used his microscope to detect these tiny particles, which he christened "spermatozoa" in 1677. He believed they contained life-giving properties necessary for conception, and that the womb, with its ovum, merely acted as a nutrient base, a temporary home. Other Dutch anatomists (especially Regnier de Graaf) concurred, as would Spallanzani almost a century later, all adhering to a view that animalculae inside the sperm were the seeds of life (i.e., they flowed in the sperm), hence the appellation *animalculists* for those who privileged sperm in reproduction.[58] But it was not long before the female gonad was also explored as a rival site, as when Danish Niels Stensen

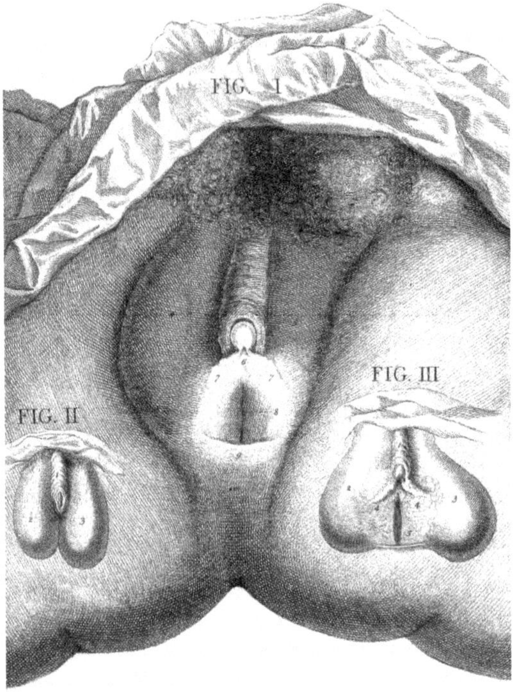

FIGURE 3.4: Hermaphrodite genitalia. In James Parsons, *A Mechanical and Critical Enquiry into the Nature of Hermaphrodites* (London: 1741). Rare Books, Wellcome Library, London.

conducted experiments and coined the word "ovary" to establish a rival camp of ovists or ovulists who attached equal importance to the female organ. Each group claimed to see a miniature organism present in the male or female genital organ and its fluids. Ultimately the microscope—in 1750 still primitive and not yet compound—would tell the truth about these rival positions.

Another context, philosophic rather than microscopic, awaited consultation: the vexed topic of preformation versus differentiation. Aristotle thought embryos developed by differentiation rather than having been preformed. In this sense, he was neither animalculist nor ovist in the eighteenth-century sense, but rather a differentiationist. William Harvey, who discovered the circulation of the blood, followed Aristotle, but by approximately 1720 a rival camp had developed—the preformationists—claiming that the preformed organism was already fully developed by the time the sperm met the egg. These preformationists could be either animalculists or ovists: they could think the fully preformed organism resided in the sperm or egg. The crucial element was opposition to *Aristotelian differentiation*, or slow development of the embryo, rather than anatomical location based on gender. Hence two concurrent debates evolved:

(a) the importance of sperm versus egg and (b) the existence or nonexistence of preformed organisms. No wonder that late historian of science Jacques Roger, who considered the matter in depth, judged it one of the most transformative moments in the history of generation, especially when viewed in the context of orgasm's role in heterosexual intercourse.[59] English microscopist and naturalist John Turberville Needham (1713–1781) was one of Roger's foci for his role in the debate: well-known to Hill and without whom *Lucina* may never have been written. Hill read Needham's treatise on generation,[60] they collaborated, and—most crucially—Hill became partial to the animalculist-preformationist position (the antithesis of epigenesis) on which he based his satire. Then, to cap this embryological welter, Hill tapped into another crest of the wave: the new, and often ridiculous, technologies of reproduction.

We have already commented on the significance of Leeuwenhoek's spermatozoa and the theories of his Dutch colleagues in the 1670s, but we have not noticed that earlier attempts than Hunter's in the 1770s also dreamed of artificially inseminating eggs.[61] As early as 1742 Ludwig Jacobi, an Italian anatomist, pondered the possibility without injecting female mammals.[62] A few decades later Spallanzani, professor of medicine in Pavia, injected toads with syringes, and again, in 1777–1780, at the time Hunter was inseminating the disappointed baker's wife, he syringed a spaniel bitch that gave birth to a litter of puppies.[63] *Lucina* was born in this welter. If its author's anxiety ever included plagiarism, it was pilfering from the contemporary ferment of reproductive knowledge rather than theft from particular writers. The Jacobis, Spallanzanis, and Hunters were earnest in their endeavor to discover an artificial insemination capable of providing offspring to those incapable of having them. Even when their experiments are less literally interpreted—as flirtations with fantasy in the laboratory—there still is no jest, or slippage, in Hill's jocular sense.[64] On the other hand, Hill intended to perpetrate a colossal prank on his Royal Society brethren who had failed, he thought, to appreciate his ability as a naturalist. In this slippage from serious attempts in the laboratory to half-baked embryo-catching machines, Hill's discovery amounted to a nervous joke: the view that asexual conception heralded the future, for reasons including the denial of sexual pleasure altogether. The further irony is that Hill, despite having discovered nothing new in the laboratory of science, and having allowed so much slippage to creep in to these ideas, turned out to be as prophetic as these figures: time has shown asexual conception to be the rival of artificial insemination in more ways than one, as contemporary commentators have commented in socioethical contexts.[65]

So far we have been silent about that *other* immaculate conception: the virgin birth of Mary's son, Jesus Christ. Mary's own immaculate conception occurred nine months before her birth (to St. Anne) and played no part in her later virgin birth of a son. Yet the two doctrines—theological and biological—were often confused in the early Christian world in light of their overlap. Still, God's implantation in her womb is germane. Despite posing problems for early churchmen, the birth of Christ to a virgin was ratified in 1483 by Pope Sextus IV and soon after sanctioned by the Council of Trent.[66] A rival alchemical tradition had also arisen, discussed by Paracelsus and Jacob Boehme, treating the Virgin Birth in very different terms, which Hill tapped into because he cites it in Roe's riposte to Johnson.[67] Hill's *Lucina* did not develop these diverse, incarnate strands, except parenthetically, but few well-read readers in the eighteenth century could have overlooked them. Biographically, Hill appears to have considered himself an Anglican without attending church, yet he tapped into this Christian context as easily as any non-Christian or non-believer. The significance for his panspermist joke is not specifically the theological confusion about Lucina's *virgin birth* but the *biological* context: that is, how it could be that reproduction occurred without the commerce of a man. So many tales about fantastical conception had arisen during Hill's formative years—not just peasant girls giving birth to rabbits, freaks, and other monsters, but the miraculous possibilities and exotic circumstances of each case—that it is unsurprising that Hill chose this specific target. In the milieu of confusion regarding generation in the 1750s anything seemed possible, even miraculous births.[68]

French cultural historian Pierre Darmon has demonstrated how the impotent were persecuted in eighteenth-century France—he called them "the damned innocent"—and elsewhere how they were scorned and marginalized.[69] These practices revealed "myths of procreation," as Darmon calls them; perceptions of fertility inherited by the *ancien regime* yet alien to our ears in an era of overpopulation: methods to propel developing nations to prosper and grow rich. Not yet Malthusians, nations suspected, and so did their ideologues, what untapped riches they had in their populations as work forces. Recently historian Lisa Cody has demonstrated that human conception was closely allied to the "birthing of nations," her work following on from Darmon's earlier studies.[70] Myths about biological conception and reproduction were deeply tied to both views—Darmon's and Cody's—but not to their views' mythic *inversion*: passing off biological virgin birth as if it *too would assist the nation.*

Hill's panspermist message to the women of the world insinuated the cheekiest of sequences: seek the wind, suck in the floating embryos from my new technology, forgo heterosexual intercourse as if you are a new Sapphic race self-sufficient on your own, desist from sexual pleasure, yet become pregnant, conceive, thrive, and enlarge the nation.[71] The turn of the screw is that the sequence pretended to act as *Christian* charity—especially the sacrifice of sexual pleasure. It was a sleight of hand: the dexterity of a deft satirical pen exploiting slippage in the joke.

FIGURE 3.5: Birth of the Virgin Mary. Line engraving by Robert van Audenaerd after Annibale Carracci, 1728. Iconographic Collections, Wellcome Library, London.

AFTERMATH: *LUCINA* AND THE SEEDS OF REVOLUTION

Lucina's afterlife was icy, as suggested already, but if frost alone were involved the post-1750 story would end here. Its immediate aftermath in England was predictable. Hill in the guise of Abraham Johnson had published *Lucina*, followed by a swift attack on his own work by the invented Richard Roe.[72] Roe and Johnson, Hill's hack puppets, are natural adversaries. This *double-ghosting* was a further facet of the panspermist joke: two authors were both sufficiently heated about the biological debates raging over generation to lash out and then hold their ground tenaciously. The rhetorical fracas was so charged the antagonists might have fought a duel. No one (not even a wife, as Hill had become widowed two years earlier, a young widower of thirty-six) urged Hill to enact it. He hit on the idea of the joke, perpetrated it, and found it redounding back to his discredit.

His intuition was tooth-sharp—he was also in the right place at the right time. Panspermism in both preformation and epigenesis swelled to major topics in Britain and France after 1750—inside and outside the Royal Society—partly owing to Hill's charades, subjects whose roots had been implanted earlier. Now the reawakened interest in fertility awaited exploitation by an opportunist-chancer-prankster. The town knew nothing. Even if it suspected Johnson to be Hill, it could not have imagined that Roe was *also Hill*—this disguise went too far. Furthermore, Roe attacked *Lucina* in the bravado of a single rhetorical display of chiasmus: not *Lucina sine concubitus* but *Concubitus sine Lucina*—"conception without fertility"—which Roe subtitled "pleasure without pain," emphasizing the idea that human life without the joy of sex was incomprehensible.[73]

Concubitus sine Lucina appeared within weeks of Johnson's original work, appearing to be a fresh reply, even composed in a different prose style. Roe's substantive arguments were technological—he argued that this machine could *not* capture the embryos in the way Johnson claimed; besides, even if it were capable of doing so, conception would not follow. Roe sounded an alarm about plagiarism: the theft of Réaumur's breeding chickens. It resonated in Hill, the notion that he had lifted the original design, which he had not. In the mid-1750s antagonists Roe and Johnson were hot properties: read and discussed, their antagonistic positions were debated and contested.

Who were they, readers asked? The Royal Society may have guessed Hill was Johnson, but few suspected he was also Roe. Then, as now, memory is short, and the temperature of debate lessened after 1751. To elevate it yet again, Hill brought out *another* edition of Johnson in 1752, again printed by Mary Cooper. After 1753, as a recently remarried father already living beyond his means, he turned to other projects. Here ended his part in the joke, and he could not have known what *Lucina*'s fate would be in Europe apart from possibly having rigged the first French translation in 1750.[74] The caveat about *possibility* must be made because evidence one way or the other has disappeared. Yet as Hill himself became increasingly well-known in the 1750s and 1760s, other publishers, such as Robert Dodsley, continued to reissue Johnson's work, often followed by Roe's reply. These were certainly not Hill's doings, even if he delighted in each appearance.[75] Nor did he play a part in the appearance of a German translation in 1751, as he then knew no one in Europe beyond the French members of Montesquieu's circle (especially his son Baptiste Secondat), befriended when they had visited Goodwood while Hill was in residence there.[76] The likelihood is that the French translators, as early as 1751, acted on their own volition, buoyed by the intensity of debates about generation across the Channel in Britain.

The aftermath would have shocked even the chancer-puffer Hill. This phase began a year after his death in 1775 (he died leaving his wife and children with debts). It was one thing for the French translation of 1751 to follow so swiftly on the heels of the English versions, especially if the jester-cum-prankster Hill had rigged its appearance, and another for a new translation to appear in 1776, long after the original rumpus fizzled and when Hill himself was dead. Controversy about generation had not, of course, diminished in that quarter century (1750–1775), as Jacques Roger, Joseph Needham, and other historians of science have shown.[77] Nor did Hill's death, so far as is known, play any part in spurring European translations: he may have been a household name in London by 1775, but he was not known abroad except among naturalists. The reasons for renewed interest are owing to the topic: conception and fertility, as the century wore on, became increasingly alluring topics. Furthermore, even in 1751 *Lucina* bore a different resonance in Catholic countries like France and Italy, where translations also followed. Controversies in France about potency and fertility were not merely related to Darmon's persecuted impotent who contributed nothing to the nation (hence "unproductive" in the double entendre that arose circa 1800) but extended to areas beyond the biological.

This transformation of a panspermist joke into something else is the facet Hill could not have predicted. While France remained in the clutches of the *ancien regime*, fascination with Hill's work was limited to realms biological. However, once the seeds of revolution—so different from the seeds blown about by Hill's benign zephyrs—took hold, *Lucina* assumed significance unthinkable earlier: an allegory of the new birthing to come that would catapult France and Europe. This is why the French translation of 1776 claimed that *Lucina* was drenched in political content, and the Italian translation of 1781 suggested that although the original *Lucina* may have been "una mera burla, un semplice scherzo" (a mere jest),[78] ultimately it was an allegory prognosticating Europe's future. Some French translators restricted their interpretations to the generative domain—for example, Jean-François Sacombe (1750–1822), a French hack who exploited the market for underbelly erotica disguised as reproductive biology—but most extracted the politico-ideological spheres from the biological, proof of Laqueur's view that the new model of female reproductive biology "solved ideological problems inherent in . . . social and political practices."[79] If these problems were not solved, they were tapped. When alive, Hill had produced moral spin every Tuesday and Friday in his "Inspector" column, printed in one of London's leading newspapers.[80] Sophisticated political allegory, of the kind promoted on the eve of revolution in 1789, lay outside his sphere.

A favorite tack of these translators was allusion to the tumultuous political events of 1637 in the Dauphine. It lent their translations a venerable antiquity, a nostalgia propelling them to glance back at a lost France. Here, in Grenoble's courts of justice, their Parliament created by Louis XIII, the birth of future political monstrosities were augured. The first anonymous translator of the

1751 French edition had included a postscript, perhaps imitative of Johnson's original, citing lists of the court's verdicts; after 1776, translators continued to include these verdicts, exploiting their ever-new relevance to a France being torn asunder by political difference, and by 1791 it was said that Hill's *Lucina* had prognosticated the Revolution itself.[81]

As the translations piled up, new prefaces replaced old, and editors claimed their versions were not merely translations but new works: original *Lucinas* or "Lusiniades," as Sacombe elegantly called his version of 1793, which he continued to reprint for a decade (1793–1803), once even in poetic form.[82] Such interest could not have developed apart from Hill's jesting reproductive technology: the original English "luciniade" of 1750 about a simple maid fertilized by the wind and now, after a half century, refashioned into a metaphor for history's most grotesque birthing: the Terror with its monstrous queen and her unnatural conceptions.[83]

This event would revolutionize the Western world. As the 1790s rolled on, editors of *Lucina* continued to note its political significance, and, truthfully or not, they documented which prior edition they had consulted and in which library. Some professed to stumble upon it accidentally in the catalog of a private library, claiming to be so intrigued by its contents that they petitioned to read it.[84] The translator of the much-cited 1786 French edition detected veiled allusions to Armand-Thomas Miroménil (1723–1796), Louis XVI's minister for whom a famous boulevard in Paris is named. His apparatus—preface to postscripts—speculates on the reproductive practices of the rich and famous of his time: Madame Pompadour, Madame Nacard,[85] a host of others.

An Italian translation, also claiming to be based on the 1786 French version, fashioned itself with a preface announcing that its "discoveries" were as important as those of "Martin Bebaimb, Christopher Columbus, and Americo Vespucci."[86] These French luciniades of the 1790s commingled eros and reproduction—Venus and Juno—with the more awesome political birthing then occurring as heads swung in the Bastille and the guillotines massacred innocents. Two generations later, another French version of 1865 claimed to be based on a translation of the 1790s edited by "St Colombe," a French author of gardening books who had become preoccupied with Hill-the-botanist.[87] Here, at least, was partial redemption.

Something in *Lucina*'s original germ sporadically haunted the Western imagination that could not be reduced to reproductive seed. Some substratum of the joke, as Freud may have argued if he could have known about Hill's bagatelle, touched the reader's unconscious, making it impossible to forget the gentleman's adolescent daughter who conceived without male sperm. The germ of Hill's idea enabled confusion and error but depended, more fundamentally, on primitive belief, miracle, and faith, on the one hand, and progressive reproductive technology on the other.

Perhaps this is why *Lucina* continued to be remembered. Still another French edition appeared in 1914 just before the events in Sarajevo plunged Europe into the conflict that became its first great war.[88] As the dark clouds gathered in that ominous Balkan landscape, one wonders whether these wartime readers construed the humble eighteenth-century maid's birthing as a metaphor foretelling the bombardment to come. After 1914 Hill's *Lucina* also took other protean shapes that restored it to its first biological impulses. Its resting place did not end in the trenches: a tour de force swelled into European conflagration.[89] Instead, its sweep over several centuries was driven by the integration of religion, reproduction, and sexual intercourse—or their lack. Over the century commencing in 1914 one can imagine Hill's reproductive technology extended to our frozen embryos and the ethical challenges they pose. It is never too late to reinvent *Lucina* one more time.

CHAPTER FOUR

Medical Knowledge

The Adventures of Mr. Machine, with Morals

JESSICA RISKIN

In 1677, Salomon Reisel, a Swabian doctor, philosophical mechanist, and personal physician to the Duke of Wirtemberg, announced that he had built an entire "artificial man" with all the internal bodily functions: circulation, respiration, and digestion. Reisel even planned to endow his creation with speech and the ability to move about on its own. In the *Journal des sçavans*, amid reports of a 300-year-old swan, a tree that grew small animals in place of leaves, a woman who vomited a cat, and a monstrous pig having thirteen feet, there appeared the following account:

> *Surprising Machine of the artificial Man of Reyselius, Esquire*. This wise *Physicien*, to demonstrate the circulation of the blood, composed a statue with such similarity & resemblance to man in all internal parts that, except for the operations of the rational soul, one sees in it all that happens in our bodies, & this by the principles of Physics-Hydrostatics. The Author hopes to perfect it to the point of giving it a voice & natural motion. This artificial man has its vessels & viscera in the same shape, structure & size as natural man, such that the water or whichever other liquid one likes, swallowed through the mouth, falling from there through a canal as through the esophagus, soon goes from the stomach to the right ventricle of the heart, after having passed through a marvelous artifice in the intestines, the Pancreas, the portal vein & all the other areas that

> we have explained in our XII Journal of this year. Nevertheless the same liqueur, descending by the emulgent vessels, & having been filtered by the artificial Kidneys of the Machine, falls into the bladder, whence it discharges itself. One notices in this Machine the natural movement of the lungs the attraction & expulsion of air, in a word all the movements of the pulse, & all the others that are natural in man.[1]

The appearance of this announcement in a philosophical journal (albeit alongside monstrous pigs and cat-vomiters) reveals that the philosophical conceit of the living body as a machine was giving a new context to old devices and, thence, to a new sort of device.

THE ENLIGHTENMENT BODY-MACHINE

The previous decade had seen the publication of Descartes's thought experiment in which he described living automata that were equivalent to human beings in every respect except that they lacked the capacity for rational thought.[2] The experiment supported Descartes's dramatic evacuation of mind from matter: the reason resided in an immaterial soul, he concluded, while all other human capacities were functions of the body-machine. Descartes had drawn on a long tradition of existing machinery for his models. Clockwork and hydraulic automata had been ubiquitous for several centuries in churches and cathedrals and as the playthings of the wealthy. In the wake of Descartes's thought experiment, automata caught the fancy of experimental philosophers, who, reinventing them, produced an entirely new category of machine: automata that reproduced as closely as possible, rather than merely depicting, aspects of their natural subjects. During the later seventeenth and eighteenth centuries, Descartes's philosophical conceit became an experimental program. Over the course of the Enlightenment, this venture would yield machines that (actually or apparently) ate, shat, bled, breathed, cavorted, walked, talked, swam, made music, drew, wrote, and played an almost unbeatable game of chess.[3] These machines constituted the experimental basis for a new and arresting figure, the Enlightenment man-machine: a personification of the hypothesis that human beings were, either completely or to some important degree, machines.

Although he cut a flamboyant, polemical figure, the Enlightenment man-machine also expressed a certain reactionary impulse: to tame the imperial, rational self unleashed by Descartes and his contemporaries, to bring it right back down to the ground. In one sense, to be sure, those who embraced and developed the man-machine idea in the eighteenth century continued what Descartes had begun: they carried the *bête-machine* to its logical extreme, encompassing even the human rational soul.[4] But in another at least as important and less explored sense, the Enlightenment authors of the man-machine reversed

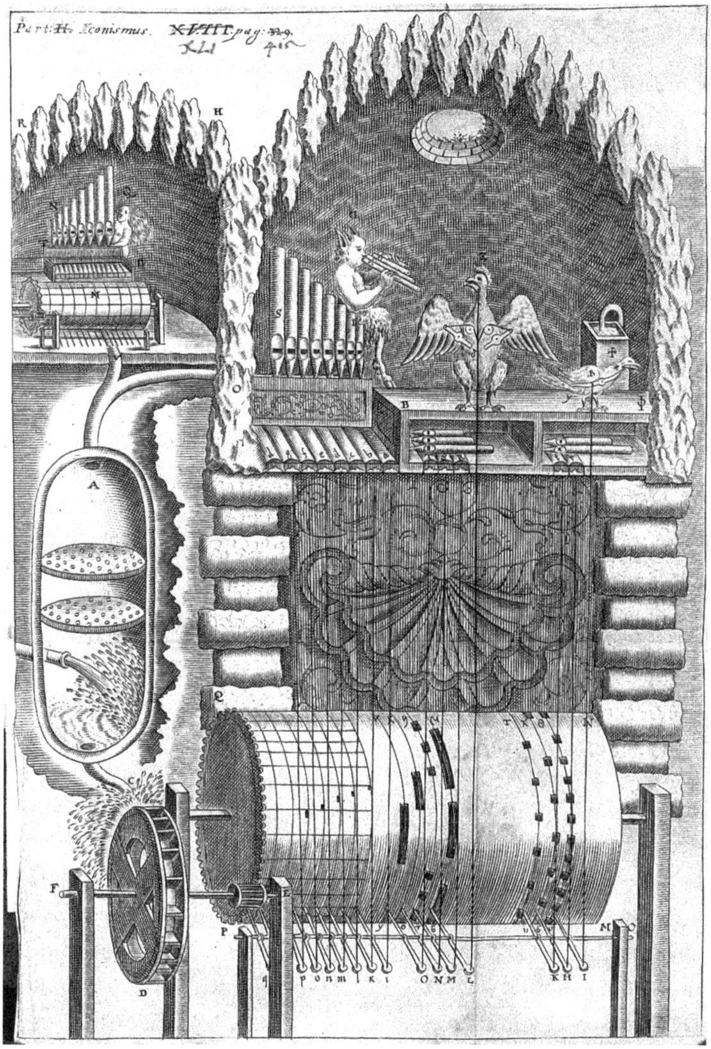

FIGURE 4.1: Mechanical toy in which a water wheel drives the music cylinder, which animates a bird puppet. In Gaspar Schott, *Mechanica Hydraulico-pneumatica* (Wurzburg, Germany: 1657). Rare Books. Wellcome Library, London.

Descartes, undoing Descartes's more dangerous idea: the stark removal of the human soul from the bodily world. Reacting against the disembodiment of the human self, Enlightenment materialists traveled to the opposite extreme. They snatched the soul from the heavens and drove it into the very earth, making of it a "*soul of mud.*"[5] Their man-machine was a rebuke to the rationalists, a denial of the infinite and omnipotent intellectual self that rationalists ascribed to humans and, above all, themselves.

Perhaps it will sound surprising to suggest that those who most urgently pressed the idea of the human-machine in the eighteenth century were driven

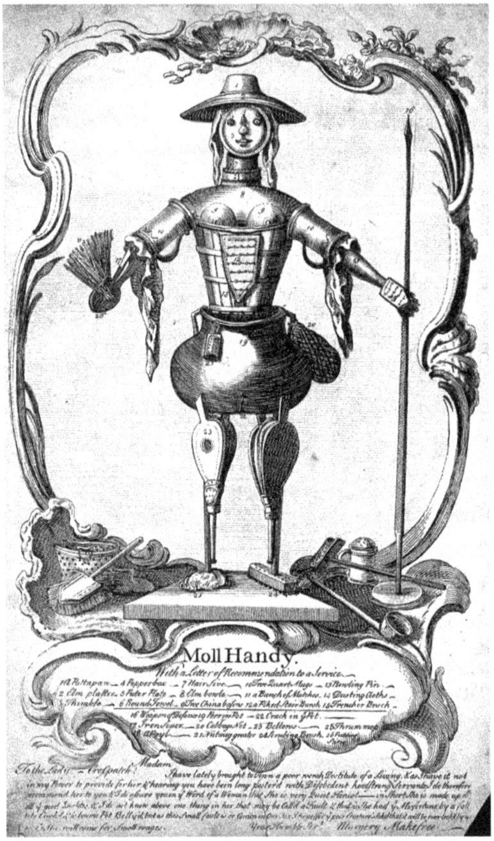

FIGURE 4.2: "Moll Handy." Advertisement for a woman-machine constructed from the implements that she uses to carry out household chores. Engraving. Iconographic Collections, Wellcome Library, London.

by a primarily moral purpose. These polemicists have appeared in historical writing not only as anticlerical, but as immoralist and anarchic hedonists who liked to celebrate in print every sexual possibility from masturbation to bestiality.[6] But neither anticlericalism nor lasciviousness, after all, actually constitutes moral anarchism, and, moreover, neither stance was specific to the materialists; deists and other theological moderates shared both. A different moral purpose did distinguish the materialist man-machine advocates, however, and that was antirationalism. They, more decisively than any of their contemporaries, were bent on humbling the Cartesian imperial self, leashing the unmoored "me," curtailing the authority of reason, and restoring to Nature her eternal mysteries. In this sense, partisans of the human-machine model in the Enlightenment were not only moralists but also proto-Romantics.

Here, then, is the perhaps counterintuitive argument of this chapter: the view of human beings encapsulated in the Enlightenment man-machine was more antirationalist than rationalist, proto-Romantic in its celebration of sensory and emotional experience and mystery, and also deeply moralized. A momentous development, moreover, rode onto the scene beneath this surprising creature's flashy cloak. Thrusting the arrogant human "I" into the material continuum of nature, the Enlightenment man-machine opened a new conversation about just where and how it fit in. Embarking on this conversation, the author of the original, definitive man-machine produced some of the first glimmers of a modern theory of species changing over time, but one that exhibited dramatic differences from its nineteenth-century evolutionist successors. Here was a radically materialist, overtly moralized, and starkly nonprogressive theory of species change. With its author's checkered tale, our story begins.

LA METTRIE'S FEVERED INVENTION

In the autumn of 1744, in the throes of the War of Austrian Succession, having entered into a secret alliance with Frederick the Great of Prussia, Louis XV laid siege to the city of Freiburg. During the siege, a young medical officer named Julien Offray de La Mettrie, who was attached to a regiment of the Gardes Françaises, contracted a raging fever. As Frederick the Great, soon to become La Mettrie's protector and patron, later told the story in his "Eulogy of La Mettrie" (1751), "[f]or a philosopher an illness is a school of physiology." Being of a philosophical bent, the delirious doctor seized the occasion to observe the effects of fever on mental function. "[H]e believed he could see clearly that thought is but a consequence of the organization of the machine, and that the disturbance of the springs has considerable influence on that part of us which the metaphysicians call the soul."[7] Frederick related, with or without irony I leave to the reader to judge, that La Mettrie had taken his intellectual direction from this early moment of delirium.[8]

Even in the pink of health, La Mettrie was apparently hotheaded. "Tumultuous and open-mouthed" is how Thomas Carlyle described him, with a "*minimum* quantity" of discretion.[9] Born and raised in the port city of Saint-Malo, Brittany, the son of a wealthy textile merchant, La Mettrie was an inveterate polemicist. During his convalescence and afterward, he pursued the idea that machinery was the basis of thought and "found only mechanism where others had supposed an essence superior to matter."[10] Persuaded that thought must be a bodily function, La Mettrie audaciously said so in his first philosophical work, *L'Histoire naturelle de l'âme* (1745). When the Paris Parlement condemned this book to be burned by the public executioner, La Mettrie prudently left for Leiden. There, he imprudently went back to work on developing his idea. Illness, after all, was not the only example of the influence of physical on mental states. La Mettrie compiled a long list that included the mind-altering power of opium, wine, coffee, sleep, pregnancy (with its "frightful schemes"), age, climate, weather, hunger ("The power of a meal!"), and of course lust, "that other frenzy of Man or Woman . . . hounded by continence and health."[11]

FIGURE 4.3: Julien Offray de La Mettrie. Stipple engraving by P. G. A. Beljambe after Claude Notté. Iconographic Collections, Wellcome Library, London.

Reciprocally, La Mettrie also considered the effects of mental states on the physical body. Why, for example, he mused, "does the sight or the mere idea of a beautiful woman cause such singular movements . . . ?" The response of "certain organs" to the simple thought of womanly beauty demonstrated beyond a doubt, La Mettrie reckoned, the intimate connection between the imagination and the muscles. The imagination had the capacity to excite a sequence of springs in the body, he supposed, and "how can this be, if not by the disorder and tumult of the blood and spirits that gallop with extraordinary promptitude and swell the hollow tubes?"[12] Seized by the idea of human machinery, La Mettrie ejected his most important work, *L'Homme-machine* (1747). No longer ill, he still had a feverish style, as even his champion acknowledged: "he wrote his *Man a Machine* or rather," amended Frederick, "he put on paper some vigorous thoughts about materialism, which he doubtless planned to rewrite."[13]

These vigorous thoughts included the announcement that soul was but "a vain word" signifying "that part that thinks," namely the brain. This organ, in turn, had "muscles for thinking as do the legs for walking."[14] With his customary mixture of serious and mischievous intentions, and to the great consternation of his beneficiary, La Mettrie dedicated his philosophical hot potato to the Swiss physiologist, poet, novelist, political theorist, and theologian Albrecht von Haller. The two were exact contemporaries and both had studied with Hermann Boerhaave, botanist, doctor, and mechanist philosopher at the University of Leiden.[15] But, in contrast with La Mettrie's ostentatious materialism, Haller was a Calvinist and a temperamental as well as a doctrinal moderate.[16] The two intellectual offspring struggled over the legacy of their carefully ambiguous father: Haller construed Boerhaave as a devout dualist, while La Mettrie made him an unflinching materialist.

At the time, Haller was poised to unleash a controversy by presenting the core idea of his physiology, first in lectures and then in print, which was an identification of two capacities of animal tissues. The first, irritability, was specific to muscle tissues: it was the capacity to contract in response to stimulus, and Haller viewed it as the basis of animal motion. The second capacity, sensibility, resided in the nerves.[17] Irritability and sensibility were versatile ideas: they lent themselves to exploitation both by vitalists, who saw evidence of a special, animating force in living matter, and by materialists, who saw support for the explanatory reach of matter alone.[18] Haller vehemently rejected both camps. Although he called himself a mechanist and saw physiology as the "description of the movements by which the animal machine is activated,"[19] he also believed in an immaterial soul and styled himself a scourge of atheists and materialists. La Mettrie, whose philosophy was a blithe combination of vitalism and materialism, invoked irritability and sensibility in the service of both at once; indeed, he anticipated Haller in his discussion of muscular irritability in *L'Homme-machine*.[20] In short, while Haller was fending off vitalists on one flank and materialists on the other, La Mettrie, with his dedication of *L'Homme-machine*, neatly planted a thorn in each.

The manifesto was too hot even for Holland, and it went the way of its elder sibling, to a ceremonial burning in the city square by the public hangman.[21] It survived this execution handily, however, provoking one of the defining controversies of the accelerating Enlightenment and becoming a fulcrum of philosophical debate.[22] The banished author, too, landed on his feet, at Frederick's court in Berlin. There he played a mixed role, officially the monarch's reader but equal parts gadfly and fool.[23] He affected a great familiarity with the emperor, not hesitating to "throw himself down and stretch out on the couches. When it was hot, he opened his collar, unbuttoned his vest, and threw his wig on the floor."[24] In his three short years at Frederick's Berlin court, La Mettrie was a whirlwind of activity. He elaborated his polemical philosophy in eight or ten more works and, at Frederick's behest and to the chagrin of many of its members, secured election to the Prussian Royal Academy of Sciences.[25] Meanwhile he cheerfully stoked the fires of controversy, joining in the fun incognito on his detractors' side as well. Samuel Christian Hollmann, a philosophy professor at Göttingen, gave La Mettrie an opening by suggesting that the author of *L'Homme-machine*, if he were correct in his claims, would himself be a machine and therefore not responsible for the gibberish he produced.[26] La Mettrie delightedly accepted the soubriquet "Mr. Machine" and dashed out an anonymous, self-satirizing pamphlet describing the life and, more saliently, the death of this personage.

Having decided that opium was the secret to a machine's happiness, Mr. Machine had met an untimely end by indulging in a good dose of rat poison. But one must not blame a creature equivalent to "the ducks of Mr. *Vaucanson* in *Paris*" (he referred to the celebrated automaton defecating duck designed and exhibited by the *mecanicien* Jacques Vaucanson during the late 1730s): "Remind yourself, if you would, that this is *Mr. Machine*. A machine does not act as it likes, but rather as it must."[27] This mischievous piece of mystification was poignantly foreshadowing: soon afterward, La Mettrie made an abrupt and suitably dramatic exit from history's stage. "Our crazy La Métrie," a jealous Voltaire wrote with ill-disguised satisfaction, "has just made up his mind to die . . . I cannot get over my astonishment."[28]

La Mettrie had gone to visit the Irish Jacobite, Lord Tyrconnel, who acted as Louis XV's ambassador in Berlin. Tyrconnel was unwell and had requested the presence of the monarch's reader-qua-fool to cheer him. Arriving just as Madame Tyrconnel was sitting down to eat, La Mettrie, according to Voltaire (in a letter to Richelieu, November 13, 1751), "eats and drinks, and talks, and laughs more than all the others; when he is full to the gills, they bring a pâté of eagle disguised as pheasant, sent from the North, and well mixed with bad lard, chopped pork and ginger. My man eats the whole pâté and dies the next day."[29] Contemporaries called it indigestion; historians have agreed on food poisoning. But if Voltaire's account is accurate, then poisoning, *tout court*, seems as good a name as any. "Voilà, my hero, one of our farces carried out," was Voltaire's punning verdict to Richelieu, a *farce* being both a farce and a stuffing.[30]

The manner of the materialist's death instantly became a test of his principles. "There is now a great dispute," reported Voltaire, "to know whether he died as a christian or as a doctor." La Mettrie had been a good atheist to the end, according to Voltaire (in his letter to Mme. Denis, November 14, 1751), begging to be buried in Tyrconnel's garden, but had been denied this final courtesy: "His body, swollen and big as a barrel, was carried, willy-nilly, into the catholic church, where it was astonished to find itself."[31] La Mettrie's death by pâté allowed his delighted enemies to equate materialism with gluttony.[32] On the other side, his supporters too were keen to attach their own moral to the parable. Frederick, in particular, scrupulously verified La Mettrie's steadfastness in extremis before undertaking to write his eulogy:

> The king inquired very exactingly about the manner of his death, whether he had gone through all the catholic rites, if he had had some sort of edification; finally he satisfied himself that the gourmand had died as a philosopher: *I am much relieved*, the king told us, *for the peace of his soul*; we started laughing, and he did too.[33]

Another chronicler of Frederick's court, the bookseller and writer Christoph Friedrich Nicolai, told the story of La Mettrie's fabled demise rather more sympathetically. According to Nicolai, Tyrconnel's chaplain, urged on by some enemies of La Mettrie who wanted to render him "contemptible" in Frederick's eyes, had "pushed into the sickroom." In this version of the story, too, the materialist moralist held fast and achieved, indeed, a form of heroism:

> La Mettrie would have nothing to do with this Priest and his talk; who, however, still sat and waited. La Mettrie, in a twinge of agony, cried out "JESUS MARIE!" "AH, VOUS VOILA ENFIN RETOURNE A CES NOMS CONSOLATEURS!" exclaimed the Irishman. To which La Mettrie answered (in polite language, to the effect), "Bother you!" and expired a few minutes later.[34]

ORGANIZED MATTER AND ITS MOTIVE PRINCIPLE

The stance to which La Mettrie apparently remained committed even to his death was not merely anti-Christian but also constituted a positive moral program. The core idea of this program was the inseparable implicitness of order, importantly including a moral order, and the individual human self to which it applied, within the material machinery of the world. An admirer of Vaucanson, La Mettrie flamboyantly described the body as "a clock," and a "machine that winds itself." He lingered over the "springs of the human Machine": the spring of the entire body backward in terror from the edge of a cliff; the blink of an eye at the threat of a blow; the expansions and contractions of the pupils, the pores of the skin, the heart, the lungs, and the sphincters of the bladder and rectum; and the heaving of the stomach when poisoned. He also considered admiringly how "the erector muscles raise up the Rod in man": "there is a singular spring in this member," La Mettrie marveled, whose study had been shamefully neglected even in the present age of enlightened anatomy.[35]

Despite this talk of winding and springs, the living machinery that La Mettrie described was importantly different from the inanimate kind. With the audience distracted by talk of clockwork and rods on springs, La Mettrie deftly inserted a special "principle," never found in any clock, and which made the whole thing go: "Grant me only that organized Matter is endowed with a motive principle," he coaxed, "eh! could one refuse that most incontestable

Observation?" This motive principle lent living matter not only movement but also "sentiment," a mix of sensation and feeling, which La Mettrie took to be essentially the same as movement.[36] This idea quickly and firmly took hold. Several decades later, Diderot affirmed in his *Eléments de physiologie* (1784) that animal machinery was of another sort altogether from artificial machinery: "the laws of motion of sensible, animated, organized, living bodies have not even been sketched," he maintained, and, above all, these relied on a property specific to animal machinery: "*sensibility.*"[37]

The man-machine was hence an erotic and passionate creature, coursing with sensations and emotions. "To be a machine," La Mettrie wrote, was "to feel, think, know how to distinguish good from evil like blue from yellow." Emotions, moral instincts, an aesthetic sense: the man-machine had all of these and also, by the same token and no less mysteriously, a sex life: "who would ever have divined a priori that a drop of the liqueur that shoots forth during coupling would make one feel divine pleasures?"[38] Even a rudimentary living machine could experience this last universal boon: La Mettrie extended the joys of sex right down to plants.[39] His libertinism was not amoral: on the contrary, it constituted a moral scheme of its own. "Natural Law" operated through the machinery as an "intimate feeling" for integrity, humanity, and virtue over their opposites. To treat others as one would want to be treated, La Mettrie claimed, was not a principle but a feeling built right into the machinery.[40] Diderot would later make a similar argument for another traditional virtue, industry: "idleness is always contrary to a living machine!"[41]

The greatest vice in La Mettrie's moral universe was therefore rational reflection: the doomed attempt to transcend one's bodily mechanism. La Mettrie's overriding project was to deflate the imperial self of Descartes and his fellow rationalists. Study, he therefore scolded, was "a catalepsy, or immobility of the Mind, so deliciously inebriated . . . that it seems detached by abstraction from its own body." Learning was an "abuse of our faculties." Philosophers trying to understand the world a priori using "the wings of the Mind" were doomed to failure. Worse, they were "Do-nothings" and "vain Pedants" whose "Balloon" brains were swollen with heaps of words and figures—which were, after all, only so much stuff, physical imprints on the "medullar canvas."[42]

The finiteness and materiality of the human-machine constituted the central moral truth in La Mettrie's philosophy: "Man in his first Principle is nothing but a Worm." Viewing the material world from within rather than above or beyond, the man-machine enjoyed the perspective of a "Mole." Therefore, La Mettrie urged, "Let us not lose ourselves in the infinite, we are not made to have the least idea of it; it is absolutely impossible for us to go back to the origins of things." With human knowledge necessarily limited and provisional, the worst pretenders were those "proud and vain beings" who claimed access to a larger, more transcendent kind of truth: rationalist philosophers and, of course, theologians. "However much they want to elevate themselves, [they] are at bottom nothing but Animals, and perpendicularly crawling Machines."[43]

While the negative component of La Mettrie's moral program was an extended rebuke to both rationalists and theologians, its positive element was a proto-Romantic celebration of feeling, emotion, experience, and mystery. Thought itself, the materialist moralist insisted, was a property of matter "like Electricity" or the "motive Faculty" and, as such, a "faculty of feeling." More-

FIGURE 4.4: The heart is depicted as a pumping machine, but the male figure nevertheless portrays the organ as the seat of human emotion. In Jacob Scheuchzer, *Physica Sacra* (Augsburg, Germany: 1731–1733). Rare Books. Wellcome Library, London.

over, neither the mechanism of thought (i.e., feeling) nor any other mechanism could ever be fully explained. The essential nature of matter and motion themselves, along with all that was not immediately visible, was ultimately "an impenetrable mystery." La Mettrie therefore ostentatiously reconciled himself to the "incomprehensible Marvels of Nature." To embrace the ultimate ignorance of an essentially material creature—a worm, a mole, a perpendicularly crawling machine—was to live a good, just, and happy life:

> What do we know of our destiny, any more than of our origin? Let us submit ourselves therefore to an invincible ignorance upon which our happiness depends. Whoever thinks this way will be wise, just, and tranquil about his fate, and consequently happy. He will await death neither fearing nor desiring It.

Such a creature would cherish life, be "full of respect for Nature; full of gratitude, attachment and tenderness." He would be grateful to partake in the "charming Spectacle of the Universe." He would "pity the vicious without hating them; in his eyes they will be but deformed Men."[44] A materialist would regard all human failings with the tolerant comprehension of a "Physician": "Do you know why I still make something of Men? It is because I seriously believe them to be *Machines*. Under the contrary hypothesis, I know few whose society I would value. Materialism is the antidote to Misanthropy."[45] To realize that humans were nothing but more or less imperfect machines, Diderot later affirmed, was to embrace the following credo: "There is only one virtue, justice; one duty, to be happy; one corollary, not to overrate life, and not to fear death."[46]

INGREDIENTS FOR A MODERN THEORY OF EVOLUTION

Through the pumplike lungs of the passionate, sensitive, moral man-machine breathed some of the first approximations of a modern theory of evolution.[47] If man was a worm and a mole, La Mettrie observed, then no sharp discontinuity separated humans from animals. Indeed, the great ape "resembles us so strongly" that it seemed perfectly possible one could teach it a language.[48] Denis Diderot, whose philosophy imbibed much of the tone and substance of *L'Homme-machine*,[49] made a similar point in relation to the orangutan at the Jardin du Roi in Paris: "Cardinal Polignac said to him one day, 'speak, and I will baptise you.' "[50]

Even more significantly, neither human—nor animal—machines were rationally designed. Flamboyantly pressing his atheist cause, La Mettrie rejected the argument from design also because of its celebration of the power of reason.

FIGURE 4.5: Orangutan, long considered to bear a disturbingly close resemblance to man. In Nicolaus Tulp, *Observationum medicarum libri tres . . .* (Amsterdam: 1641). Rare Books. Wellcome Library, London.

He ridiculed the "tedious repetitions of zealous Writers" with their endless "verbiage" and great "volume of proofs." The eye might well work like a telescope, but that did not mean someone had constructed it to do so. "Nature no more thought of making an eye to see than water to serve as a mirror for the simple Shepherd." Water just so happened to reflect images, as other substances happened to reflect sound, and likewise the eye "sees only because it happens to be organized and placed as it is." Eyes and ears required "no greater artifice" than the "the fabrication of an echo."[51] Thus the optics of vision, that most beloved example to authors of arguments from design, which would later shake the resolve of Darwin himself,[52] was to La Mettrie a bit of happenstance. Against the power of human reason to discern a machinelike order in nature, La Mettrie invoked the principle of human ignorance, which he deemed unanswerable: ultimately, "[w]e do not know anything about Nature."[53]

Scorning a rational account of living machinery, La Mettrie adopted a different approach. Human machinery had not been designed. It was the result of some other kind of process. Perhaps it had been mechanical, something like the gradual buffing of a stone: human beings were animals in whom the "raw Diamond" of the mind had been "polished" by language and culture.

Or maybe, instead, it was an organic development: the human brain was a "fertile ground perfectly seeded," yielding a hundred-fold what it received. Or, again, the working of the brain's imaginative faculty might have brought about "the generation" of mind. However it had happened, La Mettrie was sure no artifice had been involved. There was no absurdity in imagining that "an intelligent Being could come from a blind Cause," just as it required no genius on the part of parents to produce intelligent children. As a woman's womb, "from one drop of liqueur, makes a child," so a part of the machinery of human beings had simply turned out to be suited to retaining and producing ideas. "Having made, without seeing, eyes that see, [Nature] made, without thinking, a machine that thinks."[54]

Acting as a "blind Cause," Nature must have produced people and animals only "little by little," La Mettrie surmised, from the smallest and humblest beginnings. Matter would have had to pass through an "infinity of combinations" before arriving at the one that produced "a perfect Animal." Strikingly, La Mettrie anticipated a core principle of Darwinian natural selection in imagining that less perfect animals would die before reproducing, while more perfect animals would survive for longer:

> [t]he first Generations must have been very imperfect. Here the Esophagus would have been missing; there the Stomach, the Vulva, the Intestines etc. It is evident that the only Animals that could have lived, survived and perpetuated their species, would have been those finding themselves equipped with all the necessary Pieces for generation, and in which, in a word, no essential part would be missing. Reciprocally those who would have been deprived of some absolutely necessary part would be dead either soon after their birth, or at least without having reproduced. Perfection has no more been the work of a day for Nature than it is for Art.[55]

In the thought that humanness might not be constructed but rather buffed or grown or generated or otherwise arrived at over indefinitely many imperfect generations, La Mettrie found an alternative to both "Chance" and "God," namely "Nature."[56]

Here were two principal ingredients of the evolutionary theories that were to emerge over the next century: the idea that humans might be the result of a gradual process and the possibility that nature could be orderly without being designed. Both arose through La Mettrie's energetic attempts to describe a kind of machinery that was not rationally designed. In place of design or structure, accordingly, La Mettrie substituted the notion of "organization." In so doing, he picked up an Aristotelian idea with an interesting twist at its core: an "organized" body, as Aristotle had described it, was one with the potential for life (the greater the degree of organization, the higher the form of life), and one in which all the parts were arranged just as they would have been "if they had come to be for an end."[57] Organization was specific to living beings and was the quality of being arranged *as if* for a purpose: here was a kind of design that, in keeping with Aristotle's craftsman-less cosmology, invoked no demiurge designer.

The Aristotelian notion of organization, a kind of immanent purposiveness specific to living creatures, figured in some early eighteenth-century theories of life and mind.[58] Boerhaave, La Mettrie's mentor, defined an organic or organized body as one "consisting of different parts, which jointly concur to the exercise of the same function." Here again organization was an internal affair, a concurrence of parts. Boerhaave also described organization as something that became evident in the process of growth.[59] Leibniz, likewise, referred to the organization of "organic animated bodies" as a kind of unfolding. This unfolding took place across as well as within generations, since an organized being, Leibniz thought, could not intelligibly arise without having existed in simpler forms since the beginning of time.[60]

Overall, however, organization made only occasional, scattered appearances before the late 1740s,[61] and received its first extended treatment in La Mettrie's attention-grabbing manifesto. Here, organization became something new under the sun: the basis for a rigorously materialist, nonrational moral scheme. Organization in La Mettrie's usage retained its original aspects of specificity to living creatures and intrinsic purposiveness. From these ingredients, the materialist moralist derived a perfect melding of physical and moral purposes. "Organization," La Mettrie wrote, "is the first merit of Man" and the source of all the others. Rhetorically, he demanded, does organization "suffice for everything?" His answer was an emphatic "yes." Every faculty that had been attributed to mind or soul came down in the end to "Organization itself; *voilà* a well-enlightened Machine!"[62]

ORGANIC MACHINES AND SECRET MECHANICS

In the wake of La Mettrie's bestseller, and even following his controversial end, "organization" in La Mettrie's sense of the word proliferated rapidly in natural history. The concept immediately permeated another landmark bestseller, this one by the director of the Jardin du Roi in Paris, Georges Buffon. The opening volumes of his *Histoire naturelle, générale et particulière* were published in 1749, two years after La Mettrie's *L'Homme-machine*. Here, Buffon identified Nature's chief business as the production of life by "organization." Nature in general, he wrote, "seems to me to tend much more toward life than death, she seems to try to organize bodies as much as possible."[63]

Like La Mettrie, Buffon emphasized that organized bodies worked differently from artificially built contraptions. The "true springs" of animal motion, he reckoned, were not the visible muscles, veins, arteries, and nerves, but rather the "interior forces" that clearly resided in organized bodies. These forces did not follow the "laws of gross mechanics . . . to which we would like to reduce everything." After all, why must the very few properties of matter admitted by Descartes and other seventeenth-century mechanists—extension, impenetrability, movement, shape, divisibility—be the only ones? In his theory of nutrition, Buffon invoked a new sort of "active power": the "penetrating" tendency of living, organic matter to organize itself. Such forces would remain forever essentially mysterious since they acted on the interior of matter. Buffon promised, "We will never reach them by reasoning."[64]

Buffon's theory of the human constitution as an organic rather than a designed machine, like La Mettrie's, thus constituted a rebuke to rationalists, both for believing they could explain everything and, relatedly, for setting humans apart from the rest of the natural world. "The first truth that arises from this sort of serious examination of Nature is a truth perhaps humiliating for man," Buffon warned: "it is that he must place himself among the class of animals."[65]

Another person who quickly and influentially took up the concept of "organization" was a friend and protector of La Mettrie's, the French mathematician and philosopher Pierre-Louis Moreau de Maupertuis. Like La Mettrie, Maupertuis was a native of Saint-Malo and a protégé of Frederick the Great, for whom the monarch had several years earlier secured the presidency of the Prussian Royal Academy of Sciences. It was Maupertuis, indeed, who had devised the rescue plan of bringing La Mettrie to Berlin in 1748 after the publication of *L'Homme-machine*.[66] Organized bodies, Maupertuis announced in 1754, in agreement with both Buffon and La Mettrie, were inexplicable by means of the mechanical principles of the inanimate world. A "uniform and blind attraction spread through all the parts of matter," such as gravity, could never explain how the elements of a living creature came together: how, for example, some parts formed an eye while others made up an ear. To account for such a thing, Maupertuis argued, the very elements of matter must contain "some principle of intelligence . . . something similar to what we call *desire*, *aversion*, *memory*." The world of living beings could never have come together randomly out of brute and unintelligent parts; nor must God have composed it out of such parts the way an architect builds with stones. Rather, "the elements themselves, endowed with intelligence, arranged and united themselves to carry out the vision of the Creator."[67] Creatures were not made of blocks or stones but a confluence of sentient participants.

By the mid-1750s, the concept of organization was everywhere, first in France but soon proliferating in the writings of Swiss, German, and English naturalists. It was possible, for example, in a treatise on reproduction in plants and animals, to refer to the mammalian fetus as a "little organization."[68]

"Of all the modifications of nature," mused the Genevan Calvinist Charles Bonnet in 1763, "the most excellent is *organization.*"[69] Here was a principle to account for the workings of "animal machines," which, though they contained wondrous arrays of "levers, counterweights, diversely calibrated tubes, curves and bypasses," could never be explained by traditional mechanical principles.[70] Organized bodies, or "organic machines," instead contained "a secret mechanics" of their own, which allowed them to proliferate, grow, and heal.[71]

Bonnet described organized bodies both as "*looms*" that assimilated and incorporated materials into themselves and also as "cloths, networks, sorts of fabrics in which the warp itself forms the woof." Each fiber, each "fibrilla" of these loom/fabrics was itself a miniature machine, and the "entire machine is in a sense nothing but a repetition of all the *machinelets (machinules)* whose forces conspire to the same general end."[72] Extending indefinitely outward to encompass the world itself and indefinitely inward into the microscopic workings of organic bodies, organization was all-pervasive: "We do not know where organization finishes, what is its smallest term."[73] Perhaps it would be an exaggeration "to *organize* or to *animalize* everything," Bonnet allowed, but he nevertheless urged his readers not to assume that what appeared unorganized or inanimate was really so. There was "no good philosophical reason," he reckoned, to "limit the scale of animality to this or that production": the world might well be fully permeated "with life and sentiment."[74]

A sentient cosmos thrumming with feeling traveled thus arm-in-arm with Mr. Machine. This view of living machinery and its oneness with material nature was, once again, deeply moralized. Morality itself, that peculiar feature of human beings, Bonnet wrote, arose from humans' particular "organization"[75] and likewise for animal societies: beavers and "republican bees," for example, were "organized" to live in society and could not function in solitude.[76] In Diderot's view, the diaphragm was the part of the machinery responsible for moral sentiments: "the head makes men wise, the diaphragm makes them compassionate, and moral. These are the two great springs of the human machine."[77] To defy one's material organization was the greatest folly of all, and the greatest moral imperative was never to deny one's machinery or its place in the material continuum of nature. Thus, for example, there could be no purported virtue "as childish, as ridiculous, as absurd, as harmful, as contemptible" as chastity and continence.[78]

This materialist moralist impulse and the accompanying notion of "organization" found expression in the work of a growing class of people who renounced the doctrine of special creation for the conviction that species changed over time: what the anatomist and anthropologist Paul Broca would retrospectively dub "transformism."[79] (I use Broca's term here in order to avoid reading aspects of later, evolutionary theories back into these early ideas about species change.) Jean-Baptiste Lamarck, the person whom Broca, and subsequently tradition, designated as the original transformist, scrutinized the many varieties and gradations in the ongoing composition of the "animal organization" to assert that reason itself was no peculiar capacity of humans but a function of the nervous system and therefore common in lesser or greater degrees to all animals.[80]

But, as we have already seen, Lamarck was by no means the first *transformiste-avant-la-lettre.*[81] Already in 1747, La Mettrie's man-machine had been emerging "little by little" from the "smallest beginnings," and had inspired a host of transforming organic machines. Diderot reckoned there was no difference between a canary and a bird-organ, or a philosopher and a clavichord, apart from organization and degree of sensitivity.[82] Creatures were essentially keenly sensitive machines. And how silly, he observed, to suppose that animals "were originally what they are at present": "[w]e no more know what they have been than what they will become."[83] Diderot conjured a vision in which race upon race of animals came into and passed out of existence in unending succession. Our own present moment represented but an instant "in the succession of these animal generations."[84]

Diderot's eternally transforming sensitive clavichords constituted a strikingly nonprogressive view of species-change. Here is a critical way in which his and La Mettrie's transformism differed from later evolutionary ideas. "The imperceptible worm that wriggles in the mire" might be on its way to becoming a great and fearsome beast, but likewise, today's enormous and terrifying animals were likely turning into worms.[85] Diderot's vision of natural history was also startlingly indeterminist. Remote and indifferent, the sun was the cause of all. Extinguish it and everything would perish; relight it and the resulting "infinity of new generations" might never include our own plants and animals.[86] Let "the current race of existing animals pass; let the great, inert sediment act several million centuries," and there was no telling what sort of beings might result.[87] Above all, humans represented but a brief and haphazard moment, and no kind of culmination. Indeed, we might not even be the protagonists of our own moment. Being made up of infinite "animacules," themselves in constant flux, we might be but the "breeding-ground of a second generation of beings, separated from this one by an inconceivable interval of centuries and successive developments?"[88]

Another early transformist, Erasmus Darwin, represents the culmination of the trajectory that this chapter has been tracing from man-machine toward human evolution via the moral rejection of transcendent reason. Intellectual as well as biological grandfather to Charles Darwin, participant in the human-machine experimental program (he designed a machine to simulate articulate speech),[89] Erasmus Darwin, like La Mettrie, was a rhapsodist of the sex lives of plants, an enthusiastic believer in the coextension of sensation and sentiment with living matter and the material oneness of all living things.[90] Like Lamarck, he understood animals as ever-burgeoning forms of the same basic "organization" and drew the same implication too, though with greater drama.

"Go, proud reasoner," Darwin charged, "and call the worm thy sister!"[91]

CHAPTER FIVE

Popular Beliefs about the Dead Body

RUTH RICHARDSON

The cultural history of the dead body in the Enlightenment, as in other eras, offers an illuminating counterpoint to that of the living one. This chapter looks at the manner in which attitudes toward the dead body reveal themselves in this era in a number of sometimes overlapping contexts: in funerary culture, in judicial punishment, and in medicine. Toward its close, the chapter addresses the ways in which memorial artifacts surviving from the long eighteenth century indicate significant underlying changes in imaginative perceptions of the dead body, and perhaps of death itself. The main focus here is the British Isles, but much of what is said probably also applies loosely to the many places to which British culture and customs were transplanted in this period by emigration.

THE DEAD BODY IN FAMILY AND LOCAL CULTURE

The transition from life to death was much more familiar in every previous era than it is to us today. Of course we see deaths all the time on television and in film, but they are not *real* deaths: many of them are violent, make-believe, acted out, photographed, seen at a distance, or otherwise mediated. Death itself, especially natural death, and the dead human body in particular, is rarely witnessed in flesh and blood. Responsibility for the dead in our time has become institutionalized via health care systems and undertaking and to a great extent is nowadays sequestered away from the homes of the living. In the era of the Enlightenment, things were very different. People died at home, and because death usually took place in domestic settings, knowledge about it was widely shared. Deaths could be swift in disease, accident, childbirth, or epidemic. Or they could be slow, and often painful. There was much less help available for bringing recovery or relief from pain. Ambulances not yet having been invented, accidents were not rushed away to emergency rooms but were treated nearby; inquests were held in local rooms, often in an inn or public house. Death was not only more commonplace; it was more commonly witnessed and dealt with firsthand.

After the Great Plague of 1665, Britain never again suffered epidemic mortality on so extensive a scale (see Chapter 1). There were certainly scares. Defoe's *Journal of the Plague Year* was written during such a time in 1722,[1] but as the eighteenth century progressed, population growth and increasing expectation of life accompanied a developing sense of modernity. Although commerce and social change affected the form of funerary observance to an extent, the death customs and beliefs that were recorded in the Georgian era by early anthropologists (folklorists) attest to the wide distribution of long-standing beliefs and observances among the indigenous population of Britain.[2]

The deathbed was a favorite theme in moralizing tales—poetic, literary, and dramatic—and in print culture generally. The importance of words exchanged at the deathbed was emphasized in fictional and poetic deathbed scenes of the period. Even those about to die in public by hanging were permitted to speak a few words at the gallows. The fact that their "Last Dying Speeches" were commonly reported in news coverage of public executions reinforces this estimate of the importance of last words.[3]

The moments before death were considered a time of heightened importance: words spoken then had greater significance than normal speech: they were words to be remembered, repeated, and acted upon. The living made promises, and the dying made death-bed confessions or denials of guilt, revealed secrets, or expressed last dying wishes that were held to be binding on the living in some profoundly significant way. A dying curse was considered especially fearful, a malediction on the brink of eternity.

Corpse care in the interval between death and burial perhaps illustrates the central cultural importance of the dead body most clearly. In normal deaths in the community, the dead person's body would be laid out soon after death had been confirmed, that is, before rigor mortis had set in and while the body was still supple. Laying out was generally undertaken by female family members or servants, or a local handywoman (often also a midwife) would be called into the house to do these last offices.[4] The task was to wash the corpse, to plug its orifices, to dress it in its grave-clothes, and to compose its limbs and body (eyes closed, chin up) in such a way as to allow the dead person to appear decent when seen by visitors. Laying out was important for at least two reasons:

to make the body clean and decent for the afterlife and to allow it to be seen to be so prior to disposal. People would travel miles to fulfill the duty of making a last farewell to the dead. Viewing the dead was a society-wide custom that did not decline in Britain until the mid-twentieth century and that is still observed in many places. The customary farewell visit often involved touching and/or kissing the dead person and often involved the taking of ceremonial food or drink in the company of the body. Like laying out, these customs also had spiritual connotations. In many localities, drinks (usually wine) and special biscuits (sometimes proffered in traditional receptacles, such as willow baskets) were provided for visitors.[5] To be decently dressed at death was considered highly important, the favored color of the clothing or covering generally being an unsullied white. By the eighteenth century, the use of the old burial covering of the winding sheet (which had historically served instead of the coffin) had largely given way to the visible shroud as a separate garment, worn within the coffin, like night-clothes in bed. Although clean nightshirts or daywear and underclothes might be used as coffin clothes (especially perhaps where death was unexpected) many individuals commonly gave thought to preparation for death, spiritually and physically, and grave-clothes were often ready long in advance.

In the condemned prisoners' tumbrel on her way to be hanged at London's Tyburn in 1714, "Betty the Cook" passed a pawnbroker's in High Holborn where she had previously pawned the best white smock she had kept especially for use as her own shroud. She had perhaps even sewn it herself. Now she needed it for her launch into eternity at the gallows. Upon the pawnbroker refusing to return it to her, she "in very great passion swore she would plague him for it after she was hang'd."[6] That such an encounter was recorded in detail suggests that the altercation shocked or disgusted witnesses. Betty the Cook may have seemed to have more concern for her clothing than for her mortal soul, but the pawnbroker's materialism demonstrated a hard-hearted love of lucre on the brink of the grave: this was an exemplary rejection of a customary deathbed wish. Betty's fury was tantamount to a deathbed curse, and her threat of haunting exposes the means by which such maledictions were applied and reinforced.[7] Wearing their best clothes, often white, or a purpose-made shroud at the gallows was an aspiration for so many condemned prisoners that it can be said to have been general.[8] So even if those witnessing the altercation in Holborn lacked sympathy for Betty as a criminal, every observer would have understood the poor woman's last desire for a clean white shroud.

Immediately after the Great Plague, Parliament had passed the Act for Burial in Woolen (1666), designed to promote the wool industry, under which all burial clothing was to be of wool, on pain of a hefty fine. It was unpopular legislation and was so regularly circumvented that it had to be reinforced on several occasions before it was finally repealed in 1814. In response to the act, alongside (or in addition to) the dead person's own clothing there developed a new kind of burial wear: commercially manufactured, dedicated burial garments. Henri Misson, a French observer in London in the 1690s recorded: "To make these dresses [burial garments] is a particular Trade, and there are many that sell nothing else; so that these habits for the dead are always to be had ready made, of what Size or Price you please, for People of every Age and Sex."[9] Unless a family was willing to pay the fine for using linen or silk, white flannel became the fabric of choice.[10]

Recent archaeological work on the clearance of church crypts, especially that at Spitalfields (a residential district just to the east of the City of London), supports Misson's observations. Only the wealthy were buried in church vaults, so we should not read too much into the surviving evidence, yet excavations suggest that affluent eighteenth-century burials featured a mix of traditional handmade clothing, both day and nightwear, and perhaps dedicated garments, as well as commercially mass-produced undertakers' display clothing for the dead.[11] Upmarket versions of these ready-made garments were sold in what came to be called "suits" or "sets." The frilling and crimping of the prettily pierced starched fabric that surrounded the face and decorated the area of the body's upper chest was matched by similar frilling for the cuffs, and pierced repeating patterns reappeared on the coffin pillow and on wide fabric ribbons and sheets fitted inside the coffin around and along the entire length of the corpse, disguising the starkness of the bare wood and often enveloping the lower limbs. Such embellishments were not brilliantly made but probably looked fine enough starched and trim for the required week or so before burial. These important archaeological discoveries were of fabrics that had been buried for several centuries, so they were not seen at their best, but in their heyday they would have given the appearance of the lace worn to such effect by men and women of the upper classes in the Civil War era and after the Restoration, perhaps most resembling the patterns of what became known as *broderie anglaise,* though the cut edges of the patterned piercing of burial clothing were not oversewn to prevent fraying.[12] The closest contemporary garments would have been christening clothes and cradle coverings, reinforcing perhaps the affinities between birth and death discussed by Lisa Cody in Chapter 1.[13] The nearest modern equivalent to these frilled, pierced, and pinked garments is another form of substitute lace: the paper doily, associated with the display of cakes. Undertakers' wares were made for a form of temporary display, too.

An engraving of an upper class viewing taking place in 1714 shows that coffin sets had not yet become universal at that date, as the body can be seen surrounded by the bare wood of the coffin,[14] but less than twenty years later, in William Hogarth's *The Harlot's Progress* (1731), the dead woman's coffin is shown with a frill along its inner edge. This detail in Hogarth's image is significant, in that it shows the use of what looks like a coffin set in very poor domestic circumstances. The coffin frill suggests a number of possible inferences, the key one being that this is not a pauper's funeral, which would have been very plain: literally no frills. Whether the Harlot had perhaps saved up for her own funeral, or had purchased a coffin-set during better times, whether funds had perhaps been begged from a male client (in the accompanying anonymous verse, one of the men present was described as "A blade who did the funeral supply"), or whether the other unfortunate women attending the funeral had collected among themselves, remains unclear. That women of the street had a camaraderie that emerged at times of death is evident from their custom of burying one another in white, and of raising funds for one another's burial. There is every reason to believe that the group of East London women who collected money for the funeral of Polly Chapman, one of their number, exactly a century later represented a tradition of self-help probably widespread among the very poor from time immemorial.[15]

A great many customs and observances associated with the dead body—like viewing the body, touching the body, and drinking in the corpse's company—suggest that the dead person was still thought of as part of the community and that respect, care, and concern was due to the dead until burial, and beyond. The community of grief that focused around the coffin was captured in a sketch

FIGURE 5.1: Moll Hackabout in her coffin, surrounded by mourners. Engraving by William Hogarth. Iconographic Collections, Wellcome Library, London.

by Rowlandson.[16] Some customs, such as watching (keeping company with the body between death and burial) suggest that the dead body was thought of, perhaps superstitiously, as possessing some sort of awareness, or sentience, for an unspecified period after death. As we have seen, the provision of padded pillows and mattresses inside coffins supports this notion. There was of course room for doubt about the safe diagnosis of death.[17] Even today, the pronouncement of death can be uncertain. A cartoon by Wigstead from the 1780s shows a man sitting up in his coffin, shocking himself and his wife, who had been reading by candlelight while watching his corpse.

Serious metaphysical doubt also surrounded the fate of the soul. Whether the soul left the body at the time of death or remained associated with it to await the resurrection was a moot theological point, which rendered the interval between death and burial an extremely anxious time for those concerned for the dead, the soul, and for survivors. Customs such as opening windows and covering mirrors were said to owe their observance to the belief that the soul should be allowed to escape, thereby preventing haunting.[18] The prodigious number of gothic novels published during the eighteenth century suggests there was considerable scope for fearful concern associated with ghost lore.[19]

A funeral was a local festival—a somber one, but a festival nonetheless. For a significant person, or for a local disaster, entire localities would stop daily activities to walk in procession behind the dead, or stand by as the funeral passed, in an act of witnessing and communal commiseration. The body and its fate were the central focus, and those attending in procession were arrayed by their proximity of relationship to the dead person. This is still the case in present-day funerals, but we have forgotten the customary primary status of the dead that lies behind the observance and that still serves to explain why only the undertaker precedes the coffin, which for others was believed to be a position of bad luck. The location of burials, whether in churchyards, vaults, or sites of sentiment, was regarded as significant, even to eccentrics.[20]

In the seventeenth century, if not before, uncovering the causes of death was a serious matter. Public inquisitions, or inquests, served to investigate suspicious deaths, but, after the Great Plague of 1665, the new role of searcher was instituted as a public health measure, to verify that otherwise unsuspicious deaths were not due to plague, so as to isolate infection when it arose. Searchers were mature women with knowledge of the causes of death, presumably because they had experience with laying out the dead. From their observations were created the published sheets of statistics known as the *Bills of Mortality*.[21]

As the institution of searchers and *Bills of Mortality* suggests, curiosity involving the causes of death was not confined to the medical professions. The case of John Pennant, a young man who died in London in the 1630s, suggests that families could be enlightened about postmortem investigation and yet remain traditionally attached to the integrity of the body. Within hours of Pennant's death, his family, mystified by its cause, requested a full postmortem examination. The operation was supervised by a physician, Dr. Edward May, and undertaken by a surgeon, Jacob Heyden, who discovered a worm or serpent coiled up in the left ventricle of John Pennant's heart. The doctors, and family members present, investigated its "carnouse" or meaty solidity with a large pin or bodkin. We know about this case because the serpent itself became a contested specimen, which Dr. May strongly believed should have been retained, but which the young man's mother wanted buried with the body.

It was the mother's ability to resist his acquisitiveness that so riled Dr. May, who had expected to preserve the serpent as a museum specimen. Hydatid worms in the heart were a recognized cause of death for another century or more, but belief in monstrous bifurcated serpents seems to have faded as an increasingly critical spirit among medical investigators began to differentiate between genuine causes of death and simple empirical correlations, a viewpoint rooted in the careful examination and analysis both of medical museum specimens and the classification of larger series of postmortem findings.[22]

THE DISMEMBERED BODY

The solicitude toward the dead body manifest in commonly observed funerary customs was overtly and very deliberately flouted in public punishments: a deliberate transgression of funerary norms in the punishment of malefactors that is highly revealing of societal expectations. The bodies of those executed for treason were publicly maltreated: dismembered, exhibited like meat in very public places, and left to decay in public, raised aloft on spikes for all to witness. Murderers' corpses were sent to the Colleges of Physicians and Surgeons, for dismemberment by dissection. King Henry VIII originally granted four murderers' corpses a year for this purpose in 1540, and numbers had grown as medical demand had risen. During the course of the eighteenth century, government found it increasingly necessary to differentiate between murder and a rising number of other capital crimes that had been added to the statute book, mainly concerning theft of property and game.[23] In the 1750s, the bodies of *all* hanged murderers were appropriated for dissection under the Act for Better Preventing the Horrid Crime of Murder of 1752, which gave judges freedom to specify that a particular killer's corpse might be dissected or prominently hanged in chains on a gibbet near the scene of the crime.[24] The legislation specifically stated that under no circumstances was a murderer's body to be buried. Anyone attempting to rescue the body or its parts for burial was to be punished with transportation to the colonies for seven years, and with death if they attempted to return in that time. Drawing and quartering, gibbeting, and dissection were clearly designed to aggravate the death sentence, and to differentiate between crimes, by means of public corpse abuse under the protection and patronage of the state. Each ugly fate subjected the dead to public exposure in indignity, to maltreatment and dismemberment. The explicit denial of burial implicitly denied the soul a resting place.

Surviving images of British official dissections are rare. The best appears in the last of William Hogarth's series the *Four Stages of Cruelty*, dating from 1751. Hogarth intentionally echoed visual structures and features from older images of such events, such as the professor in the raised chair, the demonstrator (or "ostensor") pointing with a stick, the dissector actually doing the cutting, the presence of assistants, and the audience. Even the skeletons and the dog derive from older images of dissection, such as, for example, those by Vesalius.[25] Hogarth's image was titled "The Reward of Cruelty" and was accompanied by a verse that reinforced the implication of the image: dissection was a disgraceful fate that denied the dead rest. Their tongues are torn out at the root, their eyes wrung from their sockets. The dissectors had prying eyes and were pitiless.

FIGURE 5.2: Dissection of the body of Tom Nero. Etching by William Hogarth, 1751. Iconographic Collections, Wellcome Library, London.

Hogarth was not alone in portraying public dissections as beastly affairs, as is revealed in a contemporary engraving showing animals dissecting a human corpse. We know very little about this print or its history, not even the artist's name. It is certainly British, and is believed to predate Hogarth by perhaps a couple of decades. It may possibly belong to the same era (the 1750s), just when the new Murder Act extended the punishment of dissection to all murderers. The caption reads:

Behold how in the Colledge hall
The Surgeons and Doctors all
Are met in Consultation wise,
A Carcase to Anatomize.

The Master there displays his art,
Sagely discants on every part,
And that with Ears & eyes & nose
We hear and See and Smell, he Shows.[26]

Considering these two images of official dissections, one is reminded of Oscar Wilde's unspeakable and uneatable: the dissectors are presented hardly more sympathetically than the dissected. The last verse quoted here implies that the entire event stinks of bestiality.

The gallows never yielded sufficient corpses to supply the needs of anatomists, surgeons, or the teachers of human anatomy. During the Enlightenment, a growing number of anatomy schools opened their doors, turning their backs on the old manner of teaching shown by Hogarth in favor of newer, hands-on teaching methods, whereby students learned anatomy by doing dissection themselves.[27] Such methods demanded considerable numbers of dead bodies, which the gallows could not furnish. Grave-robbers provided for the shortfall, and body snatching developed into a thriving, secret, and highly profitable business. Behind closed doors, corpses were bought and sold; they were touted, priced, haggled over, negotiated for, discussed in terms of supply and demand, delivered, imported, exported, and transported. Human bodies were dismembered and sold in pieces, or measured and sold by the inch.[28]

Despite widespread public and individual efforts to thwart the commodification of the human body and defend the sanctity of the grave, much of the increased understanding of human anatomy, which developed apace during the Enlightenment, was accomplished on bodies provided by body snatchers.[29] Anatomical practitioners and teachers created medical museums of anatomical and pathological specimens, generating esoteric skills of preservation and display, as well as sales and exchanges, and a growing interest in the comparative anatomy of the animal kingdom. In these places the dead body, skeletonized

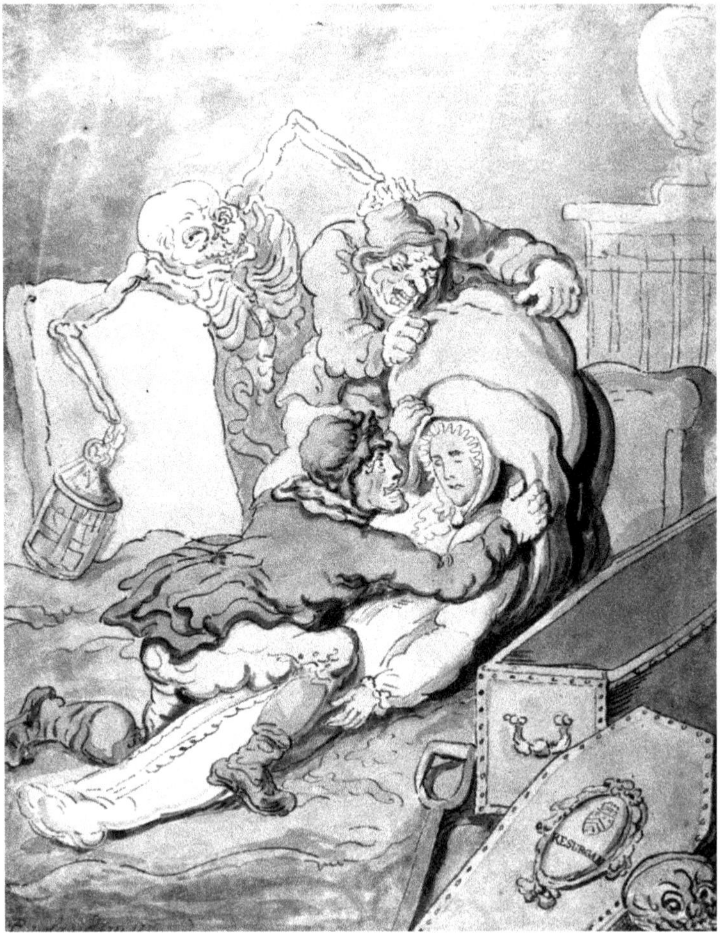

FIGURE 5.3: Grave robbers placing a disinterred, shrouded body into a sack. Pen and watercolor by Thomas Rowlandson, 1775. Iconographic Collections, Wellcome Library, London.

or dismembered into its constituent parts, became a privately owned exhibit.[30] Anatomical museums were used primarily for teaching but served a wider educative and public relations function: selected gentlemen and ladies, enlightened (and sympathetic) members of the elite lay public, were invited visitors.[31]

Book production in anatomy flourished in this period: extraordinary volumes appeared, such as William Cheselden's grand work on the bones of the skeleton, *Osteographia,* and William Hunter's great atlas of the gravid (pregnant) uterus.[32] Both works took years to assemble, not only because the dissections and exquisite illustrations took time to create and engrave but, certainly in Hunter's case, because of the difficulty of obtaining human material on which to work. William Hunter, whose dissecting rooms were near Leicester Square, in the heart of the West End of London, had probably let it be known that he would pay a high price if a grave robber brought him the body of a pregnant woman. Even so, it took him twenty-five years to acquire the fourteen bodies on which his great book was based. Hunter's difficulty in acquiring such bodies, given the high rate of deaths in childbirth, provides a measure of the customary care that was taken to protect the bodies of the dead. The casts of the dead bodies of these women, with their babies in situ, survive in the Hunterian Museum in Glasgow. Observance of the customary sorrowful interval between death and burial is manifest in the softness of the tissues on the day they were cast, still evident in the plaster: these bodies were decomposing when they were delivered to Hunter's dissecting room.

The bodies shown in Hunter's great book, and cast by him, have thighs and chests sawn across, their layers of flesh and fat revealed like butchered meat: headless, armless, legless mothers. Such apparently brutalized images of human body parts would have been shocking to contemporaries (many people still find them so today) but of course they were not intended for public view, since they appeared inside enormous, very expensively printed and bound volumes. The decision to remove extremities may have been taken because the bodies were in a bad state or because human arms, legs, and heads possessed economic value and were useful to students and fellow anatomists. Hunter needed access only to the lower abdomen in these bodies, to try to comprehend and demonstrate the causes of such calamitous deaths.

Jan Van Rymsdyk's meticulous drawings show the structures of gestation from full-grown death in the womb, right back to the bubble of a five weeks' conceptus. He recorded knotted or torn umbilical cords, placentas in the wrong place, ruptured wombs, infants lodged in breach, and other impossible positions. We can only guess at the hope extinguished, the pain, the exhaustion, and the sorrow that involved and surrounded these women's bodies before they were obtained by Hunter. Rymsdyk's work mitigates the tragedy to some extent by the precision and beauty of its representation. He rendered immortal the careful process of excavation of the grim double mortality of these natural childbirths, down to the smallest humane detail: the particularity of a baby's finger-crease, the tenderness of tiny testicles, even the reflection of the bright window (above the artist and the corpse) in the waters inside the soft translucent membranes.[33] Both anatomist and artist were looking directly at the close-up detail of the causes of death, and both were endeavoring to convey to the few who might access their book (like the select visitors to John Hunter's museum) the intimate information borne within the corpse itself.

Enlightenment anatomy schools and private medical museums served as research resources (it was William Hunter that demonstrated the separate blood circulations of mother and child). They were crucial to a developing understanding of anatomy (including the developing science of biological understanding via comparative anatomy of animals), which allowed new approaches to surgery and was fundamental to the developing study of disease by the close examination and comparison of so-called diseased appearances, which came to be known as morbid anatomy and, later still, pathology.

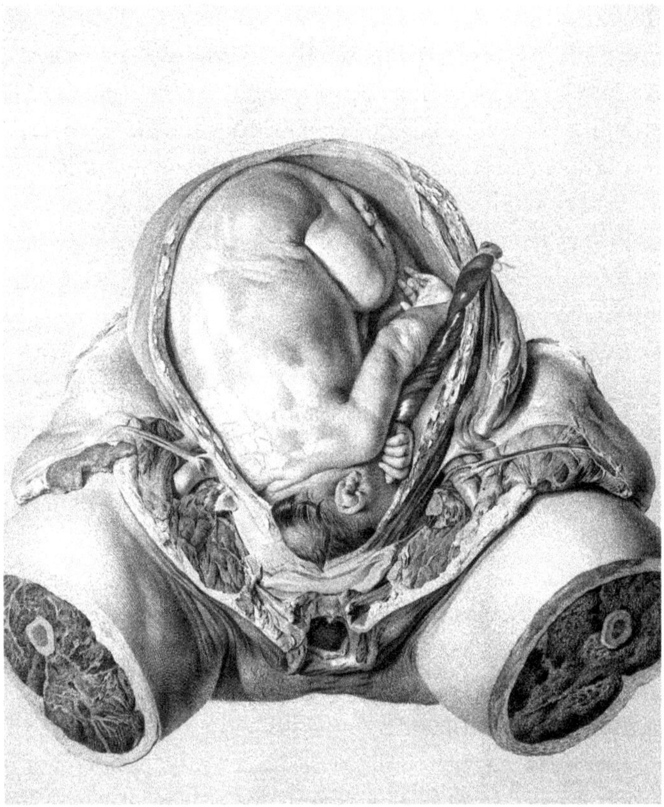

FIGURE 5.4: Full-term fetus in birth position. In William Hunter, *Anatomia uteri humani gravidi tabulis illustrata . . .* (London: 1774). Rare Books, Wellcome Library, London.

The meaty qualities of human flesh, and the apparent cruelty of dismembered display in Hunter's *Atlas,* were taken further in a number of other works on human anatomy, which one might say almost glory in depicting the infliction of pain and indignity on the dead.[34] The bodies of the dead appear in these anatomical works as profoundly vulnerable: in undignified postures, with their hands tied behind them or their half dissected limbs suspended by ropes, their heads resting awkwardly on hard blocks, body parts left hanging or exposed, in positions of unwarranted intimacy and pain, skin flaps held back with sharp hooks and chains, as excavated vestigial but still recognizably human remains, or with identifiable, sad, undignified, or mere remnants of facial expressions. In this sense, Hogarth's *Fourth Stage of Cruelty* does not exaggerate contemporary dissecting room attitudes toward the dead. To anatomists themselves, such disrespectful treatment of the dead was perhaps justified by the fact that dissection was regarded as a fate worse than death for the worst of malefactors. But there is a troubling sense of premeditated cruelty in the picturing of this kind of treatment of the dead, a sort of static sadism that remains a strand in

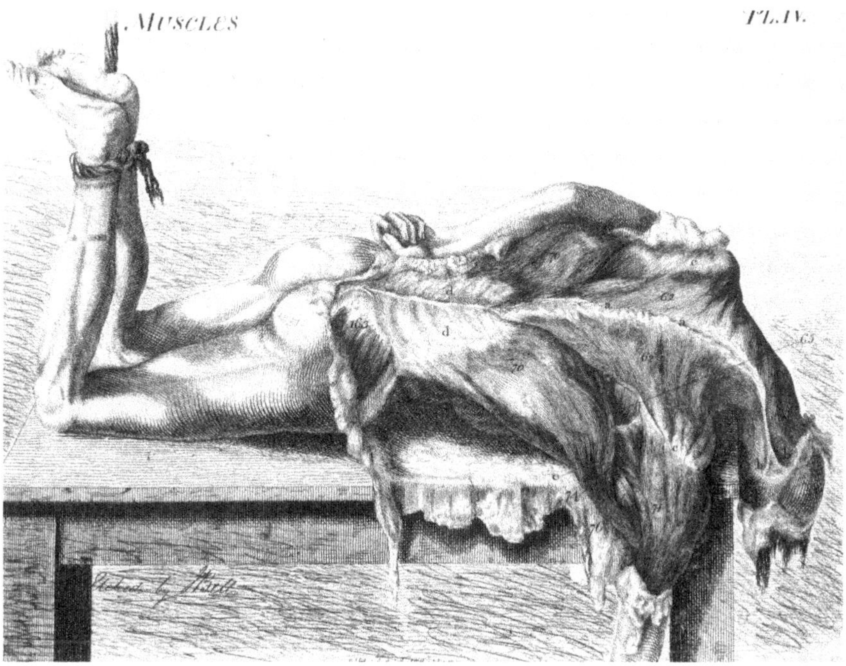

FIGURE 5.5: Flayed and roped corpse showing musculature. In John Bell, *Engravings, Explaining the Anatomy of the Bones, Muscles and Joints* (Edinburgh: 1794). Rare Books. Wellcome Library, London.

anatomical illustration that infected even student texts well into the Victorian era. Images of this kind, which present dissection almost as a ritual of predation, might possibly have had some contemporary justification when directed at a malefactor (who "to Pity has no claim," according to an accompanying verse) but not when inflicted on the defenseless dead who had done nothing to warrant such treatment. Such images appear to represent a knowing and calculated spurning and overturning of traditional care for the dead, closely akin to the values of judicial punishment as presented by Hogarth. The anatomist John Bell, some of whose illustrations are memorably on par with Goya's *Disasters of War* in their exhibition of mangled human bodies and body parts, asserted that works by other anatomists—in presenting views of the body that did not accord to the reality of the dissection room—were less honest than his own.[35] But it is unclear whether mastery of a subject and honesty about one's discipline necessarily entailed the unashamed demonstration of the power to inflict the degradation some of Bell's images portray. The profound disrespect they demonstrate toward the dead helps explain why dissection was equated with butchery, just as a commentator on Hogarth described it as being "butcher'd like the carcass of a Hog!"[36]

The reasons for these behaviors by some anatomists (Bell was not unique) probably cannot now be ascertained. But one cannot help but ponder what relationship they might have had with the treatment of the living, and whether such treatment articulates an assertion of religious disbelief or more nearly expresses the strain of bearing the taint of complicity with grave-robbers that anatomists and artists such as Bell carried with them. Were these images a sort of exhibitionistic display of transgression against the dead to assert the stupidity of sentimental attachment to the corpse, the absurdity of religion, and the fundamental materiality of the body? Or were they a sort of public denial of association with body snatching: a play-acting of punishment so as to reassert the legality of dissection and celebrate its protection by the law? The latter cannot be a sufficient explanation, insofar as the brutal honesty of Van Rymsdyk's work was done entirely on the bodies of pregnant women, and this was a constituency of the condemned prisoner that was never hanged: pregnant prisoners could plead their "belly," and execution would be delayed until after the birth of the child. Were these images, then, an attempt to assert the fact of insensate death? William Hunter had described clinical detachment as "a necessary Inhumanity," so perhaps it was rather the insensate anatomist that these illustrations reveal: the hard-earned detachment acquired in the dissecting room by both dissector and artist, masked behind or manifested in a kind of unflinching bravado.[37]

Toward the end of the Enlightenment, a lively medical debate erupted concerning the location of the soul during life, which eventually generated such disharmony and disfavor (and branded the entire anatomist profession with the disgrace of atheism) that a long silence of self-censorship descended on all sides.[38] That there was an element of medical bravado in the abuse of the dead seems likely; breaking the taboo on maltreatment of the dead was an individual and collective achievement of no small dimension. But the hardness of heart generated under the circumstances of body snatching and the commodification of the dead body in this era is not merely conjecture, as it is a matter of historic record that it resulted in the terrible series of murders for anatomy (in Edinburgh and London, 1828–1831) which marked the end of the road for this kind of enlightenment.[39] Jeremy Bentham's exemplary gesture bequeathing his own body for public dissection occurred in 1832,[40] the same year as the passage of the Anatomy Act, which Bentham himself had drafted. The act effectively undercut the corpse market by institutionalizing body procurement from workhouses and other places in which the poor died, establishing a cheap source of bodies and parts that endured until the rise of donation after 1948.[41]

GOTHIC AND CLASSIC IN THE GRAVEYARD

The arts often serve to reveal or emphasize undercurrents in a culture, and in this period they reflect imaginative anxieties among the public concerning the disposal of the dead. The illustrations to this chapter demonstrate how close to the surface the subject of death and the dead body was in the visual arts. Disturbed remains, it was popularly believed, caused haunting, and haunt society they did. On the one hand the so-called Graveyard Poets—Young, Blair, and Gray—pondered the human remains they took to be decently sleeping in rural churchyards (1742–1745, 1743, and 1750, respectively), while on the other, the ghosts of the dead shrieked and gibbered through the fiction of the period in hundreds of Gothic novels, culminating in the worm-eaten corpse of Ann Radcliffe's *Mysteries of Udolpho* (1794), John William Polidori's blood-sucking *Vampyre* (1816–1819), and in the murderous monster—a living assemblage of grave-robbed body parts—of Mary Shelley's *Frankenstein* (1816–1818).[42]

The emergence of these themes in poetry and fiction present a literary parallel to the fierce defense of the grave that was occurring in reality, in graveyards all over the country.[43] Every buried body was in danger of predation. Cities and towns had the most densely packed burial grounds, where pit burials for the very poor offered easy pickings, but even burial vaults were defenseless where sextons could be bribed. River transport, sea routes, and especially post roads rendered many other localities at risk. Suburban graveyards were in easy reach, and snatching in quiet country places had advantages of safety for those involved if activity went unsuspected for a considerable period. All attempts were dangerous: ferocious riots could follow discovery.[44] As I have shown elsewhere, the danger involved in body snatching promoted great ingenuity in the transportation of human corpses for the dissecting rooms, and doubtless in thousands of cases as the disguised parcels and packages reached their destinations. But there was also a considerable number of cases in which human remains were discovered, to the shock of many, in unexpected places and at unexpected times. Human bodies were compressed into boxes; packed in sawdust; packed in hay; trussed up in sacks; roped up like hams; sewn in canvas; packed in cases, casks, barrels, crates, and hampers; and salted, pickled, or injected with preservative. They were carried in carts, barrows, and wagons and were manhandled, damaged in transit, and hidden under loads of vegetables. They were stored in cellars and on quays.[45]

High levels of public anxiety about body snatching served to make the business of undertaking highly profitable. Considerable effort, ingenuity, and money were invested in burial security. This was the era in which secure burial vaults became popular and when tomb railings and strong monuments were erected in churchyards; it was the era of the triple coffin, the lead coffin, the mortsafe, and, eventually, the patent coffin. I have written elsewhere about the new lamps erected in churchyards, the watchmen's huts, the heavy body stones, the exclusive security features such as catacombs that were developed by private cemetery proprietors to attract clients, and the shifts to which poor people unable to purchase any of the above were driven by fear of their vulnerability to body snatchers in the Georgian era.[46]

In this process, funerals gained superadded importance. Undertakers set up shop as a distinct occupation in centers of population during the late seventeenth century, purveying funerals to suit the pockets and aspirations of their clients. Advantage was there to be taken of public anxieties, as it was of the new semi-industrial processes of mass production. Coffin plates and ornamentation, coffin nails, and, as we have seen, coffin linings and shrouds were being produced in industrial numbers by Hogarth's day. Appropriate levels of expenditure could purchase various permutations of coffin strength and durability, grave or vault size, security, commemoration, and funerary display.

The eighteenth century also saw a highly significant shift in monumental imagery: a remarkable transition in which older skeletal forms of imagery were forsaken and a new cluster of imagery was embraced. Between the seventeenth and the early nineteenth century, the skull, and its associated imagery of skeletal bones and burial paraphernalia, disappeared from British monumental imagery where it had been ubiquitous for centuries, part of a long continuity from medieval times. These older emblems of mortality would still have been visible, just as they are visible today, since old tombstones survived in churches and churchyards from previous generations. But enlightened Georgians preferred monuments of often severely simple neoclassical design, featuring such elements as the down-turned torch, the broken column, the obelisk, the pedimented sarcophagus, the weeping willow, and the grieving figure.

The new-style monuments exhibited more flesh, more light, more smooth and unadorned surfaces. As the skeletal figure of Death receded, in its place the more kindly (and more fleshy) bodily figure of Old Father Time became more prominent: alive, active, and, over time, less gaunt. The shreds of Death's shroud became less ragged as Time's robe, as the neoclassical angel's garb, and as the draped fabric on the urn or sarcophagus. Central features were invariably shown raised off another surface, either by plinth or pedestal, or, in the case of the sarcophagus, often by small feet, so that there was a layer of air, even of light, below. The imagery aspired to a lightness quite foreign to what I have called Old Mortality, often visually assisted by the use of a pyramidal formation of subject matter, in which raised architectural features, flames rising upward from the urn's neck, or occasionally upturned wisps of drapery or floating angelic figures, formed the apex. The eye is held between this tendency toward ascension and the gravitational pull of mourning figures' downturned faces and gestures, and their echoes in the pendulous foliage and branches of the willow. The more successful designs resolved the tension by setting the visual focus on the inscription, itself often pyramidal in form.[47]

The preeminence of neoclassicism in the commemorative monumental sculpture of the Enlightenment was synchronous with that in architecture. The celebration of clean lines, and the new ascendancy of pattern books, developed alongside the adoption of more industrialized approaches to the working of stone and a concomitant decline in traditional monumental carving styles in Britain. Along with this decline went the traditional subject matter. Marble, some of it ready-prepared, was increasingly being imported from Italy for use in the erection of fashionable monuments. Stone surfaces were smooth, cleanly cut, and finely polished. At the same time, the clean lines, and the emphasis on whiteness, smoothness, and upward movement, served to deflect thoughts of earthiness and physical decay: a direct counter to the older tradition of darkness, decay, and decomposed remains.

The rejection of the imagery of Old Mortality in favor of the new designs took place slowly and unevenly and over a long period of time. Traditional skull and bone imagery survived longer out of doors than on the more expensive monuments inside churches, and in rural churchyards later than in urban ones. Both developments suggest that less exalted folk, and those further away from urban influences, held on to the old traditions longer.

Although all of its features were not entirely new, the imagery that replaced Old Mortality cannot really be understood as traditional or indigenous to Britain. Excepting for the willow (which arrived by a different route), they derived from a reworking of themes from other traditions and other times: from the pre-Christian civilizations of classical Greece and Rome. Probably the most omnipresent symbol of the newer imagery was the urn, a curious choice for a symbol of Christian burial. As an enclosure for ashes, the urn referred back to the cremation ritual of pre-Christian and classical times, and to the ancients, but in fact cremation was not a disposal option in eighteenth-century Britain. Cremation was not revived there until the late Victorian era, and cannot be said to have come into general use until after World War II. The draped urn served ubiquitously as an analogue for the body—a pale vertical container upright in the light, arms akimbo—in contrast to reminders of a dark box of decomposing remains, and as a standing emblem of euphemism.

A similar transition took place in print culture, particularly in the printed funerary ephemera created for the expanding constituency of the financially fortunate. Like tombs of the pre-Enlightenment era, older funeral invitations often featured skulls and crossbones, the figure of Death as an animated skeleton with his hourglass and scythe or dart, and views of the coffined body surrounded by friends, all these surmounted with the caption "Memento Mori" or, in English, "Remember Death" (or sometimes, more curiously, "Remember to Die"). Margins often featured earthy reminders: blackened coffins, bones, skulls, open graves, grave-digging tools. These older examples tend to be dark in character—black lettering, darkly ornamented borders, dark scenes.

The sense of lightness, even of weightlessness, in the newer style was materially assisted by new techniques of illustrative printing. During the eighteenth century, up-market printers moved away from the dark crudity of the woodcut, employing instead copper (and later steel) engraving, which allowed a thinner and more delicate line, and hence more whiteness on the page. Printed designs bear strong similarities to those on tombstones and church monuments: classical architectural features and floating figures. It has proved difficult to date the transition precisely in printed ephemera, however, because of the circulation and re-use of engraved printing plates.[48] None of the old memento mori imagery survives much beyond 1800 in funerary ephemera. Reminders of mortality gave way to reminders of memory, crudity to delicacy, and darkness to light. Once again, the rejection of the imagery of Old Mortality proved permanent: the skull and the animated skeleton were not revived with the Victorian gothic revival and have never undergone a revival in the field of funerary ephemera. They remain completely absent from modern examples.

Georgian mourning jewelry provides a clearer chronology that allows more precise dating and a better understanding of the cultural change at which we are looking. British mourning jewelry surviving in museum collections is often of very fine quality and known provenance, designed to commemorate wealthy members of society: the aristocracy and gentry, and the higher reaches of the commercial classes. Here, in affluent society, the chronology is much more easily ascertainable because the marking of precious metals is often date-specific, either by jewelers' marks, or because the date of the death commemorated is known. Change seems to have occurred in three phases. First, the tiny ivory skulls (so beautifully carved that they seem like miniature realities), skeletons, and memento mori inscriptions, which had been in vogue since at least the 1500s and possibly much earlier, seem to have disappeared between about 1720–1740. They were not replaced by a new visual vocabulary immediately. Instead there followed an intermediate phase in which mourning jewelry featured elegant monograms and inscriptions: mainly in black enamel on gold, or formed in gold wire protected by crystal. There appears to have been a genuine appreciation in the mid-eighteenth century of the beauty of calligraphy or typography, and an increasing emphasis was placed on significant words or alphabetical letters, particularly initial letters signifying names. Letters of the alphabet were used in symbolic ways, such as intertwining to signify loving relationships.

The neoclassical imagery of pedestalled urns, willows, and melancholy figures seems to have come into vogue in jewelry in the 1770s, when crystal was used increasingly to protect delicate samples of human hair, presented decoratively in twirls, plaits, or other shapes. The skull, which probably more than any other emblem represents human bodily decay, was thus replaced by other imagery, and by the *least* corruptible part of the body, the hair.[49] The jeweler's art therefore reveals a chronology that subverts David Irwin's notion that the antimacabre ideas of Winkelmann and Lessing influenced taste in Britain.[50] It suggests, rather, that both writers reflected a significant cultural change already very well established among the British public *decades* before either composed his thoughts in print. The process of change was slow. Elements of both styles were present in the seventeenth century, but the newer style predominated as the eighteenth century progressed. It is not all that difficult to see how the old crossed bones or the crossed gravediggers' implements could be superseded by crossed, downturned torches, or the winged skull by the winged cherub's head or the anthemion. Nor is it surprising how King Death's crown could transmute into the incandescent Heavenly Crown of Everlasting Life.[51] Skeletal King Death, an image that dates back half a millennium to before the Black Death, is first accompanied by, then usurped by, Old Father Time. Indeed, tussles between these two emblematic figures appear occasionally on mid-eighteenth-century tombstones. One of the most dramatic moments in this struggle is of Death's overthrow by Time, which can actually be witnessed taking place in 1757 on an extremely fine wall monument, by the sculptor Roubiliac, in Westminster Abbey.

Old Father Time seems to have inherited Death's props of scythe and hourglass, and he increased in benignity as the eighteenth century progressed. He looked better fed, his wings became larger and more noticeable, and the angels who in turn supplanted *him* became lighter, more ethereal, and more often caught in flight. By the later eighteenth century, the imagery seems to have undergone another process: feminization. Veiled or hooded figures in mournful poses (often genuine portraits of mourners) usually female, occasionally men in classical robes, became an adjunct of the urn or the sarcophagus and later seemed to merge with, or contribute toward the development of, the Victorian angel, also swathed in generous yardages of cloth.

A common monumental image, the broken column, relates to another preoccupation of the age: the love of ruins. The Italian artist Gian-Battista Piranesi (1729–1778) found a receptive audience in Britain for his magnificent prints of architectural decay, and views of ivy-clad ruins graced the pages of popular books. Ready-made ruins—gothic or classical—were constructed in the gardens and estates of the wealthy, with ready-weathered stone, even planted with appropriate mosses and ferns so as to hasten the appearance of overgrowth. The love of ruins also articulated an attitude toward death. Architectural ruination is emblematic of transience: transience of human effort and its artifacts, and also of the transience of the human body. This may perhaps help explain why the broken column developed into such a potent monumental symbol in this period, especially since individuals were spoken of as pillars of church, family, or community. There is a very real sense in which the architectural ruin came to stand in as an analogue or euphemism for the human skeleton, as a symbol of the vanity of human wishes.

A great deal more might be said about this cluster of changes involving the contemplation of the dead body in the culture of this era: the exploration of the deepest recesses and secrets of the body in the dissection room, the confrontation with traditional corpse care in anatomical illustration, and the disappearance of skeletal imagery from jewelry, funerary ephemera, and monumental sculpture. There is a sense of darkness lifting, especially in the remarkable process of lightening in the monuments and the funerary ephemera of the era, which really does feel like a process of enlightenment. Yet while both processes cast light in dark places, the anatomists were confronting mortality by looking more closely at the corpse, and the public was looking away.

CHAPTER SIX

The Body Beautiful

DAVID M. TURNER

What is beauty? Why have certain bodies been idealized while others have been stigmatized as ugly or deformed? Such questions are of key importance to the history of the body.[1] Yet beauty is a slippery subject for the historian. As Samuel Johnson wrote in 1751, "the idea of beauty is vague and undefined, different in different minds, and diversified by time and place."[2] Faced with the prospect of analyzing something so abstract and elusive, one may be tempted to follow the example of Voltaire's philosopher who concluded that "beauty is very relative; that what is decent at Japan is indecent at Rome, and what is fashionable at Paris is otherwise at Pekin; and thus he saved himself the trouble of composing a long treatise on the beautiful."[3] In spite of these difficulties, the study of ideals of physical beauty and their influence on society is worth pursuing since it may tell us much about past understandings of body image and the ways in which corporeal ideals are used to inscribe gender and other social differences.

Beauty was a key topic of cultural debate in the long eighteenth century. The critical appreciation of beauty in all its manifestations was central to philosophical and aesthetic discussions of taste and refinement.[4] Body image and human appearance were recurring themes in a variety of periodicals, ballads, chapbooks, moral tracts, and didactic texts prescribing ways of beautifying the body or advising on the art of bringing forth beautiful children.[5] This chapter examines the attributes and qualities that defined a body as beautiful in this increasingly body-conscious age and analyzes the ways in which beauty was viewed as a source of both power and anxiety. It also explores the ways in which the myth of physical perfection was sold as an attainable ideal in an increasingly commercialized society and the various methods for achieving the beautiful body. In the process, it analyzes the social value of personal appearance. Indices of beauty and deformity were allied to broader systems of social inequality and played a critical role in establishing systems of power and hierarchy, inclusion and exclusion.

DEFINING BEAUTY: BODY, POWER, AND ANXIETY

The term "beauty" had a variety of uses. First and foremost it denoted "a sense of whether, and how, a particular object or person gratifies the viewer."[6] Adopting a Platonic view of beauty common to many writings on the subject, the popular conduct book *The Ladies Dictionary* (1694) defined the beautiful body as "humane splendour, lovely in its own Nature, and which hath the force to ravish the Spirit with the Eyes."[7] The question of what constituted human perfection was of as much interest to people in the early modern period as it is to us today. In May 1691 the coffee house periodical the *Athenian Mercury* answered a query from a young gentlewoman who asked "Whether beauty be real or imaginary?" "We dare *almost* venture to affirm 'tis both," replied the paper's editors. Although concepts of beauty and deformity were subject to the vagaries of "*Custom* and *opinion,*" the law of nature dictated that "there must be a *best* somewhere," for there were some things "being shaped more neat, cleaverly and handsome than others, as we may grossly see in a Horse compared with an Elephant, a Gray hound with a Swine or Cow." Human beauty could be gauged objectively and measured by taking account of a variety of physical and mental attributes.[8]

Although the shape of the body's features as a whole contributed to the estimation of beauty, these were often augmented by other characteristics. Very important was symmetry or proportion. Popular culture mirrored classical ideas about the harmonious body that defined the rules of beauty in aesthetic theory.[9] The *Athenian Mercury* emphasized the importance of a clear complexion and a "good meine and air." Mien related to the body's deportment. Identified in conduct literature as an important indicator of good breeding in men and women, it combined dignity, ease, grace, and a lack of affectation.[10] Mien was perfected through "Education and Converse," but a good air was harder to learn and define and was to be seen in the countenance and the face's natural expressions. An air might determine whether a person was beautiful or merely pretty. Sometimes:

> we meet with such an *incomparable sweetness* (mostly residing in the Mouth and Eyes, tho' the whole *turn* of the *Face* contributes something to't) that it charms all that see it, and those who have it, we rather call

pretty than *beautiful*, since 'tis often found where there's hardly *one good feature.*[11]

The *Athenian Mercury* rated "good Air" the highest characteristic of beauty, "because when good Mien and Complexion fails, when there is sometimes little that we can like either in Feature or Proportion, *this* always lasts, and nothing but *Death* . . . can alter or destroy it."[12] Sixty years later, Joseph Spence also placed greater emphasis on the body's demeanor than on complexion or form. Of chief importance was a sense of grace, which "often depends on some very little Incidents in a fine Face; and in Action . . . consist[s] more in the manner of doing things, than in the things themselves."[13] This combination of body and spirit was also evident in visual representations of beauty. In Isaac Fuller's *Iconologia* (1709), "Beauty" is pictured as a nude young woman surrounded by a radiant light carrying in one hand a lily, symbolizing the senses, and in the other a ball and compass, symbolizing beauty as consisting of measure and proportion.[14]

Qualities of beauty differed in men and women. According to Spence, "[t]he distinguishing Character of Beauty in the Female Form, is Delicacy and Softness; and in the Male, either apparent Strength or Agility."[15] A "majestical

FIGURE 6.1: Personification of Beauty. In Isaac Fuller, *Iconologia* (London: 1709). Rare Books, Wellcome Library, London.

face" in a man like "that of *Caesar's* or *Gustavus Adolphus* dazzles all that behold it" and embodied the qualities of leadership.[16] But discussions of beauty were gendered in more fundamental ways. As Robert W. Jones has demonstrated, the term "beauty" was most frequently deployed in eighteenth-century cultural debate in relation to the attributes and behavior of women.[17] Although essayists and authors of conduct literature did not deny beauty to men, and purveyors of beauty products aimed their wares at both sexes, when the body beautiful was discussed it was mostly the female body under scrutiny.[18] Beauty had long been associated with women's power over men. "A Beautiful Woman fixes her unshaken Empire in the hearts of her Admirers," noted *The Ladies Dictionary*,[19] while the *Gentleman's Magazine* observed that "wherever Beauty appears, like a sovereign, it exacts Awe and Obedience."[20] Yet this power was viewed with distrust by male commentators. Seventeenth-century ballads often emphasized beauty's cruelty and its tyranny over those whom it ensnared.[21] Beauty's conquest of men's hearts was described in military metaphors, reducing men to servitude:

> Oh! Her power's too great for his armes to withstand,
> His conquests are yielded up at her command,
> She rules his dread fury and wildness a war
> And wins all his tropheys without sword or spear,
> While tamely he bows nor has power to resist
> When such pointed beauty does enter the list.[22]

Beauty was regarded ambiguously in popular literature of the late seventeenth and eighteenth centuries, which inherited the Platonic tradition that equated beauty with goodness and deformity with sinfulness and evil.[23] For instance, a late seventeenth-century chapbook, *The Merry Dutch Miller* (1672), told of a magical mill that transformed ugly, scolding, and adulterous wives into "handsome" spouses, easier on the eye and easier to govern. The miller

> undertaketh to grind all Sorts of Women, as the Old, Decreped, Wrinkled, Blear-ey'd, Long-Nos'd, Blind, Lame, Scolds, Jealous, Angry, Poor, Drunkerds, Whores, Sluts; or all others whatsoever. They shall come out of his Mill Young, Active, Pleasant, Handsome, Wise, Loving, Vertuous and Rich; Without any Deformity, and just suteable to their Husbands Humours.[24]

Sermons and moral advice literature also equated deformity of features with sexual sin. Readers of *Onania*, the popular antimasturbation tract, were warned that "licentious masturbators" could be recognized by their "meager

FIGURE 6.2: The magical mill that transforms ugly wives into beautiful and obedient ones. Engraving by Paulus Fürst, ca. 1650. Iconographic Collections, Wellcome Library, London.

Jaws and pale looks" or (if women) by their "swarthy and haggard" faces.[25] On the other hand, it was believed that physical beauty could be deceptive and the source of man's downfall, as attested by the stories of Eve's temptations and the "War and Confusion" occasioned by the beautiful Helen of Troy.[26] Contradictory messages might be found in the same text. The anonymous misogynist tract, *Look e're You Leap: Or, A History of the Lives and Intrigues of Lewd Women* (ca. 1760), advised its male readership to choose a "moderately fair and beautiful" wife, since "a lovely and fair Face does generally prove the Index of a fairer Mind." Elsewhere it urged men to beware of some dissimulating women who hide "the inward Vices of their Minds, under the Veil of fair Appearance."[27]

Female beauty was thus often associated with inconstancy and lewdness. "Beauty and chastity (honestly) seldom meet," ran a seventeenth-century proverb.[28] Francis Osborne advised his son that if he wed a "celebrated beauty," his house would become "as populous as a confectioners shop; to which the gaudy wasps, no lesse than the liquorish Flyes make it their businesse to resort, in hope of obtaining a lick of your honey-pot."[29] The author of a marital advice book, *Learn to Lye Warm* (1672), counseled young men to favor age before "youth and beauty," for any man "after a weeks repose in the bosom of a lusty and youthful beauty" will "become a Victim to Venus," enslaved, emasculated, and cuckolded.[30] While mutual "good liking" was deemed essential to a happy marriage, moralists advised that "he that marries for a face, marries for a year."[31] "Thousands are daily ruined by a handsome Person," warned the hunchbacked MP William Hay in his autobiographical plea for acceptance of human variety, *Deformity: An Essay* (1754), "for Beauty is a Flower, that everyone wants to gather in its Bloom, and spare no Pains or stratagem to reach it."[32] As the *New Spectator* argued in 1784, the jealousy and unease caused by marrying a beautiful woman was just punishment for favoring looks over more durable qualities.[33]

As well as causing tension between the sexes, female beauty was apt to provoke rivalry between women themselves. A beautiful young woman "raises the envy of the world," remarked the *Female Tatler* in October 1709, and suffers a "constant prying into her conduct" by jealous women who are "as earnest in railing at those that have more beauty than themselves, as men are in hating those that have more wit."[34] Beauty was frequently associated with pride, affectation, and vanity. The *Spectator* (1711) related a story of two sisters, the beautiful Laetitia, and Daphne who was in "no way remarkable for any Charms in her Person." Yet for Laetitia, "consciousness of her charms has render'd her insupportably Vain and Insolent, towards all who have to do with her." Being "confident of favour" on account of her beauty, she had "studied no Arts to please," and thus conversed with others with a "haughty impertinence," whereas her plainer sister was much more agreeable, depending much more on her "merit." The lesson was that just as "it is an Argument of a light Mind, to think the worst of our selves for the Imperfections of our Persons," so too was it "equally below us to value our selves upon the Advantage of them. The Female World seems to be almost incorrigibly gone astray on this Particular."[35]

Women's beauty was thus seen as a dubious property unless allied with virtue and morality.[36] Robert W. Jones has drawn attention to the "profoundly ethical nature of most eighteenth-century accounts of the beautiful."[37] As the poem *Beauty: Or the Art of Charming* (1725) put it:

> Where lies this charm?—Alas, not in the Skin;
> The Life of Beauty rises from within;
> Flows from the Soul, and animates the Breast,
> In words and Actions, Looks and smiles exprest.[38]

The right balance of "Virtue, Modesty and Beauty," advised the *Universal Spectator*, is the "Foundation of a Woman's claim to Love and Respect."[39] *The Matrimonial Preceptor* (1759) noted that beauty was "in some degree dependent upon SENTIMENT and MANNERS," because the "finest features, ranged in the most exact symmetry, and heightened by the most blooming complexion, must be animated before they can strike."[40] The winning combination of fair features and inner virtue had the power to act as the "sweet reformer of mankind," and was very important to the cultivation of politeness and sensibility.[41] Writing on "The Empire of Beauty" in the *Tatler* (1709), "Jenny Distaff" remarked that "Every Temper, except downright insipid, is to be animated and softened by the Influence of Beauty."[42] Beauty of this kind regulated the passions and might inspire men to improve their conduct. "As it excites a purer passion," observed the *Lady's Miscellany*:

> it also more forcibly engages to fidelity; every man finds himself more powerfully restrained from giving pain to goodness than to beauty; and every look of a countenance in which they are blended, in which beauty is the expression of goodness, is a silent reproach to the first irregular wish.[43]

IDEALS OF BEAUTY

For all the emphasis placed on inner "beauty of virtue," physical appearance mattered. From quasi-medical texts on the "art of preserving beauty," to poetical tributes to the society beauties—the "toasts" of the town—preferred body image was anatomized from forehead to feet. The face was regarded as "the chief seat of beauty" and a woman's principal weapon in conquering men's hearts.[44] The aesthetic ideal, deriving from classical notions of beauty, was that the face should be symmetrical with a high forehead and an aquiline nose.[45] Advice literature on personal appearance emphasized the importance of an unblemished complexion and offered recipes to cure the deformities of redness, freckles, and spotted or dry skin. According to *Artificiall Embellishments. Or Arts Best Directions How to Preserve Beauty or Procure It* (1665), "[t]he smiling glories of beauties spring are often ript with an early autumne; when sharp sith'd time cuts those flowry graces down, and shrouds them in the furrows of a wrinkled face." Wrinkles were signs of mortality and also "humane frailty"—the embodiment of declining powers.[46] The forehead, described as "the Ivory throne where Beauty sits in state," should be smooth and of a "decent height" in order to inspire reverence and respect.[47]

A comely face was not simply one that was elegantly proportioned or blemish free, but also one that reflected the deeper sentiments of the soul. As the French surgeon Nicolas Andry explained in *Orthopaedia: Or, the Art of Correcting and Preventing Deformities in Children*,[48] the face "takes the features of the Soul, and moulds itself by them. The soul's "habitual sentiments such as are contracted in a good or bad Education, by good or bad Habitudes" serve to "imprint upon the Face such deep Characters, that they are never afterwards to be defaced."[49] The power of female beauty resided above all in the eyes. *Artificiall Embellishments* described the eyes as "Cupids chrystal burning glasse, to kindle devotion in your Captives hearts."[50] Beauty manuals prescribed remedies for inflamed, bloodshot, or spotted eyes and yellowness of the eyelids. Andry listed fourteen conditions "necessary to handsome eye brows," nine deformities of the eyelids, and seven of the eyes themselves.[51] Society beauties were lauded for their "piercing" eyes, or those giving a "soft engaging look."[52] Eyes that were "soft" or "languishing" conveyed an appearance that was sensual or alluring but also indicated women's passivity and reinforced notions of female weakness and delicacy. Andry remarked that the "arch of the eyebrows too high raised" in women was a great deformity since it gave them "too much of a bold air."[53]

One of the "chiefest ornaments of a woman," according to *Art's Master-Piece: Or, the Beautifying Part of Physick,* was a "long Train of dishevelled Hair."[54] Remedies were prescribed to make hair soft and not "like thatch on a country cottage," to curl it, cure baldness and hair loss, remove dandruff, and remedy split ends.[55] Blond or black hair was considered most favorable.[56] Despite the historical example of illustrious redheads such as Elizabeth I, red hair was seen as a deformity by some writers.[57] In July 1705, a young man wrote to Daniel Defoe's *Review* periodical asking whether it was a "real disadvantage" to his reputation if he married a "pretty agreeable Young Woman" whose only "fault" was her red hair. His concerns were raised by the behavior of the mob at the recent Middlesex election who had chanted "No carrots, no carrots" at one of the candidates, Mr. Barker, in reference to his ginger-haired children. While the *Review* counseled him to ignore the opinion of the world and marry his sweetheart, its reasoning that "A man certainly lies under less disadvantage with such a Wife, than with a Black or Fair [headed] one, because he is in less danger . . . of losing her love by the Temptations of other men," did little to challenge prevailing negative stereotypes about red hair.[58]

The ideal body shape was classically proportioned and should be "five times as tall as it is broad."[59] Posture was significant in evaluating the body's beauty. "When the spine is strait, well set, and freely turned, it makes a handsome body, and when it is crooked and ill turned the Body is deformed."[60] Physicians warned against tight swaddling, which might prevent the babies' limbs from straightening, and urged parents to take special care to teach their children to hold their bodies upright, especially young girls, who, according to Andry:

> ought not to be allowed to sew or read, except in an erect Posture; they should hold their work or their Book to their Eyes, and not their Eyes to their work or book, without which their Body will infallibly become crooked.[61]

Raised haunches made for a fine or slender waist, which was valued as a sign of femininity. The size, shape, and appearance of the breasts received much discussion in the later seventeenth and early eighteenth centuries when female fashions emphasized the cleavage. *Art's Master-Piece* contained advice on reducing breast size, hindering their growth, and recipes for hardening the

FIGURE 6.3: Woman wearing curled hair pieces attached to her natural hair. Engraving by James Delegal, ca. 1750. Iconographic Collections, Wellcome Library, London.

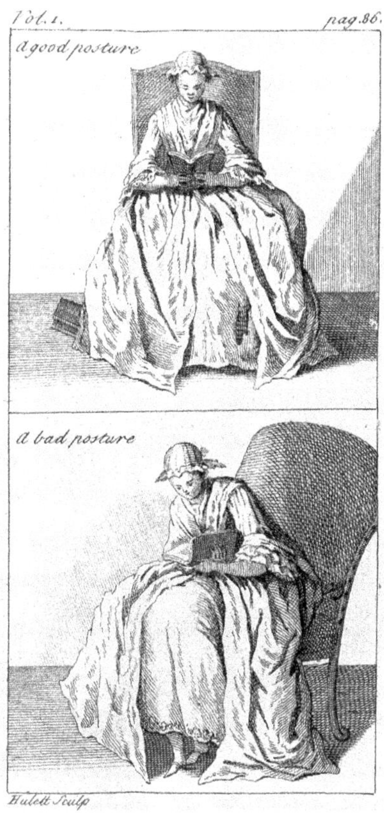

FIGURE 6.4: Good and bad posture. In Nicolas Andry, *Orthopaedia* . . . (London: 1743). Rare Books, Wellcome Library, London.

breasts and making them appear smooth.[62] *The Ladies Dictionary* remarked that breasts that "hang loose" or were of "an extraordinary largeness, lose their Charms, and have their Beauty buried in the grave of Uncomliness." Ideally, breasts should be "small," but also "plump and round," being "like two Ivory Globes, or little Worlds of Beauty." In order to improve the roundness of the breasts women were advised to ' "Bind them up close" with "Caps or Bags that will just sit them, and so let them continue for some Nights."[63]

For much of the Enlightenment era, a certain plumpness of figure was regarded as desirable. Stoutness generally was esteemed as a sign of well-being and an attribute of the wealthy. Ample flesh gave a woman allure and hinted at fertility.[64] However, the truly obese body was viewed with contempt. Rotundity might suit the contented country squire but was not for the young woman of fashion. The Rubenesque female form was fleshiness taken too far for some eighteenth-century writers. Rubens, noted Joseph Spence, "delights so much in Plumpness" that "[i]t seems as if nobody could be a Beauty with him, under Two hundred Weight. His very Graces are all fat."[65] The supposed corpulence of the women of the painter's Dutch homeland was a frequent source of ridicule. Remarking of the women of Tonga in 1773, Captain Cook's astronomer, William Wales, observed that they were "rather too fat to be esteemed beauties any where but Holland."[66] Body size was indicative of moral qualities. "When ere the carcase swells itself into a bulk too voluminous," remarked *Artificiall Embellishments*, "idlenesse is there describ'd in folio."[67] *Abdeker: or, the Art of Preserving Beauty* (1754) went further: fatness destroyed all semblance of grace, and caused "great Sloth and Carelessness" in the actions of those afflicted by it. Contradicting the association between corpulence and a robust, healthy constitution, fat people were

> stupid; their apprehension is not so lively as it naturally ought to be; they breathe with Difficulty, and are subject to frequent Distempers. They have also a certain Incapacity of Breeding, so that Women of that Complexion are commonly barren. The Soul is overwhelm'd with the Weight of a huge Lump of Matter, and all the Functions of the Understanding are in such a languishing Condition, that it can shew no Marks of its former brightness.[68]

Obesity not only caused physical problems but encumbered and trapped the spirit.[69] While it was acknowledged that obesity might be caused by medical and environmental factors, it was inextricably linked with "idleness and luxury."[70] Weight loss was recommended as a means to "regaine your credit and your beautie too." *Artificiall Embellishments* recommended that the obese rise early in the mornings and "use some violent exercise to sweat often," and prescribed a special diet of "Oily and fattie things that the appetite may soon be satiated, and the body kept soluble," followed by "sharp, salt and bitter things," advising readers to "eat all your meats with vinegar, pepper, mustard, juice of oranges and limmons." If that failed, purges and bleedings might help, or simply fasting often and rising from one's meals "halfe satisfied."[71]

Weight loss, as Roy Porter has shown, became something of an obsession in the eighteenth century, particularly following the publication of George Cheyne's *The English Malady* (1733), which explained how obesity threatened both physical health and mental stability.[72] Yet it was the skinny, rather than the corpulent, body that aroused most concern. Before the later decades of the eighteenth century, when fashions began to favor the lean, youthful body, and slimness came to represent the delicacy and fineness of sensibility, excessively thin bodies were deemed inconsistent with health or beauty and, like the corpulent, carried the stigma of moral defect.[73] The thin were no more than "breathing skeletons" whose bones look "as if they had eaten up the flesh, and were readie to leap of[f] the skin to fall upon others."[74] The emaciated body was indicative of poor physical health, poverty, and a guilty conscience.[75] A correspondent to *The Ladies Mercury* periodical in 1693 described how her beauty had been destroyed by an all-consuming guilt for a secret premarital love affair in which she had lost her virginity, for her "shame-sick" condition had "almost worn [her] to a skeleton."[76] Also prone to excessive thinness were young men "that are too much given to Women" and those generally who were "too passionate, or who have too lively an Imagination."[77] As already noted, eighteenth-century masturbators were warned that they might be given away by their gaunt, haggard appearances.[78] One beauty manual painted a morbid picture of the skinny body:

> When the body is lean, the Face grows long, the Eyes are sunk into the Head, the Mouth enlarges, the Cheeks are Hollow, the Face is pale, often yellow, and sometimes of the Colour of Lead; the Bones are prominent and seem to be almost out of Joint; the Breast exhibits a dismal Representation of a Vault, wherein one may count all the Arches; and the spindle shanks seem hardly able to sustain the Bones of this walking Skeleton. It is a spectacle that strikes all that look on it with Horror.[79]

Thinness was blamed on a number of factors: bad diet and digestion, "immoderate use of spirituous liquors," too violent exercise and excessive excretion as well as a lack of a sufficient quantity of nourishing particles in the blood.[80] *Artificiall Embellishments* prescribed a diet of fresh eggs, mutton, veal, capons, and moderate recreations such as "dancing, singing [and] discoursing" as a means of regaining one's looks.[81] Such foods were considered particularly nourishing and easy to digest and also generated heat, which was deemed important to fattening the body.[82]

BEAUTY AND DIFFERENCE

Notions of physical beauty in the long eighteenth century were comparative and evaluative. They established differences not simply between individuals considered more or less favored in terms of their looks, but between classes, races, and nations. The body beautiful, as Porter noted, was "indexed onto the 'higher' or ornamental ranks" of society, and prescriptions for beauty were not just about preferred body image but about preserving the status quo.[83] Claude Quillet's *Callipaedia: or the Art of Getting Beautiful Children*—written in 1655 but subject to a number of popular English translations in the early eighteenth century—was critical of wet-nursing using poorer women since the beautiful bodies of children of the social elite might be corrupted and defiled with the "meaner Passions of the nurse," transmitted through breast milk.[84] *Artificiall Embellishments* was aimed at women of the social elite, and its author apologized for including a discussion of remedies for "an itchy or scabby skin," since this was the complaint of "kitchen maids" or "such as are forced to content themselves with courser concerns."[85] The body beautiful prescribed in this and many other similar texts was essentially decorative, geared to attracting the admiring looks of "inamorato's" rather than performing manual labor.[86] The fear that a servant might be esteemed more beautiful than her mistress, stealing her husband's affections and usurping her in the household order, was a potential source of anxiety. *Artificiall Embellishments* recommended that houses were well-lit since in the gloom of the "sooty night" chambermaids might appear "as handsome" as their mistresses.

Some writers explicitly denied beauty to the lower orders. Since, as we have seen, facial beauty was gauged by the quality of the sentiments of the soul, it was argued that people of "low birth" lacked the ennobling passions and grace necessary to be beautiful.[87] It was only those exposed, by dint of their manners, education, and breeding, to the improving "arts of polished life," maintained Samuel Stanhope Smith in *An Essay on the Causes of the Variety of Complexion and Figure in the Human Species* (1789), who might aspire to beauty, both physical and moral:

> What a difference is there between the soft and elegant tints of complexion in genteel life, and the coarse ruddiness of the vulgar!—between the uncouth features and unpliant limbs of an unpolished rustic, and the complacency of countenance, the graceful and easy air and figure of an improved citizen!—between the shaped and meaning face of a well bred lady, and the soft and plump simplicity of a country girl![88]

The poor diet and hard-working lives of the "poor and labouring" sort also supposedly damaged their looks. They were commonly "more swarthy and squalid in their complexion," more "hard in their features" and more "coarse and ill-formed in their limbs" than people of "better fortune, and more liberal means of subsistence." Such characteristics were believed to be hereditary—a clear warning against interclass breeding.[89]

Prescriptions for beauty also divided peoples of different countries and races. As Londa Schiebinger has observed, beauty deeply influenced the assessment of the world's peoples by anthropologists, travelers, and other commentators.[90] Writers were fascinated by the varieties of bodily form and physical adornment deemed beautiful in different parts of the world. "[A]ll People do not agree upon that which constitutes the Beauty of the Body," observed Andry. The "Tartars will not allow any Person to be handsome, unless the Eyes are little and hollow, the Nose large and flat, the Face hollow, and the Waist thick, especially in the Women." Among the "Moors," flat noses and thick Lips were marks of beauty, whereas "it is reckoned a Beauty amongst the Ladies in *China* to have the Feet less than the natural size." Although Andry maintained that his purpose in pointing out these differences was not to condemn the tastes of other peoples, the fact that such physical characteristics would be stigmatized as deformities among his genteel French and English audiences implicitly suggested that the beauty of some peoples was valued over that of others.[91]

Other writers were more explicit in arguing that some nations were more favored in terms of physical appearance than were others. The perceived ugliness of Dutch women has already been noted. The avowedly superior beauty of "Britannia's fair Daughters," became a commonplace of nationalist boasting. "Of all the nations in Europe, the English females are acknowledged to possess the pre-eminence in beauty and chastity."[92] Johann Reinhold Forster noted in his journal made during the voyage of the *Resolution* that although the women of Tonga were "all well proportionate," of "regular" features, and with becoming "brown complexions" and "fine brisk and lively eyes," their charms failed to match "those of our European fair ones . . . among which the British ones, no doubt, deserve the first rank." When one considered the "innocence and chastity of our British maids, their improved minds and all the other accomplishments they so commonly are masters of," then "not the least remembrance is left of all the charms of these copper beauties."[93]

Discourses of beauty played a key role in early modern fictions of racial superiority. White skin was deemed an essential attribute of the body beautiful. Restoration ladies were advised on how to "whiten a tan'd visage and to keep the face from sunburn," and protect their "Alabaster Armes and Hands."[94] A "fair" complexion, enlivened with a rosy hue, was esteemed as a sign of good health.[95] Fairness was also a sign of exalted feelings and refined sentiments—the term "fair sex" was coined during this period to convey the idea that women were naturally more refined in their passions.[96] Moreover, only in a fair complexion might the blush of modesty be seen.[97] Whatever the varying standards of beauty among different nations of the world, argued the *Athenian Mercury*, it was a principle of the "*law of Nature*" that "*White* is *Lovely* and *Black horrid*, one resembling the *light* the other *darkness*." While the striking physical presence of certain noble savages was admired by some writers, the majority of texts looked on blackness with disgust.[98] *Callipaedia* contrasted "Fair harmony and order" with "Black Deformity" and recoiled from the "horrid Fronts" and "thick Lips" of the inhabitants of the "sultry Regions of the *Torrid Zone*."[99] Whereas "fair" skin denoted superior qualities of the mind and exalted feelings, blackness was associated by eighteenth-century writers with "idiotism."[100]

Attempting to give the racial politics of beauty the stamp of scientific authority, Stanhope Smith argued that physical form was the product of both environment and civilization. "Coarse and deformed features are the necessary production of the climate," he explained. The skin darkened "in a torrid climate, where the inhabitants are naked [and] . . . the ardour of the sun is both more constant and more intense." Heat also relaxed the nervous system, disordering the body's bile and causing a darkening of the skin. Moreover, in extreme climates of hot and cold, the eyes appeared to sink into the head, whereas "in the temperate zone . . . the agreeable warmth of the air, disposing the nerves to the most free and easy expression, will open the features and increase the orb of the eye," presenting a more pleasing countenance. "Nakedness, exposure, negligence of appearance, want of cleanliness, and lodging, and meager diet" serve to "so discolor and injure their form" that, even without the influence of the environment, "render it impossible that a savage should ever be fair." Just as cultivated manners supposedly placed patricians above plebeians in the hierarchy of beauty, so the inhabitants of self-declared civilized nations were believed to surpass those of so-called savage states in terms of appearance. Thus, "the Europeans and Americans are the most beautiful people in the world, chiefly because their state of society is the most improved."[101]

THE PURSUIT OF BEAUTY

Achieving the body beautiful became a marked cultural (and commercial) obsession during the long eighteenth century. The starting point for beauty prescriptions was advice to parents on bringing forth fair and handsome children.[102] Concerns about birth irregularities were a recurring theme in early modern medical writings, midwifery texts, ballads, and pamphlet accounts of monstrous births.[103] The popular belief that the maternal imagination might affect fetal characteristics persisted into the eighteenth century, despite increasing scientific skepticism.[104] Sexual immorality, "too hasty" love-making, and astrological factors were also blamed for the birth of deformed children.[105]

The beauty of children was not merely an aesthetic issue but also guaranteed the nation's strength and moral well-being. Writing against the backdrop of heightened concerns about the growth of sodomy, one early eighteenth-century writer looked back nostalgically to an age when boys' education was governed by manly principles that favored martial exercises over book learning. Healthy and vigorous, such a man would be blessed with a "dutiful, beautiful Offspring" capable of serving "his King, his Country, and his Family." In contrast, the coddled effeminate youths of the present era were capable of fathering only a "feeble, unhealthy Infant, scarce worth the rearing."[106] In this pre-eugenics century, there was concern over fitness to breed. *Callipaedia,* for example, suggested that it was proper to prevent the disabled from reproducing:

> Not every woman was design'd
> To propagate and multiply their kind
> Forbid we rightly the Deform'd and Foul,
> To clothe with ill-shap'd limbs the heavenly soul
> . . .
>
> But he who judges right of what is fair
> With healthy sons will healthy Daughters pair.[107]

Procreation Refin'd: A New Method for the Begetting of Children with Handsome Faces similarly argued that couples should be matched in "personal perfections" and cited the example of a landowner who refused to allow his daughter to marry an ugly man, being "resolved to take all the Care [he could] from so fair a stock that the Breed shall not degenerate" and thus "hazard" his hopes of a "fair Race of Grand-Children."[108]

The connection between health and beauty was also apparent in advice literature aimed at procuring and preserving beauty in adults. *Artificiall Embellishments* employed the Galenic model of the nonnaturals in its general prescriptions for beauty, in which moderation was the key. A "comely face" and complexion was best achieved through protection against extremes of hot and cold air, through moderation in sleep and waking, regular bowel movements, avoidance of "perturbations" and dangerous passions of the mind, and temperance in eating and drinking. The body should be purged to rid itself of superfluous humors and then replenished "by food of light concoction and good nourishment."[109] Over a century later the advice was very similar. "A healthful constitution, and temperate habit of body, is the very ground-work of beauty," wrote the author of *The Art of Preserving Beauty*. Dancing was particularly useful since it "sets off beauty to advantage," being good for the posture, the muscles, and helping to lift the spirits. In contrast, beauty was endangered by fashionable pursuits that heightened the passions: "nothing has a greater influence on the loveliness of the human frame than those violent agitations of the mind which frequently attend the ill success of a gaming table—*Quadrille* will murder beauty."[110]

Where willpower and healthful remedies failed in improving appearance, "art" might assist. Attitudes toward cosmetics and physical beautification underwent a significant change during the long eighteenth century. Seventeenth-century puritans were deeply opposed to cosmetics or the application of spots or patches to cover blemishes. Proceeding from the assertion in Proverbs 31:30 that "beauty is vain," religious moralists inveighed against the "monstrous pride" of artificial beautification that deigned to improve the work of God's creation.[111] They argued that, like Jezebel and the Whore of Babylon, the painted face symbolized falseness and wantonness.[112] According to John Bulwer, painting and spotting among English ladies made them akin to barbarous savages.[113] Beauty spots (renamed "Beastly Spots" by the puritan Thomas Hall) came in for particular condemnation.[114] "Whereas you suppose that applying such spots and patches to your faces ye beautifie and adorn them," thundered one typical invective, "ye indeed make them visibly deformed and abominable in the eyes of all beholders."[115] Aside from supposedly terrifying children,[116] the "monstrous and prodigious custom" of wearing spots and patches was condemned as a foreign affectation used principally to hide the disfiguring effects of venereal disease—"French plaister" to conceal "French-pimples," as Hannah Wolley put it.[117]

This distrust of cosmetics persisted long into the eighteenth century. In 1754 the *Gentlemen's Magazine* criticized *Abdeker, or the Art of Preserving Beauty* as consisting mainly of "directions to paint and patch, to make washes, pomatums, lip salves and constringents, which are not necessary but to prostitutes, as they give an appearance of transient beauty, when the reality is lost by intemperance and disease."[118] However, by this time moral arguments against beautification were losing their force. Moderate Protestant opinion held that there was no harm in the "use of *colour* and *complexion* to the face," provided that "no *adulterous*, *wanton*, or *evill* purpose is harboured in the soul of those that use it."[119] *The Ladies Dictionary* agreed that there was a world of difference between the "harlot" who beautified herself to "allure and ensnare the unwary into her Embraces," and the "virtuous Lady" who does so for "Decency and the Credit of her family," or to please her husband. It was more becoming, therefore, for "grave Divines" to "meddle more with Ladies Hearts, and less with their Faces."[120] As Puritanism lost its influence after the Restoration, cosmetics became increasingly widely embraced by fashionable society. Samuel Pepys noted the ambition of his patron, the Earl of Sandwich, to have his wife and daughter wear patches as a symbol of their acceptance into courtly society.[121] The diarist's wife also employed a variety of beauty aids, from washing her face in dog's urine ("puppy-dog water")—criticized by Samuel as somewhat faddish—to rising early to collect May dew to rejuvenate her skin.[122]

By the eighteenth century, a wide variety of washes, tinctures, creams, powders, fragrances, waters, and oils were available to rejuvenate, revitalize the complexion, and heighten sexual attractiveness. Selling beauty products be-

FIGURE 6.5: Fashionable patches on a white face and those on a black "savage." In John Bulwer, *Anthropometamorphosis . . .* (London: 1653). Rare Books, Wellcome Library, London.

came a lucrative commercial venture for quacks and quasi-medical entrepreneurs.[123] *A Guide to Health, Beauty, Riches and Honour* (1785) advertised the services of Mr. Patience, dentist and dancing master of 8 Bolt Court, Fleet Street, who offered for sale "Rose Powder for preserving the teeth," "universal pills" for all ailments affecting the complexion, and cosmetic dentistry to ladies whose teeth had rotted. Elsewhere it advertised Dr. James Graham's infamous "celestial bed," whose electrical vibrations could be used for preserving health, "exalting personal beauty and loveliness," and for help in the "propagating of Beings rational, and far stronger and more beautiful in mental as well as in bodily endowments" than the current "puny, feeble, and nonsensical race of probationary immortals." Those willing to pay the shilling entrance fee to see it demonstrated at Graham's "Temple of Health and Hymen" would receive gratis "A very large printed pamphlet, containing advices of the highest importance to health, beauty, vigour, happiness and serenity of the human species."[124] Some products were aimed at both sexes and all ages, such as hair restorers or the miracle skin cream that was advertised in 1719 to take away

"Redness, Pimples, Roughness, Worms," and other imperfections.[125] However, most products were aimed at women. The proliferation of cosmetics was part of the development of a cult of femininity in the eighteenth century, emphasizing fundamental physical differences between the bodies of men and women. Cosmetics placed further emphasis on the qualities of "softness" and delicacy, apparent in so much prescriptive writing on beauty, and associated femininity with being alluring, youthful, and sexually attractive.[126]

THE VALUE OF BEAUTY

Discourses of the body beautiful were far reaching. Prescriptions for body shape were not simply a reflection of prevailing fashions or tastes but also a potent tool of differentiation between social classes, races, and nations. Standards of beauty were never just relative: writers sought to establish objective norms of bodily appearance that could be allied to broader systems of power and hierarchy. Most importantly, beauty emerged as a significant means of evaluating both the appearance and moral conduct of women, and of developing concepts of femininity. Beauty was an ambivalent quality: while the connection between the beautiful and the good was evident in many writings, beauty was also seen as a source of corruption and vanity, viewed by many with distrust. Although beauty represented as a source of power for women over men, providing both a means to conquer their hearts and reform their manners, prescriptions for beauty emphasized simultaneously female qualities of modesty and passivity that were key to women's subordination.

The perceived value of physical appearance to eighteenth-century women is revealed most vividly in stories of beauty's loss. The fear that loss of beauty would mean social death was a recurring theme at a time when frequent smallpox outbreaks threatened death or disfigurement.[127] In 1751, the *Gentleman's Magazine* serialized the story of "Victoria," a society beauty who from an early age had received "praise and admiration" for her looks. She had been encouraged by her mother, "whose face had luckily advanced her to a condition above her birth" and who "thought no evil so great as deformity," to take every opportunity to profit from her appearance. Courted by eligible bachelors and being daily "followed by gazers in the mall," "celebrated in the papers of the day," and "imitated by all who endeavored to rise into fashion," she experienced the adoration enjoyed by those few who "stood in the first rank of beauty." But at the height of her powers, she was "seized by that dreadful malady which has so often put a sudden end to the tyranny of beauty." Despite returning to health, her face was forever disfigured by smallpox scars, leaving her to lament that

> The condition of a young woman who has never thought or heard of any other excellence than beauty, and whom the sudden blast of dis-

> ease wrinkles in her bosom, is, indeed, sufficiently calamitous. She is at once deprived of all that gave her eminence or power, of all that elated her pride, or animated her activity, all that filled her days with pleasure, and her nights with hope, all that gave gladness to the present hour, or brighted her prospect of futurity.[128]

Beauty and appearance were seen as increasingly important, not just as a means of judging women, but as an indicator of how women saw themselves. The psychological benefits of beautifying the body as a means of procuring "love, affection and esteem" were promised by eighteenth-century purveyors of beauty advice or products.[129] The ways in which body image might affect a woman's feelings of self-worth are hinted at in the words of an anonymous female correspondent to the *Athenian Mercury* in 1692, devastated by her husband's infidelity:

> I am a young woman, not having the vanity to think myself beautiful enough to be ranked amongst the first Rate, but those of a lower degree, yet having as I hop'd just so much as to have kept my husband to my self.[130]

Further research is needed into how concerns about appearance were articulated by women at all social ranks. While the value of beauty was supposedly felt most strongly by elite women, there are signs that by the end of the eighteenth century, as the commercial appeal of beauty was becoming apparent, plebeian women might also experience anxieties about their appearance. A milliner wrote to the *Ladies Monthly Museum* in 1799 under the pseudonym "Martha Plainface" to complain that she was unable to get work because she was "ugly," for

> it is all the fashion now among the masters and mistresses to hire none but pretty journeywomen; so they place them in a row at the end of their shops, putting the handsomest most in sight, to entice the gentlemen into the shops to buy things.[131]

Pressure to conform to a prescribed body image—and the association of success and acceptance with a particular shape or appearance—is often seen as a condition of the modern age,[132] but, as this chapter has shown, its antecedents may be located in the long eighteenth century.

CHAPTER SEVEN

Marked Bodies and Social Meanings

LAURA GOWING

Amid the announcements of deaths, auctions, elections, bankruptcies, and exhibitions, the newspapers of Enlightenment England advertised for missing persons.

> Morgan Auberry, a black down lookt Fellow, of a middle Stature, thick Lipt, black short bushy Hair, went away from his Master, Mr Thomas Bartlet, at the Crown-Tavern without Aldgate, on the first Instant, with Eighty odd Pounds.[1]
>
> Went away from her Friends on Tuesday the 29th day of July last *Martha Thedham*, a Girl of about 15 years of age, she went from Mrs *Turners* School, a Plain worker in *Sherborn lane*; when she went away she had on a plain Muslin Cap and an old black hood, a yellow Stuff Gown and Petticoat, with red stripes, pretty old, her under Petticoat of old Silk of Gold colour and black, and old pair of Stays on, white sticht with red, the Stomacher with some Gold in it; she is not very tall of her age, thin, dark ey'd, and of a brown complexion.[2]
>
> Whereas Robert Fenton absented himself from his Service on Wednesday Evening, with a considerable sum of French and Spanish pistoles; he is about 23 Years of Age, about 5 Foot 4 Inches high, much pitted with the small Pox, the lower Part of his face thin, a thick Nose, black Eyes inclinable to be Sore, and black Eyebrows, a large Mouth, down Look,

> takes a great Quantity of Snuff; had on when he went away a yellowish Olive Colour old Drugget suit of Cloaths, and black Stockings, usually wears a light bob Wig, sometimes a black bob Wig, a Silver Pair of large wrought Buckles, a sore Finger on his Left Hand.[3]

The descriptions of those missing, usually servants, trace a history of distinguishing marks. Their anatomies of height, hair, dress, scars, and complexion recorded what employers or passersby were expected to notice about the strangers in their midst. Height is calibrated with increasing precision. Wigs, by the early eighteenth century, are as significant as hair. The fabric, weight, and color of clothes are recorded in ever-increasing detail, despite the sensible note in the last of these advertisements that "It is supposed they have changed their Cloaths." Alongside the indelible marks of nature (a thick nose) and disease (smallpox scars, sore eyes) stand those of character—for example, a "down look," denoting evasiveness or shiftiness.

The relationship between the visible body and the intangible self in the Enlightenment was changing. One strand of recent historiography has identified the period between 1650 and 1800 with a transition from bodies that were flexible, with both sexual and racial difference somewhat unfixed, and externalized, to a newly circumscribed corporeality, eventually experienced internally.[4] The evidence for this transition has been seen in new narratives, textual and visual, of sexual difference, race, and class. For gender historians, in particular, the mid-eighteenth century has come to represent the point at which corporeal sexual difference began to stand more solidly for the social meanings associated with it, with absolute differences in physiology replacing inflections of humors. Yet the historical purchase of this narrative remains debatable. The idea of male and female as fundamentally physically different was hardly a phenomenon of the Enlightenment, although it was certainly fleshed out with new details about anatomy and physiology. Perhaps the most striking change from representations of the body in the seventeenth century, to those of the eighteenth, is the proliferation of different kinds of sources, which enabled distinctive expressions of subjectivity: family portraits, memoirs, fiction, and journalism.

The relationship between those representations and popular experience is, as always, elusive. A consensus in print does not rule out a less visible and more complex mass of popular ideas. If medical or literary ideas presented bodies as less flexible, the ideas themselves remained highly negotiable; if anything, by the late seventeenth century there were more ideas in circulation, and more debate, about how bodies worked and what they meant.[5] As popular interest in anatomy, physiognomy, astrology, and the humors waxed and waned, the body's external marks remained a testament to what lay under the skin.[6] Alongside a story of the changing meanings of the human body, I want to explore here the multiplicity of ways in which bodies were understood to be marked, and the stories those marks told.

NARRATIVES OF THE BODY

Print culture facilitated a close attention to the visible evidence of difference on the body. A growing literature of manners encoded performances of gentility, offering guidelines for the aspiring young.[7] Popular medical books displayed the female and the male body to lay readers, demonstrating the foundations of sexual difference and the mechanics of reproduction. Travel literature introduced readers to African and Indian customs, exploring the meaning of racial difference through anecdote; science developed its own narratives of race. But histories of the embodied self have typically focused on the subjectivities imagined in elite culture, not those of lower cultures like cheap print, street talk, or pauper petitions.[8] If the eighteenth century is understood as an age of emerging individualisms, it has not yet become apparent how historians are to recover the individuality of those who were most often characterized as a mass of stereotypical figures: laborers, paupers, whores, or thieves.

This chapter traces some of those lost understandings through the records of a legal system that reached deep into everyday life. Legal records constitute something of a meeting point for the different kinds of story and memory that ran through early modern culture. The plots of early fiction often took criminality, immorality, and deviance as their subject. Moral crises that would later become private, like those of adultery and illicit sex, were still matters of judicial concern and neighborly intervention. Witchcraft and murder trials were already being printed as cautionary tales or providential narratives in the sixteenth century; audiences at assizes heard stories of murder, rape, and arson rehearsed as cases were tried. By the late seventeenth century, a regular series of London's trials was in print: the records of the Old Bailey. Together, these forms constituted a corpus of narratives in which old and new plots, readings of character, and notions of the self could be developed.

Legal records are a kind of fiction, mediated by memory, time, and clerical interventions. They are also very diverse in nature: any reading of them as historical sources must attend to their variability across place, time, and form. The records of the *Old Bailey Proceedings* (*OBP*), use a different style of narrative from the manuscript records of quarter sessions and assizes, or the Catholic or Protestant church courts across Europe. By the early eighteenth century, the Old Bailey records present witness statements in the first person, with an increasing amount of detail and complex narratives. Yet by the late seventeenth century, legal records such as these also all existed in a more homogenous context than had earlier texts (like those of witchcraft trials, for example). Pamphlets outlining trials, and case summaries in newssheets, were widely available throughout Europe. In legal documents, lawyers, clients, and their audiences developed a new concern with the representation of states of mind: legal responsibility was described and explored in relation to new ideas of selfhood and subjectivity.[9] The wider textual context for these explorations included print journalism and, of course, the early novel.[10] Printed legal records and criminal biographies constituted a discourse that encouraged the exploration of morality, sin, character, and reform. Those explorations concerned not only ideas about the mind, but about the body. The art of persuasion used with such craft in legal cases was also used to develop character and plot in the early novel. And in plots, or cases, about such matters as sex, romance, infanticide, or rape, the intimate relationship between body and mind was crucial.

In the late seventeenth and eighteenth centuries, the body was a source of knowledge. Its expressiveness was traced by professionals and amateurs, through the study of physiognomy and pathognomy: facial and cranial structures demonstrated the mind's character, while momentary or habitual expressions were testimony to passing states of mind. Artists and authors mined the stock of character knowledge this provided. In the novel *Clarissa*, the heroine's body is minutely legible, as she raises her eyebrows, or shrinks from her seducer; in ribald contrast is the world of Tristram Shandy, where a whole series of ways of reading the body authoritatively are tried out, and rejected.[11] In Aphra Behn's *Love Letters from a Nobleman to His Sister* (1684–1687), whose plot is based on the cause célèbre of Henrietta Berkeley's elopement with her sister's husband, the loss of the heroine's virtue is traced through both physical and mental characterizations. Once Silvia's chastity is gone, her body is irretrievably marked, and she becomes incapable of the kind of continence owing to a husband. Instead, she becomes the model of a female rake. In the first seduction, she is the trembling, innocent victim of her brother-in-law; in the last, her own seduction of her lover's servant, she masquerades innocence with shocking ease.[12] The loss of virginity is at once moral and corporeal.

Behn's portrayal of this transformation is a sympathetic and brutally telling account of how Restoration culture perceived the slip into whoredom. The ability to trace this kind of story came with the development of the novel and its associated narratives. It was, of course, not entirely new. Sixteenth- and seventeenth-century commentators on the body knew how to read its signs as proof of certain physical conditions, like virginity, promiscuity, or pregnancy, all of which, it was asserted in medical texts, could be detected from facial features and expressions. But one of the shifts traced by historians of gender has been in the relationship between women's bodies, sex, and sexuality: over the course of the eighteenth century the female body began to be seen as naturally chaste, rather than predictably sinful. At the same time, the developing narratives of fiction, drama, and journalism made it newly possible to follow stories about the body's decay into immorality.

The centrality of female subjects and tales of seduction to popular narratives means that this was a story most powerfully told about women's bodies. The stories of male corruption are less conclusively about the body or about sex. Hogarth's Tom Idle, even his Rake dying in Bedlam, bear barely any visible marks of their crimes; it is the women and children around them who most often carry the evidence of poverty (the emaciated, once plump wife of the Rake) and promiscuity (the signs of syphilis in the spot and calipers on the child in the final scene of *Marriage à la Mode*).[13] Female promiscuity was, of course, understood to have significant social implications; the increasingly publicly available records of adultery, as well as stories about illegitimacy and infanticide, made such implications abundantly clear. Popular culture also embodied a long history of interrogating, and drawing conclusions from, the female body: virginity, promiscuity, and pregnancy were all supposed to be visible in the lines of the face and the comportment of the body. But other sins followed the same path of lighting on the female body most harshly; the hopeless mother at the center of Hogarth's *Gin Lane* crystallizes a longer history of representations of female addiction.[14]

FIGURE 7.1: *Marriage à la Mode* (detail). The countess bids farewell to her child, whose facial spot and callipered legs suggests congenital syphilis. In J. Heath, *The Works of William Hogarth* (London: 1837). Wellcome Library, London.

FIGURE 7.2: A drunken, sore-infested mother drops her baby while preoccupied with taking snuff. Engraving after William Hogarth, ca. 1751. Iconographic Collections, Wellcome Library, London.

The seventeenth and eighteenth centuries inherited the idea of a humoral body—one in which a person's nature was determined by the corporeal balance of wet and dry, hot and cold elements. That vision remained powerful at the same time that medicine, science, and philosophy were reimagining the relationship between body and mind. One of the manifestations of this was the transformation of public policy toward the poor and criminal. From the later seventeenth century, both state discipline and private philanthropy were apt to assume that personal character was changeable through bodily work. In hospitals and later in workhouses, hard labor was construed as the cure for slack minds. Across Europe, disciplines of the flesh ranged from whipping and branding to workhouses and imprisonment; in all of them, it was assumed that the dangerous character of the shiftless, poor, or criminal could be reformed through the mortification of the body. In the seventeenth and eighteenth centuries, criminals were still being marked by punishments, publicly whipped, mutilated, or branded on the hand or on the forehead as robbers or vagrants; complaints were sometimes made that the brandings were not strong or lasting enough.

What people saw when they looked at the urban population, then, was shaped by the fantasy that bodies told truths—about morality, constitution, character, and history. Eighteenth-century bodies, as bodies always had, bore the marks of their social origins, of poverty and of disease. In expanding cities, with threatening numbers of indigents and vagrants, those marks perhaps demanded a sharper attention. In the moments of social drama recorded in legal records, the body's marks helped witnesses order the world of the streets.

BODIES ON THE STREETS

In 1700, Tom Brown's *Amusements Serious and Comical* imagined an Indian visitor "dropped from the clouds" to London's streets. His imaginary alien was struck first by the amazing variety of vehicles and people crowding the streets: "half a Dozen Brawny Bulk-begotten Footman," a "Gouty-Leg'd Chairman," "an old Burly Drab," a "Fat Greasie Porter."[15] His vision, of course, is not that of an Indian, but a native Londoner: he knows how to read the corporeal signs of bulk, brawn, gout, and grease. Ned Ward's *The London Spy* amassed these physical stereotypes to even greater effect, listing in each episode a whole crowd of stock characters distinguished with grotesque precision. His Billingsgate fishwives are

> a tattered assembly of fat motherly flat-caps, with their fish-baskets hanging upon their heads instead of riding-hoods, with every one her nipperkin of warm ale and brandy, and as many rings upon their thumbs as belongs to a suit of bed curtains; every one as slender in the waist as a Dutch skipper in the buttocks, and looked together like a litter of squab elephants.[16]

Fishwives are the consummate disorderly urban female, represented as bawdy, unwholesome, and dishonest traders. Here they are also the archetypal female drunkards, their cheeks crimson with the evidence that "brandy is a nobler dye than claret." Traversing the Royal Exchange, his narrator observes Spaniards, "lank-haired formalists," and French who "talk more with the heads and hands than with their tongues." The "bum-firking Italians" are followed by the "water-rats of Europe," "with huge logger heads, effeminate waists, and buttocks like a Flanders mare, with slovenly mien and swinish looks" and a "compound of Scotch and Irish," a "kind of lean carrionly creatures with reddish hair and freckly faces, being very much given to scratching and shrugging, as if they held lousiness no shame and the itch no scandal."[17] The topography of immigration, where new and old arrivals clustered together, as well as the business practices of a trading port, may have encouraged perceptions of the distinctiveness of Scotch, Irish, French, or Flemings; whether people really observed each other on such terms is another question.

Urban life was conventionally imagined to encompass a dazzling variety of people; anatomizing the marks that characterized character, occupation, and morals provided both a guide and a spectacle in itself. Image and fiction characterized the poor as shorter, uglier, and grotesque; the self-contained, classical body was an elite one.[18] It was well-established, too, that the difference between city and countryside was marked on the body. The widening gulf between city and country life had been noted in the sixteenth and seventeenth centuries, in comments on behavior, manners, civility, and clothes; for balladeers and journalists of the eighteenth century, the very bodies of country folk and city dwellers looked different. Some saw Londoners as brisk, animated, and peculiarly alert; others noted their shortness, paleness, and "languid, sallow" looks, the results of polluted air and food.[19] A conviction

FIGURE 7.3: Fishwives in the market with their tankards of ale (detail). Engraving ca. 1751 by William Hogarth. Iconographic Collections, Wellcome Library, London.

that bodies told truths about environment, personal history, and character, and a disagreement about how those truths were manifested, shaped what people saw.

Some of the ways that the urban environment affected bodies are possible to reconstruct. The Georgian period saw the average heights of Londoners beginning to decrease; by the nineteenth century men's average heights were down by about two inches, and women's by one, from the Tudor and Stuart periods.[20] Poor diet stunted growth; injured limbs were withered or amputated; deafness and blindness, disease, and disability shortened the lives of the poor. By the late eighteenth century, malnutrition, smoke, and pollution were leaving children of all classes with rickets, and the damage that resulted caused bent limbs and narrow pelvises that made childbirth riskier. Contemporary observers noted some horrific visions of urban destitution and disease. A witness described his sight of a prisoner who, he believed, was being starved to death in the house of his neighbor in Holborn: "he saw a poor Wretch who had the Small-pox that run through his Cloaths, that they stuck to his Back like a Searcloath, and as black in the Face as Leather."[21] The vision of pox as a mobile, active agent, piercing through clothes, gives some hint of the peculiar understandings of disease that governed early modern thought. In 1732 another suspected victim of starvation, Mary Vezey of Stepney, was described by a beadle who met her in the street:

> Her Body was all black, and her Legs were perfectly covered with a white Mold. She had on a thin old Crape-Gown, and a Bit of a red Petticoat, but no Shift nor Stockings. . . . Her flesh was all over wasted.

Another witness, her cousin, said "She shew'd me her Legs and Breast; she was nothing but Skin and Bones; her Skin was all black, and she look'd just like an Anatomy that I have seen at a Surgeon's."[22] A surgeon judged her death to be, in fact, from consumption. City life gave people plenty of damaged and diseased bodies from which to build a language of comparison and judgment. Hospitals, prisons, and street beggars were part of the urban exhibition (see Chapter 9); "anatomy" here probably means a skeleton or a surgeon's model, but it could conceivably also suggest a dissection.

Smallpox was a familiar hazard, particularly for city dwellers, and was likely to scar its victim for life. Old Bailey witnesses often commented on smallpox scars, but they were so common they were generally simply another feature. A soldier in 1711 was noted to have defected "with the small pox fresh upon him."[23] The significance of smallpox might be gendered (see Chapter 6): Fanny Hill, heroine of John Cleland's *Memoirs of a Woman of Pleasure*, is grateful to escape unscarred; but she also notes that one of her male clients is marked in a way that merely made him more masculine, adding "a grace of more manliness to features rather turned to softness and delicacy."[24]

As the idea evolved that moral ills could be physically manifested on the body, the example of venereal disease dominated. From its first representations in Western Europe at the end of the fifteenth century, the pox was understood to mark not just the sexual organs but other, more visible parts, too: it functioned as a kind of literal proof of the indelible effects of promiscuity on both women and men (see Chapter 8). Damaged noses were read as syphilitic and were metaphorically related to genitals. Proof of infection with venereal disease was one of the few means of prosecuting rape, providing otherwise problematic evidence of physical damage. Even then, though, jurisdictions were readier to attend to underage girls than to adult women, whose infection was less incontrovertibly proof of rape.[25] Knowledge of the ways that venereal infection marked the body did not, either, necessarily determine treatment. Medical practitioners through the seventeenth and eighteenth centuries treated men without informing their wives; such practitioners may have even given an incorrect diagnosis to pregnant women rather than reveal the misdeeds of their husbands.[26] Signs of the pox were not as reliable as people imagined them to be: the belief that promiscuity left tangible evidence was a fantasy projected onto both male and female bodies.

Age changed bodies quickly and visibly. For women, the phrase "aged" was too often coupled with "poor," and the dependence of old poor women put them at the parish's disposition for, for example, the dangerous work of searching the dead for plague signs.[27] Without poverty, though, age brought a different status: the thief of a green silk petticoat in 1678 was described as "an elderly comely woman."[28] Perhaps particularly for older women, the marks of old age were hard to mask: osteoporosis and curvature of the spine, lost teeth, and the accumulated injuries of a working life made up a picture of postmenopausal women that was familiar from seventeenth-century representations of witchcraft.[29]

Observers knew how to read class from costume: in the series of images that depicted street sellers in Marcellus Laroon's *Cryes of London* (1687), the poorest hawkers were dressed in shreds of clothes; the slightly better-off in patched but intact garments; and the dignified workers in plain, tidy clothes.[30] These are conventional images; their relationship to actual appearance is complex. The patchwork appearance of the begging poor, Tim Hitchcock suggests, was itself a response to the cultural representations of pauper archetypes: torn, dirty clothes, rags bound under the feet, and sticks as crutches were the beggar's only tools (see Chapter 9).[31] Despite a small revolution in consumption, in the late seventeenth century clothing still clearly marked status, and for many people their daily garb made them as recognizable as the hair or their faces.

Criminals, it was alleged, took advantage of this, resorting to disguises that were almost as criminal as the acts they concealed. In 1732 John Griffin ("King John"), a blacksmith accused of burglary, was described as follows:

> a short, square, well-set Man, about Thirty-five Years of Age, much pitted with the Small-pox, round Vizage, his Hair short, thick, and curling, of a very dark Brown, his common Apparel a striped Flanel Waistcoat, and a Leather-Apron; if he wears a Coat, which is seldom, 'tis generally a Snuff-colour'd Camblet, and for the most Part open and unbutton'd, but he has several Disguises; sometimes he appears like a Countryman, a Sailor, a Butcher, &c.[32]

Expectations of criminal appearance, too, might be frustrated. George Gallimore, robbed by two women in a brandy-house in 1733, swore he would recognize the one who picked his pocket; but the next day, dressed in different clothes, she turned out to be unrecognizable to him, and he told her the thief was "lustier than you, and pitted with the Small-pox, and was in a black gown." By then she was dressed in a "green gown and flower'd robings," and looked quite different.[33] Still, in the face of multiple disguises, witnesses continued to describe what they saw as if clothes were likely to be as unchanging as features. This was particularly so where livery was concerned, since it provided an outfit that would not be readily discarded. A man living in the Fleet, accused of robbery, was freed from suspicion when it emerged that he had been wearing a livery of gray stockings for the last eleven months; the robbers were seen in white ones.[34]

At the other end of the social scale, aspiration encouraged, perhaps required, self-improvement: a refined body demanded better skin, teeth, and hair. Dyed hair, reddened faces, and mouse-skin eyebrows were accompanied by miracle cures to make an old complexion young.[35] Wigs were available in every possible cut to echo or enhance social status; many men in the 1700s had their heads shaved for them. By the 1710s, advertisements for missing men described both their own hair and their wigs. The expense of wigs made them perhaps even more reliable than original hair as an identifier.

Mouths were particularly important: smiles eased social mobility. The aspiring urban public was expected to want to cure "cankers," "ulcers," "scurvy in the mouth," and loose and blackened teeth. In 1707 an Aldgate practitioner was advertising a "certain and infallible Cure for the Tooth Ach, without Drawing, and that so effectually, that the Pain will never return; and not only so, but makes Teeth as white as Ivory, and fastens those that are loose to Admiration."[36] The public smile was sufficiently significant that people would buy false teeth that were unusable for practical purposes, but impressive in appearance.

FIGURE 7.4: An elderly woman surrounded by the trappings of youth, attends to her toilette. Only her breasts bespeak her gender. Etching, early eighteenth century. Iconographic Collections, Wellcome Library, London.

SKIN AND RACE

In newspaper descriptions of runaways and deserters, details of complexion record the spectrum of what was perceived as white, from pale and fresh-complexioned, through sandy, russet, and reddish, to swarthy. Skin color mattered partly because it had implications for the temperament: a red face meant healthy blood and a sanguine character, while dark skin suggested the presence of black bile. The vocabulary of whiteness was rich and complex. Blackness, at least at the start of our period, was a vaguer concept: in 1700 it could mean anything from an African to someone with dark hair.

European communities were beginning to rethink the meaning of skin color, and to crystallize the prejudices associated with ideas about race. Greater immigration made for more ethnically diverse communities, particularly in ports and large urban areas. The fear of racial contamination voiced by some eighteenth-century commentators, of a "morisco taint" and a "little race of mulattoes," was offset by the necessary daily acceptance of racially mixed marriages and relationships among the working poor.[37] The practice of reading race in the body involved constant transformations of language. The word "black," long-used to describe the unusual dark-haired British, did not yet definitively mean either African, or even the dark-skinned; there was an overlap in usage for at least a century. The man who, trying to trace a theft of clothes in 1726 said "a black Man may turn grey, and so may a black waistcoat," meant a man with dark hair.[38] Twenty years earlier, Mary Harris, accused of stealing sheets and smocks, was described as a "Black-woman": this time, the speaker seems to have meant that Harris was African.[39]

All this suggests a lack of differentiation at the turn of the eighteenth century between the meanings of "black." If dark-haired Europeans, Africans, or Indians, and those (rarely identified) of mixed race, could all be black, perhaps the flexibility of terminology reflected a popular culture with no need for an established taxonomy of racial classifications. Visitors from Europe to the Indies and Africa were sending back reports of racial characteristics and their origins that were highly individual and often contradictory. Popular texts published in the late seventeenth century were still suggesting that a white couple could, by means of the maternal imagination, produce a black child and that black people might become white after a long stay in northern climes.[40]

However, there is also evidence of a use of more apparently specific terms, at least by the mid-eighteenth century. "Mulatto" was used in court starting in 1726. John Sutton, of Clerkenwell, accused of rape in 1745, was described in court as "a black," and called by neighbors "Black John"; his alleged victim was described as "a mulatto."[41] Over the course of the eighteenth century, blackness was identified and anatomized more precisely. It was not until 1751 that someone was recorded at the Old Bailey using the phrase "black as a Negro."[42] Along with the development of racial terminologies came, eventually, another set of criminal identities and an intensification of pejorative language. A black gardener in 1780s London was called "you sulky negro."[43]

A violent fight ending in death in 1731 suggests some of the complexities of the visible signs of race, and their use in urban life. Robert Ormes was killed in an alehouse argument in Chelsea in 1731, in the course of which his attacker, Francis Woodmash, had called Ormes's wife, Elizabeth, "Brandy-Face." She replied, she told the Old Bailey, "that hers was no more a Brandy-Face, than his was a Jew's-Face, for he look'd like a Mulatto (he being of a swarthy Complexion)." Another of the company added "Jews were circumcis'd." The defendant, though, said only that he had been called "Irish rogue," and witnesses reported that the original fight began with him quoting Latin, claiming to be a scholar, and calling the others blockheads, tradesmen, and "scrubs."[44] This is a war of gendered as well as racialized images: "brandy-faced" evoked all the stock images of female drunkards, and the eventual addition of circumcision

FIGURE 7.5: Female Hottentot as observed by an eighteenth-century visitor to southern Africa. Iconographic Collections, Wellcome Library, London.

to her conflation of Jews, mulattos, and swarthiness was surely meant to undermine his manhood.

The otherness of the Jewish body was defined in different ways than in later discourses. There was a seventeenth-century debate about whether, and which, Jews could be defined as black.[45] Other writers looked for a distinctive voice, gesture, and looks, a peculiar gait, and crooked feet; Jewishness was associated with swarthy skin, poverty, and plague (but not, apparently, with a particular shape of nose).[46] The mention of circumcision raises another interesting point, for the first generation of London's new Jewish community in the 1660s, being mostly conversos (forced converts) from Spain and Portugal, were often uncircumcised, to the unease of rabbis.[47] Precisely what made Jews physically different was, then, in dispute. Occasionally trials shed light on the popular grasp of traditional Jewish bodily practices. One witness had told another that Aaron Polack was at dinner in a tavern on a Friday night but would eat no bacon; in the process of questioning Polack about where he had been on a Saturday, the judge asked him if his religion allowed him to walk as far as Highgate on his Sabbath. (Polack's response was that there were good Jews and some bad ones, but there are "some good Jews that walk out of the Sabbath.")[48] A theft case reveals Solomon Isaacs asking a shopkeeper for a boy to come home with him to collect payment, as it was close to 7 P.M. on a Friday and "we Jews don't touch Money after our Sabbath is begun."[49]

By 1700 there was also an interest in the foreign voice. The eighteenth-century proceedings of the Old Bailey attempted to transcribe depositions by Jews and other foreigners phonetically, leaving an extraordinary record of the (perceived) speech patterns of immigrant voices. A French woman accused of bigamy was recorded as saying that her second husband "Monsieur Benchain, he no do vid me in de Bed, he do nothing in de Varld;—noting but de Ceremonie;—vy den dat be no Husband."[50] A Jewish man testifying to a theft from his shop in 1732 was quoted in phrases that would suggest to the reader not just a recognizable voice, but a way of storytelling:

> I vas but a very leetel vile gon out of de Shop, and ven I vas come back to go into de Shop again, dare I see dis Man, de Precsonare, vas in de Shop-Antry, and he vant to come out; so I aska to him, vel Friend, vat vas it you vas please to have? And he say, noting at all, he vant noting in de Varld, not he. Vel den I say, Vat you do here? He make a me no answer to dat; but he creep, creep along by de Side of de Vall, and sleep out of de Door, and run avay.[51]

MARKING GENDERS

Narratives of criminality are rarely explicit about gendered bodies. The boundaries are marked implicitly. Useful comparative descriptors of strength and body type like "lusty" and "stout" must have meant different things for men and women; the evidence of drunkenness and pox certainly did. When the lines of gender were crossed, rules became more explicit. The cross-dressed woman was almost a stock character of sixteenth-century popular literature, heroic in her ability to perform masculinity, even in conjunction with crime. Contemporary images of Moll Cutpurse show her looking convincingly male, but by the time her story was retold in the eighteenth century, she was posing like a woman in men's clothes. Dror Wahrman has pointed out a transformation in the telling of such stories: by the eighteenth century, cross-dressed women's ability to appear male looked seriously threatening.[52]

In the Old Bailey records, a few women are distinguished particularly as masculine. Until the end of the eighteenth century, this did not necessarily mean manlike, or unwomanly; it was still being used in its older sense of strength. But it was also coming to mean the opposite of womanly. In 1747,

FIGURE 7.6: Moll Cutpurse (Mary Frith) as depicted in 1662. In James Caulfield and R. S. Kirby, *Collection of Four Hundred Portraits of Remarkable, Eccentric and Notorious Personages Printed from the Original Copper Plates of Caulfield's Remarkable Characters, Grainger, and Kirby's Wonderful Museum* (London: Reeves and Turner, 1880). General Collections, Wellcome Library, London.

a woman who intervened in a robbery in Shoe Lane was described as "a masculine Creature." Allegedly, she told the (male) victim "Damn you, why don't you seize and charge?"[53] Nothing, other than her forwardness, is advanced as further evidence for her supposedly masculine nature. A much later case conflated masculinity more specifically with violence. Ann Fell, accused in 1790 of stealing a gown and poor relief money from Rachel Keen, a widow who shared her room in a poor-farm in Mile End, was described by her victim as masculine and strong. In contrast, Rachel Keen and her witnesses described herself as old and weak: Keen's story told how Ann Fell:

> laid me down in her lap, and took my property; and then she pushed the candle out; it was the 28th of June: I go to bed between seven and eight: I am sure it was candle-light; for she pushed the candle out of my hand; and I screamed out at the instant; and I said, she has got my tin box and my money; she beat me on the hips; so I let it alone; I had no stays on; and she is a great powerful woman.

Other witnesses, she said, did not intervene because they were frightened of Fell, because she was "such a masculine strong woman." Fell, in return, brought witnesses alleging Rachel Keen had been drunk.[54]

In 1799, Jane Gibbs, describing her robbery by Jeremiah Beck, found herself counteraccused, based on her alleged demeanor and looks, of being a "girl of the town." In her story, she was an honest girl robbed by a strange man; in the stories of witnesses, she was a woman (not a girl) of suspicious manner, overly tall and masculine looking. A girl would have been entitled to some assumption of innocence, and to protection. Instead, one of the victims, Reverend Dr. Ford, testified that it was she who did the robbing:

> I was a little alarmed, I confess, because the woman was taller apparently than myself; I looked up and down the street, and saw no creature walking but myself; she then cried, watch; she took me by one hand, and put her hand to my pocket to feel for the money which she said I had taken from her, and said, you have it here; I immediately, with my right-hand, seized her by the right-wrist, in this manner, (describing it); and being left-handed, and pretty strong in my arm, I said, if you do not let me go, I will certainly knock you down; and I was going to knock her down, for I really thought, from the face and the manner, that it was a man in woman's clothes.[55]

Likewise, Gerald Fitzgerald deposed that he had met Gibbs late at night, and taken her arm; though she had protested she was "not a bad girl," he said:

> she gave me several squeezes of the hand while she was saying she was not a bad girl, but having seen her face, I wished to quit her as soon as I could, as she was such an extraordinary looking animal as I had never seen before.[56]

If Gibbs did not present exactly as a "girl of the town," it was also possible to describe her as a strange creature, an "extraordinary looking animal." This suggested a peculiar, unfeminine middle ground, by which women could be described as neither whores nor virtuous, but simply wrong-looking and hence dangerous or suspicious.

In earlier criminal records, what caused women to be read as masculine or male was generally their clothes; for women, cross-dressing had always been, and continued to be, a means of disguise for both criminality and personal safety.[57] But in these cases, it was physiognomy, not clothes, that marked the extremes as well as the norms of gender: the face, the height, and the general physical affect, were either feminine or masculine. This kind of face reading seems to be quite absent from earlier attempts to characterize female criminality; it suggests a culture in which masculinity and femininity could be seen to be composites of recognizable attributes, not inflexions of humors. The naming of some women as "masculine" might, then, be a further marker of the redefinition, and straitening, of femininity.

If women's strength made them masculine, weakness in men was more likely to be identified with poverty or old age than with effeminacy. By the early eighteenth century, too, a sufficient concept of sexuality had developed such that when men were identified as female-looking, they were likely to also be called mollies or sodomites. In 1738 John Stevens was sentenced to transportation for stealing five silver seals to the value of seven shillings from a shop in Bishopsgate; he defended himself, saying he was too drunk to know what he was doing. To the witnesses, who knew him from Borough Market and said he was of no trade, and not much reputation, he protested: "they call me a Molly, and say I am more like a Woman than a Man, and how can I help my Face?"[58] Ideas like this were part of a cultural landscape of sex and gender that is rarely fully visible. In it, gestures, features, and manners were combining to make an urban sexual subculture, recognizable at least to those who searched for it, and sometimes to those who feared it. The same development has been noted in trials of sodomites in eighteenth-century Amsterdam.[59]

The Old Bailey trials also picked up men who regularly passed as female. The most notorious was John Cooper, known in his Chelsea neighborhood in the 1730s as "Princess Seraphina." A female neighbor explained:

> she commonly used to wear a white gown, and a scarlet Cloak, with her Hair frizzled and curl'd all round her Forehead, and then she would so flutter her Fan, and make such fine Curtsies, that you would not have known her from a Woman.[60]

There is a careful attention here to the mechanics of making a woman: clothes, hair, and gesture are artfully combined, and the result is someone who, while still not a woman, cannot be distinguished from one, and who demands female pronouns as she transforms herself. Princess Seraphina also worked at the roughest end of women's work, going out to nurse those having mercury treatment for the pox, and ran "sodomiting errands" between gentlemen. John Cooper also had some good men's clothes. At the Old Bailey in July 1732, Cooper claimed that he had been robbed of his good (men's) clothes at knife-point by Thomas Gordon, who had threatened to swear Cooper had given him the clothes "to let you bugger me." At the center of all the testimonies was the absolute power of dress. When John Cooper appeared in an alehouse in his black coat, white waistcoat, and black breeches, and returned the next morning in "a dirty ragged Suit of Cloaths, and a speckled Shirt," a witness said he had "never set his Eyes on a Man so metamorphos'd." When a friend saw Thomas Gordon in his new, better clothes, he said he hardly knew him. Clothes, whether men's or women's, were the currency of sex and advancement, and the means of almost complete disguise: no one expressed any surprise at the suggestion that Thomas Gordon had accepted a set of fine clothes in exchange for a sexual act that was punishable by death.[61]

Most trials that revolved (as this one did not) around sodomy made no suggestion that sodomites or mollies were in any way distinguishable; a culture of cross-dressing, secret signs of recognition, and bawdy nights in secret alehouses only became public with the exposure of molly houses like that of Mother Clap.[62] But there was already both a language in which to talk about the disguise or loss of masculinity and an established connection with sodomy.

Physiognomy was achieving a new seriousness in the eighteenth century, becoming a mainstay of lay as well as medical discourse. In criminal records, as in novels, the shape of the face and the arrangement of features were coming to denote not just a means of recognition but a sign of temperament. In 1739 Arthur Jackson recognized the man who had tried to press him into enlisting by "his Phisiognomy, and the deep Scar in his Forehead."[63] In 1678, the judge's summation after a case of coin clipping proclaimed to the young man convicted:

> I am sorry, heartily sorry, and very much lament to see a Youth, in whom there seems to be so much modesty, far from persuading any one to believe, that any manner of Villany should lurk underneath so promising and so good a Face, come under the guilt of so great an Offence. But the truth of it is, the Apprentices of London have got such a Trade of abusing their Masters by Clipping, and such tricks, which they are encouraged to by a pack of Goldsmiths Men, who are fit for their purpose, that if some of them be not made Examples, it will be the ruine of many . . . And I am sorry to see you the first sad lamentable instance of that Justice, which must pass against Offenders of that kind, whose modesty should have prevail'd upon you, not onely to look like a vertuous Boy, but so to have acted.[64]

In this period, when adolescent boys were likely to be dependent on masters, "modesty" denoted the kind of subject carefulness that would also be expected of women. To be "virtuous" and "modest" was a young man's role, as it was a woman's, and the most "promising" and "good" faces suggested as much. Renaissance literature made much of the disjuncture between women's faces and their characters; the physiognomy craze of a century later encouraged similarly moral reflections about men.

The body marked by gender also registered the social order. Subordination was registered in physical gestures, in degrees of privacy, in public manners, and in bodily integrity. In households, in marriages, and on the streets, the young, the poor, and particularly women found their bodies might be public property. The authority that maintained early modern hierarchy was a physical one. The formal context of this—judicial punishment—was more visible than the informal side; in a female-dominated world of sexual and reproductive control, women's bodies were subject to the authority of mistresses, matrons, and midwives.

In the eighteenth century, touch and intuition remained one of the powerful ways in which women were regulated by women. The formal role of so-called juries of matrons gave some women the work of examining those who claimed pregnancy to escape the gallows. Women suspected of infanticide were likely to have their breasts searched to check for evidence of milk. In 1712 Anne Nichols's mistress (also a midwife) followed up her suspicions of her condition by testing her breasts: finding Anne up earlier than usual in the kitchen, she saw "an Alteration in her face, and having a Suspicion, felt her Breasts, and told her she was sure she had been newly deliver'd."[65] In another case in 1729, a midwife checked a woman's breasts and found plasters on them to dry the milk out.[66] In Newcastle in 1753, Ann Milbourne was targeted by her young cousin, who said, "Nanny you are cock-bellied I think"; she defended herself by saying "her stays were bad." The girl was reproved by her mother for "speaking so forwardly," but the substance of the accusation was a familiar one. Another woman observed that Milbourne "seemed to be thick about the waist and go dirty in her clothes."[67]

The eighteenth century saw this habit of physical surveillance begin to change, becoming more formal and less female focused. The role of midwives and lay married women in policing other women's bodies was in decline, less because of the incursions of male midwives and surgeons than because of the transformation of justice, bureaucracy, and the poor law. Across much of late seventeenth- and eighteenth-century Europe, the place that women had traditionally, if informally, held in various parts of the public sphere was diminishing as citizenship was formally defined as masculine. In England, the formal, judicial role that had previously required women to testify to paternity, or report illegitimacy, was much less prominent, as parishes managed their poor differently, and workhouses came to embody a new kind of institutional supervision. This seems to have affected the disciplines to which women's bodies were subject. In the seventeenth century, both neighbors and constables instigated searches, and women performed them as part of the process of neighborhood monitoring of sexual and social norms that regulated early modern communities. Eighteenth-century records make much less note of such encounters; the job of interrogating for paternity was no longer midwives' responsibility, and perhaps with that, other women's functions in the regulation of reproduction declined. Instead, the role of male experts took on a powerful, politically charged place.[68]

These kinds of stories make women's breasts the object of regulation, a target for control. But there were other ways of using them. Mary Cut and Come Again, accused of highway robbery in 1745, was said to have used her breasts to assault the men who came to arrest her: "she pulled her breasts out, and spurted the milk in the fellows faces, and said, damn your eyes, what do you want to take my life away?"[69] Mary was a ballad singer, accused of stealing from another woman by cutting her apron strings (as a witness noted, "I have seen many a ballad singer, and I never saw one with two aprons").[70] She was also described as a spinster, though that may have been a legal fiction. There is no mention in the printed records of any children or recent pregnancies she may have had; the presence of milk (and enough of a supply to squirt it at someone) was presented as just one of the fluids at her command. There is a long history of the "leaky" female body, bawdy and grotesque; to find a woman using her body like this on the streets might attest to a continued popular subculture that subverted it.[71]

Culture and experience marked bodies; people worked their own way with them. Beggars wore false beards or faked pregnancy; the toothless bought better mouths; the fashionable affected beauty spots. In response to the state's marks, criminals and prisoners inscribed their own tattoos. In their daily lives, people learned both the potential and the limits of transformation. But it was the language of bodily marks that made them legible. Early modern people had well-established physical traditions and languages of the body to work with; they saw women's chastity marked out in their faces, people's temperaments in their complexions, gender roles in gestures. They had new ideas too. Changing ideas about race, gender, and disease expanded the vocabulary of difference and simultaneously contracted it: if the eighteenth century's conception of physical difference looked, eventually, more rigid and constrained, it was also distinguished by a depth and variety of interpretation that must have left much room for flexibility. Alongside the new discourses that marked the bodies represented in elite culture stood a popular world where the everyday observer used a rich, complex language to read the meaning of the scars, complexion, dress, and demeanor that defined people's places and told their stories.

CHAPTER EIGHT

The Puzzle of the Pox-Marked Body

SUSAN STAVES

Eighteenth-century medical men and laypeople alike puzzled over questions of how to read what seemed to be signs of a disease they knew by many names, including the "pox," "lues" or "lues venerea," "venereal disease," the "French disease," the "loathsome disease," or "syphilis." Most obviously, pox seemed to produce a set of visible symptoms, marks on the body, which were difficult for a sufferer to conceal, though many tried to conceal them. There were skin lesions ranging from generalized but superficial rashes to deep, necrotic ulcers, sometimes exuding foul-smelling discharges. Notoriously, a poxed nose, from which disease had eaten away the cartilage and bone, no longer looked like a human nose. Perhaps worse, a palate so eroded that the passages of the nose and mouth were no longer separate could make normal speaking and eating impossible. Patches, wigs, and masks might cover some lesions, and prosthetic devices like silver palates might alleviate some of the disagreeable signs, but such coverings and devices themselves could also be viewed as marks of the disease. Endless jokes played with anxieties people felt in a society where venereal disease was endemic and where sufferers, motivated by shame or sometimes more simply by a wish not to frighten off potential sexual partners, tried to hide its marks.

While a few bold libertines might boast of pox as a badge of honor in the so-called wars of Venus, most Restoration and eighteenth-century sufferers felt its marks were shameful. Medical writers reported patients' distress at the social stigma of bearing visible signs of the pox and warned fellow practitioners that patients were frequently not truthful about their sexual and medical histories because they tried to avoid a diagnosis of pox. Pioneers of medical demography in the period complained that the *Bills of Mortality* failed to provide useful statistics of the numbers of deaths from the pox because friends and families of the afflicted conspired with officials to create false cover diagnoses. John Graunt in 1662 remarked that since it was common knowledge that "a great part of men have at one time or other had some *species* of this Disease," he wondered why so few deaths from pox were registered in the *Bills of Mortality*. Upon investigation, he determined that only in two London parishes, where there were notorious houses of prostitution, were deaths from pox typically reported; elsewhere, the cause of death was given simply as "*Ulcers* and *Sores*": "from whence I concluded, that onley *hated* persons, and such, whose very *Noses* were eaten of, were reported by the *Searchers* to have died of this too frequent *Maladie*."[1] Similarly, at the end of the eighteenth century, an Irish physician practicing in London, William Black, in *An Arithmetical and Medical Analysis of the Diseases and Mortality of the Human Species* commented that those reported as dying from the venereal disease were "in all probability" mostly prostitutes and lower-class men, because among those "of less straitened circumstances, a small bribe to the searchers would conjure venereal mortality into sores or ulcers, or into consumptions."[2]

Contemporary conventions, especially in satire, comedy, jokes, and satirical prints, recognized a set of marks as visible signs of the pox. These conventions differed somewhat from later conventions. Among the most common signs was the patch worn to conceal a skin lesion, the sufferer hoping that the patch would be read simply as a fashionable beauty patch. Other common conventional signs were associated with the mercury treatment often recommended for the pox: ointments, sweatings, confinement to one's room, and loss of teeth. Thus in William Hogarth's moral series, *The Harlot's Progress*, we see Moll Hackabout in the second plate, as mistress of a wealthy merchant and wearing a beauty patch, but in the third and fourth plates she has additional patches and medicine by her bedside. In the fifth plate, weak and near death, she sits in a chair wrapped in sweating blankets.[3] More directly, in a wonderful pastel-colored print, "Bawd on Her Last Legs," Thomas Rowlandson shows a pretty young prostitute dressed in pink watching a bewigged medical man lean over a plump older woman with a big pocky ulcer on her leg. Eighteenth-century representations of the pox often did not include what was later called the chancre of primary syphilis, recognized as a sign of the pox by some but not much focused on either by sufferers or medical men since it was of short duration, typically appeared on the genitals, and was not sufficiently bothersome to motivate most people—rightly suspicious of medical treatment as fraught with iatrogenic peril—to seek treatment. Nor did conventional signs of pox include the dreaded neurosyphilis of Henrik Ibsen's *Ghosts*, since most contemporaries did not think of its symptoms as a possible form of tertiary syphilis.[4]

FIGURE 8.1: Moll Hackabout being treated for the pox. Engraving by William Hogarth, ca. 1750, plate 3 of *The Harlot's Progress*. In J. Heath, *The Works of William Hogarth* (London: 1837).

On the other hand, in addition to the skin lesions found to be hallmarks of syphilis, eighteenth-century people were especially concerned by bone pain. This was often the symptom that drove patients to consult the medical men who specialized in venereal disease. Even in the absence of external lesions, bone pain, frequently in the shins or forehead, could be severe enough to drive patients to risk treatments that were themselves toxic and painful. Tobias Smollett, both medical man and novelist, in *Roderick Random* makes the first evidence of the presence of his prostitute character suffering from the pox "a groan" heard through the wall of a garret room.[5] In works of Hogarth, Rowlandson, and Smollett, it is usually the body of the female prostitute that is represented as afflicted with the signs of the pox, and it is usually women who die of the pox; male characters catch the pox, but their bodies are rarely represented as marked by it and, after an embarrassing and usually comic interlude of treatment, they recover.

Despite the contemporary sense that the pox often revealed itself with all-too-distressing visible signs, contemporaries were also aware that signs of the pox might be difficult to read. In this chapter, I want to consider some representations of the signs of the pox in the medical literature and associated debates over how to read such bodily signs. For the purposes of this exploration,

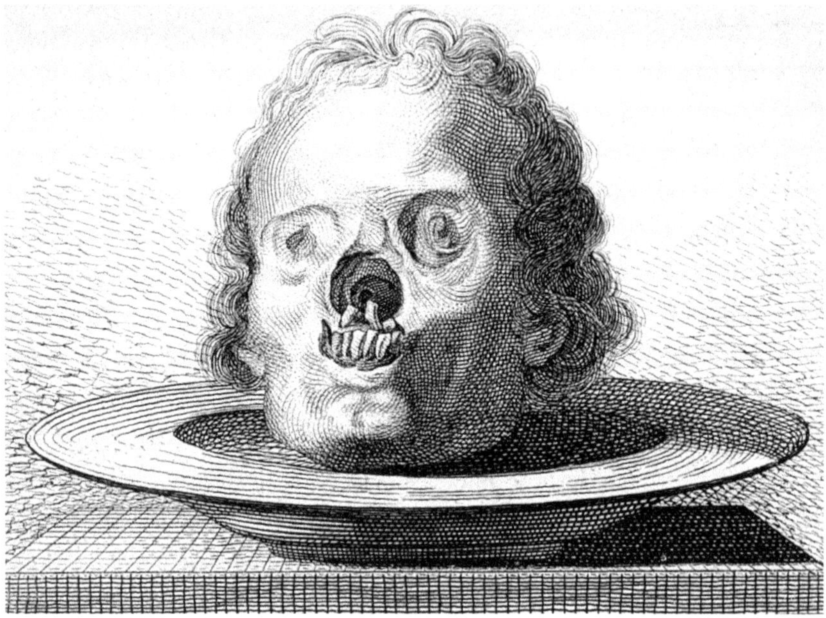

FIGURE 8.2: Preserved head of a poxed prostitute. The nasal cartilage and soft palate have been totally destroyed. Line engraving, 1796. Iconographic Collections, Wellcome Library, London.

I am going to focus on the pox rather than on the clap or gonorrhea, despite the fact that the most common eighteenth-century opinion was that clap and pox were stages of a single disease, the "lues venerea," the clap being an early stage, which, if not aggressively treated, might develop into a pox.[6] For patients, the pox was normally a distressing, even frightening affliction, but for medical men it could be a significant source of income, of professional prestige, and an intellectual puzzle to be solved.

I will draw most of my evidence from the many books written by contemporary medical men engaged in treating patients they diagnosed as suffering from venereal disease.[7] Authors typically published to take positions on contested diagnostic or therapeutic questions and to state that their understandings of the disease led to superior therapeutic outcomes. Some were university-educated physicians and fellows of the Royal College of Physicians, a number had foreign medical degrees, some were licentiates of the Royal College, others were surgeons, and still others lacked these official qualifications. While distinctions have often been made between orthodox and fringe or quack practitioners in this period, with respect to the understanding of venereal disease such distinctions do not seem to me particularly useful.[8] Thus, here I would prefer to consider all these authors in the single category of *medical men*. Some authors were in private practice; others held positions that gave them access to patients with venereal disease—for example, appointments as military surgeons or physicians and surgeons appointed to the Lock hospitals, which specialized in venereal disease. Virtually all these medical authors published with an interest in enhancing their professional reputations and hence their practices. Authors typically expressed their therapeutic optimism in a kind of romance formula: a patient comes to the author in distress (sometimes after having been misdiagnosed or mistreated by a rival), the author makes a correct diagnosis, persuades the patient to follow his recommended treatment, treats him or her, and the narrative ends triumphantly with the vindication of the author's ideas about the disease and its treatment, and the restoration of the patient to good health. Remarkably, once under the author's treatment, patients' bodies—ravaged by deep ulcers, suppurating nodes, and even the destruction of cartilage and bone—are transformed into healthy, apparently, unmarked bodies. In contrast to the representations of marked bodies in the satires, the majority of poxed bodies represented in this medical literature are male.

CONTESTS OF INTERPRETATION

These books reveal contests of interpretation of bodily signs between medical men and between medical men and patients. Unsurprisingly, the typical medical author was convinced that he could correctly diagnose and treat the pox, although he was concerned that other practitioners could not. For example, Daniel Turner, the surgeon author of one of the more important treatises on the disease, describes the case of a young married woman who, having complained of "great Pain in her Head" and fevers, had been diagnosed by a physician as having "an *anomalous Intermittent*" fever and treated with decoctions of bark, bleeding, blistering, and gargles for her sore throat, all to no avail. Turner immediately perceives "a *Serpigo* [a creeping skin lesion] on the Eye-brow," finds that the headaches are worse at night, and pronounces the symptoms, including the sore throat, as signs "of the unsuspected Disease she laboured under."[9] His mercury treatment cures her. John Sintelaer, an antimercurialist, presents this case of a husband and wife who come to him after having been treated elsewhere with mercury (which produced salivation):

> the Gentleman after all the Miseries of these repeated Salivations, found neither himself nor his Lady a jot the better, but on the contrary, in a much worse condition than before, for he was afflicted with such tormenting Pains in his Head, Shoulders, Arms and Legs, that he could scarce take any Rest in his Bed: Besides which, his Legs were beset down to the Shin-bones, with hard and painful *Nodes*. The poor Lady was still in a more doleful Condition than her Husband . . . not . . . able to stand without Help, she was most miserably afflicted with the most tormenting Pains in the World, not only in her Head, Arms and Legs, but also in her

FIGURE 8.3: Painful pox treatments including sweating, salivating, scarification, and cauterization. In John Sintelaer, *The Scourge of Venus and Mercury . . .* (London: 1709). Rare Books, Wellcome Library, London.

> Shoulders, Back and Back-bones, she had also divers very deep Ulcers or running Sores on her Back; . . . and out of her right Ear, issued such a stinking putrid Matter, as enough to strike any Body down that came near her.[10]

Sintelear's nonmercurial treatment restores this couple to health. Since many patients had previously been treated for venereal disease by someone other than the author, there was often the problem of deciding whether a particular sign was a sign of disease or, instead, a sign of some previous treatment.

Many authors also report contests of interpretation between themselves and patients. The medical author is convinced that he has correctly diagnosed and successfully treated a case of pox, and a patient, who agrees that he or she has had the pox and been treated, continues to experience signs and symptoms that he or she interprets as treatment failure. Some authors describe cases in which they believe that patients have misread their own bodies, erroneously concluding that they are afflicted with the pox, when, in the opinion of the medical man, they are not. In what was an age of new interest in psychology, some medical men interpret some cases in which the patient insists he or she is poxed as psychological, arising not from venereal infection but rather from causes such as the patient's guilt over having engaged in illicit sex acts.

Turner offers an account of one pious tradesman so distressed over his premarital fornication with a servant that, despite an absence of any symptoms and repeated reassurances that he is free of disease, the tradesman continues to insist that he is afflicted. Because one unusual, if disputed, feature of the venereal disease was that it might be symptomatic, then latent, then symptomatic again, an absence of detectable signs was not necessarily entirely reassuring. The tradesman's melancholy situation is worsened by another medical man's having previously told the patient he suffered from an "an inward Pox" and, Turner says, by the man's obsession with "the *Bills* distributed by *Quacks* about the Streets . . . as well as by their Advertisments in the *News Papers*." Turner manages to get the patient better for a while, sometimes, against his better judgment, treating him with various medicines, yet in the end reports that his "*Hypochondraicism*" is recalcitrant. The patient finally has to retire from his business to the country, where he lives without society, sometimes weeps, and sometimes "feeling of his Nose, he will run to the Glass, that he may have both Senses to ascertain he has not lost it."[11] Similarly, Charles Hales writes of patients who have no physical signs or only slight signs, "yet in their own phrase, are *rotting gradually* with a dry pox." Advising medical men not to administer drugs to such people, Hales recommends instead attentive listening to "their tedious tales" and describes one case of a fifty-year-old man cured by Hales patiently spending two years giving him "the fullest indulgence in the recital of his dismal and tedious narrative."[12]

More commonly, patients who are supposed to have been successfully treated return to complain of recurring pox symptoms. On occasion, the medical man concludes that his treatment had been successful but that a wife, having been infected by her husband and been treated, continues to complain of symptoms, not because she is still afflicted with the disease, but because she refuses to forgive her husband. William Buchan worried that too many afflicted married women refused to believe that they had been "radically cured":

> Under this persuasion they impute every ache, pain, or pimple to the venereal poison lurking in the system, and lead both themselves and husbands most miserable lives. Humanity should induce medical men to do everything in their power to eradicate these hurtful notions. But this is, to my knowledge, not an easy task.[13]

Apparent recurrences often inspired a patient to seek a new practitioner, but some returned to the man who had treated them before.[14] Under such circumstances, the medical man usually resorted to one of the following options: he tried to convince the patient that the posttreatment signs were not venereal, he accused the patient of noncompliance with the prescribed treatment regimen, which had resulted in failure to cure, or he claimed that some constitutional idiosyncrasy of the patient had caused a generally successful treatment to fail. He was unlikely to say that his initial diagnosis and treatment were mistaken. John Hunter, with a candor not typical of these authors, includes in his *Treatise* a section titled, "Of the Effects Remaining after the Disease is Cured, and of the Disease Sometimes Produced by the Cure," acknowledging, "These cases puzzle considerably; for it is difficult to say when the venereal disease is absolutely gone."[15] Most frequently, medical men seem to have resorted to the first option of trying to convince patients returning after treatment that the present symptoms were not venereal.

It was widely acknowledged that the pox was contagious, and there was less reliance on the older humoral paradigms of bodily constitutions in the eighteenth-century venereal disease literature; nevertheless, especially under circumstances of apparent treatment failure, medical men did sometimes resort to these older paradigms. As scholars of Renaissance medicine have argued, knowledge that the plague and the pox were contagious diseases had significantly challenged existing Galenic medical ideas and contributed to weakening and changing them.[16] Still, these older humoral paradigms provided a model for the common eighteenth-century idea that a local gonorrhea might somehow get deep into the patient's constitution in the form of a "confirmed pox." Indeed, those who thought that clap and pox were one disease sometimes wrote of late-stage pox as a constitutional pox.[17] Moreover, humoral medicine provided a model for worry that mistaken treatment could worsen a disease by driving the offending humors to a more vulnerable part of the body. Mercury treatment also rested on a humoral model of expelling bodily fluids through the mouth by salivation.[18] In any case, confronted with physical signs that seemed to suggest treatment failure in a patient, eighteenth-century medical men could still invoke the idea that that particular patient possessed a noncompliant constitution, a move that helped preserve their confidence in the efficacy of their treatment regimens.

SOME EFFORTS TO RESOLVE AMBIGUITIES

Recognizing that it could be difficult to decide whether a particular visible sign on the body was or was not a sign of pox, medical men tried a variety of ways to resolve ambiguities. Jean Astruc, professor of medicine first at the University of Montpelier and then at the University of Paris, in a widely cited treatise, attempted to distinguish between equivocal and "univocal" or "pathognomic" signs. The equivocal signs included obstinate headaches and deafness; the pathognomic signs included "communicating infection to a sound person," ulcers of the palate and inside of the nose, and bone caries.[19]

Most medical men appreciated that the differential diagnosis of skin lesions posed problems. Books published in the late seventeenth and early eighteenth centuries typically contained brief descriptions of lesions that would have been relatively difficult to match to a particular lesion on a body. Early writers, though, did often try to give some definitions for their terms, defining words like "Chancre" or "Bubo" that might not be equally familiar to every reader. Thus, Richard Bunworth explains that buboes "are small tumors in the arm-holes or groyn and privie parts."[20] Much more rarely, some tried to be more specific about the appearances of the skin lesions. One Restoration writer, for instance, tried an analogy:

> When the Body is covered with Blisters, and those Blisters are turned into Chancres, that is to say, round Ulcers, the sides whereof are hard, and the middle somewhat white, or (as they say) resembling the Eyes of a Partridge, there needs no other proofs to assure us of the Pox.[21]

In the second half of the century, however, a few surgeons offered more elaborate, more precise descriptions. John Andree was especially detailed, invoking color, shape, appearance under magnification, texture, characteristic locations on the body, and change over time:

> A Chancre is an eroding ulcer: it usually begins from a minute inflamed spot, much resembling a part which has been burnt by a spark of fire. Within the space of one week after this appearance, the centre of the inflammation ulcerates. The part on which the ulcer is seated is always hardened . . . The colour of a chancre is a pale-red, sometimes inclining to a cream colour; but this, in different cases, varies from a deep-red to a dingy-cream colour: and, if examined through a magnifying-glass, will be seen to be studded over with numerous small hillocks of flesh. The surrounding diseased part is most commonly swelled in a circular form, which sometimes is not raised higher than the sound part, but is always hardened, as may be perceived by gently pressing the part between the finger and thumb.[22]

The early books were not illustrated. When a few illustrations did begin to appear in the late eighteenth century, they were typically black-and-white engravings of particular pieces of apparatus rather than of a body. A few very striking, even oddly beautiful, engravings of bones eaten away by the pox appeared about mid-century, but such bones would only have been fully visible after autopsy; these images were not very useful for diagnosis. What was really needed to help with differential diagnosis was something like Anthony

Wisdom's *Color Atlas of Venereology*, a book with good color photographs of skin lesions (1973). Not until quite late in the eighteenth century, in France, did medical science and printing technology combine to produce a harbinger of the twentieth-century color atlas in Jacques Gautier D'Agoty's *Exposition anatomique des maux vénériens sur les parties de l'homme et de femme, et les remèdes les plus usités.*[23]

Eighteenth-century physicians and, especially, surgeons who cut into patients' bodies in operative procedures or autopsies could report on signs of the disease not otherwise manifest. There were significant differences of opinion about whether mercury treatments should be supplemented with various operative procedures—like incising the lymph gland tumors in trying to drain them or cutting away necrotic tissue from a deepening ulcer in trying to control it. But medical men might resort to such procedures for a variety of reasons, including hope of palliating a patient's pain, and surgery sometimes seemed to shed light on diagnostic conundrums or to reveal truths about the disease within the body. For example, Turner reports a case of one gentleman still suffering from dreadful headaches, despite an avalanche of medicines, including liberal doses of opium, from Turner and other practitioners. Finding a bad

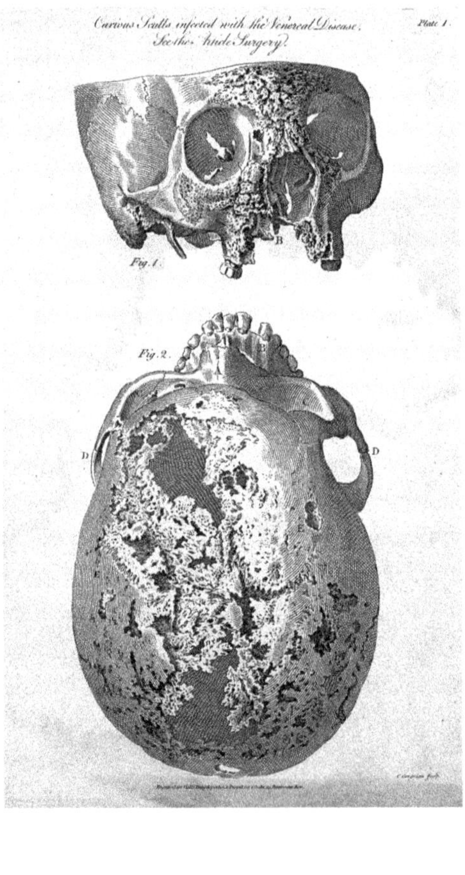

FIGURE 8.4: Skull with the bone caries typical of tertiary syphilis. Line engraving by Charles Grignion, ca. 1788. Iconographic Collections, Wellcome Library, London.

lesion on the man's scalp, they apply a caustic, wait for a scar to form and separate, and "perceiv'd the Skull rotten, the Compass of an half Crown." Further observing the caries in the skull bone, Turner reports, "I was once by when the Surgeon passing up his Instrument several Inches, found a Communication with that superior protuberating Fullness formerly remarked, the Bone being discovered porous and rugged all the Way."[24] Giovanni Battista Morgagni's *De Sedibus et Causis Morborum* (1761), a pioneering and well-illustrated work of morbid anatomy, aimed to correlate symptoms of disease reported in patient's histories with pathological changes in particular organs visible at autopsy.[25] Morgagni's autopsies revealed the syphilitic aneurysm; he also pointed out that people afflicted with gonorrhea rarely had urethral ulcers. This later observation strengthened the case for considering that gonorrhea was an independent

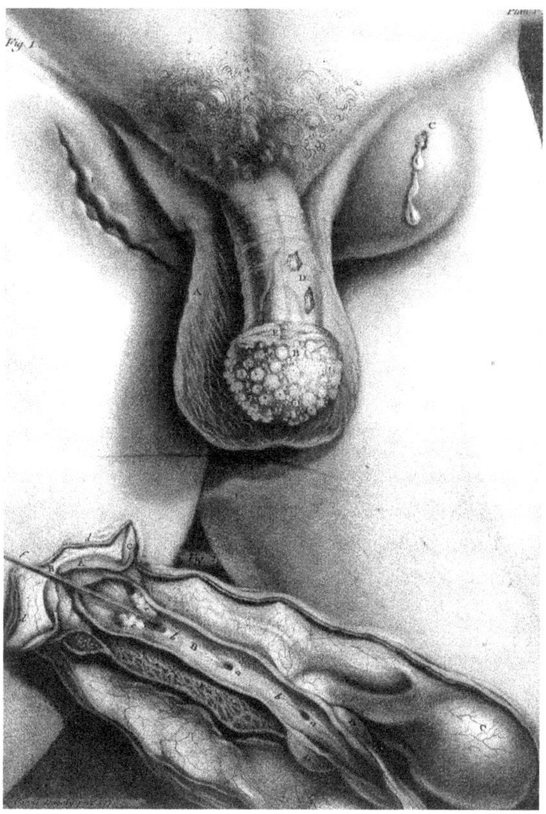

FIGURE 8.5: Male genitals with syphilitic sores and enlarged, infected inguinal glands. A cross-section (presumably after death) shows lesions in the urethra. In Jacques Gautier d'Agoty, *Exposition Anatomique des Maux Vénériens* . . . (Paris: 1773). Rare Books, Wellcome Library, London.

disease rather than an early stage of the pox and became part of the evidence used by Francis Balfour in his 1767 Edinburgh dissertation arguing that these were two distinct diseases.

A more crucial way in which eighteenth-century medical men tried to resolve diagnostic ambiguities in pox was to focus attention on the body over time. Pox was harder to understand than plague not only because its symptoms were more various but also because its course could run over decades and because it sometimes seemed asymptomatic. Eighteenth-century medical men carefully elicited histories from their patients, and they attempted to develop good accounts of the natural history of the disease. The hope was that a bodily sign that was ambiguous when viewed at a single point in time could be better read in the context of an extended temporal perspective. Confronted with the case of a married woman who had debilitating headaches and skin lesions including an ulcer—"very extensive, reaching from one mamma to the other, and almost from the inferior point of the sternum to the top of it"—and who others had treated with "blood-letting, blisters, bark, cold-bathing, and a variety of nervous medicines," all to no avail, Benjamin Bell queried her husband. He discovered that the husband had a past venereal infection and that the wife had then "complained of a soreness in the pudendum about the same period," although another surgeon had not then thought the wife infected.[26] Convinced that accurate patient histories could be crucial aids to interpret equivocal bodily signs, medical men report attempting to pierce the lies told them by embarrassed or frightened patients.

PATHWAYS OF CONTAGION

A related crucial method for resolving the ambiguity of bodily signs was to study pathways of contagion. Since pox was understood to be contagious, evidence that a person suspected of having the disease had communicated identifiable pox symptoms to another person added to the weight of evidence that the original patient was infected. A number of the most thoughtful writers, including Hunter, tried to isolate precisely what matter from an infected person had the power to communicate the disease to another, thus refining the natural history of the disease. Some pathways of contagion were almost universally accepted, most importantly, sexual intercourse with an infected person. Others were generally accepted, but instead were subjects of significant debate. Two debates concerned transmission between an infant and a wet-nurse and transmission in the womb to a fetus. Some considered that a fetus acquired venereal disease at the moment of conception. Others suggested a postconception infection in the uterus of an affected mother. A few argued that infection occurred only during the birth process, as the infant passed through the birth canal. Still others denied that an infant might be born with venereal disease. Many commented on infants infected by wet-nurses shortly after birth. One of the most famous and oft-translated eighteenth-century descriptions of the pocky infant was Astruc's: "[They will be] strumous [scrophulous], rickety, humped, hectical, and emaciated, die miserably before their Time; or, if they live, are short, broken-backed, large headed, crooked, bandy-legged, variously distorted, and thick-jointed."[27]

Walter Harris's *Treatise of the Acute Diseases of Infants*, first published in Latin in 1689, then twice translated into English and several times reprinted, offers a reasonably representative late-seventeenth-century English physician's view. Harris was Oxford educated, took a French medical degree, and then was awarded a Cambridge MD; he became physician-in-ordinary to King William III and Queen Mary II. The *Oxford Dictionary of National Biography* says that Harris's *Treatise* "became the standard pediatric monograph for a century."[28] For the most part, Harris still maintained the Galenic and Renaissance concept of disease as essentially constitutional. As a learned physician, he considered himself obliged to provide some theoretical framework for his therapies and to say something about the etiology of disease. Yet his big idea that acid is "the Cause" of "all the Diseases of Infants" does not fit very well with his account of pocky infants.[29]

Harris locates the origin of venereal disease itself in "the impure Wombs of common Prostitutes" where the seeds or semen of many different men are mixed and corrupted. This process of the corruption of healthy male seed he likens to that by which "Lice and Fleas are produced from Filth and Uncleanliness."[30] The older medical histories of pediatrics and even the new *Oxford Dictionary of National Biography* say that Harris is significant in part for "estimat[ing] the effect of heredity on disease in children." Yet Harris's discussion of a general mechanism by which disease is inherited from parent to child omits pox and is at a considerable distance from his much later discussion of venereal disease. When he does get to venereal disease in children, he warns that remnants of the pox may lie hidden in a parent, presumably invoking a concept of latent disease. A hidden pox can not only make parents ill, but also cause "some tralatitious" (which can mean "common" and, here, also, I think, "inherited") and "almost incurable diseases of Children." When he comes to his actual cases of infants with venereal diseases—a small number—these are cases not of infants with hereditary venereal disease but of infants infected by their wet-nurses. These infants he represents as "all covered with Spots, Pustules, and Ulcers, and . . . tortured with cruel nocturnal Pains." In contrast to the pessimism Harris expresses about the curability of children born with hereditary disease, he optimistically reports solving these infants' problems by removing them immediately from "the poisonous Nurse, as from a Plague" and offering them a treatment using sarsaparilla mixed with their pap.[31]

Much has been written about the struggles for professional authority between physicians like Harris and other contemporary medical men, including the surgeons who were sometimes said to have made treatment of venereal disease their specialty. W. F. Bynum rightly argues that in the eighteenth century both physicians and surgeons treated the disease, sometimes in collaboration with one another, and that a number of surgical writers borrowed some of the hallmarks of the learned discourse supposed to be characteristic of the university-educated physicians.[32] Nevertheless, several of the books written by Restoration and eighteenth-century surgeons with extensive practices do seem more independent from late Renaissance medical theory than Harris is, more engaged with more precise observation of their patients' bodies and the natural history of their diseases, and more apt to raise questions about the adequacy of current conceptions of the disease. In one of the most interesting late seventeenth-century surgeon's books, *Eight Chirurgical Treatises* (first published in 1676, with a sixth edition in 1734), Richard Wiseman offers case histories that are unusually detailed for the period. Wiseman was a surgeon who rose from treating Royalist troops in the Civil War to becoming Charles II's personal surgeon, and eventually the King's "Principal Surgeon and Serjeant Surgeon." He had extensive experience in military and private practice. He says he wrote his *Treatises* for young surgeons, hoping to give them better information "than a meer Academick could have done."[33] Unusually for the day, Wiseman urged medical writers to describe both successful and unsuccessful treatments, "thereby increasing knowledge in our Profession, and leaving Sea-marks for the discovery of such Rocks as they themselves have split upon before . . . there being more of instructiveness often in an unfortunate case than in a fortunate one."[34]

Wiseman is not concerned with high theory, but he is interested in trying to understand what the existence of pocky children means about the pathways of venereal disease contagion. He finds that the infections of children are "more obscure" than those of adults and observes that although many children are born pocky, others come

> into the World very clear, and with the appearance of very sound Bodies, which notwithstanding have been certainly infected and have Dyed of it, or at least have been very infirm; and though they were brought to their ends by some other apparent Disease, yet the Pox hath been judged the foundation of all.[35]

As his cases involving mothers, children, and wet-nurses reveal, he tried to make differential diagnoses distinguishing the pox from the many other diseases that may have been confused with it by combining patient histories with his observations of the physical signs. He looked for additional signs beyond skin lesions—signs including buboes or bone caries—before he was willing to diagnose a confirmed pox.

Wiseman's case narratives do not pose explicit questions and answers about paths of transmission, but several seem implied. For example, this case seems to ask: does the birth of a pocky child to an apparently disease-free mother demonstrate that the mother was, in fact, diseased?

> A Woman who suspected herself Venereal was brought to bed of a seeming healthful Child; it thrived very well for some months, and the skin was clear . . . This Infant being about four months old brake out in the Head, afterwards in the Face, and most parts of the Body with such eruptions as shewed it to be Venereal, yet the Mother seemed untainted.[36]

Wiseman reports his successful treatment of the child and its cure, though he adds that it "died afterwards of the Small Pox." Wiseman concluded from his experience and reasoning that a pocky child could infect a healthy wet-nurse and that a pocky wet-nurse could infect a healthy child. He gives cases in which he attempts to establish whether the infection began with the infant or with the nurse. Indeed, as his case of the apparently healthy mother with the pocky infant suggests, he finds that the sudden appearance of pox in a healthy wet-nurse who has been suckling an apparently healthy child is evidence that the child was infected.[37] In the following case, he seems to decide that an infant must have infected its wet-nurse rather than, as the infants' parents' actions suggest, that the nurse infected the child:

> A Nurse by giving suck to a diseased Child was infected with great ulceration and chops with *verrucae* on the Nipples and parts about the Breasts, upon which account the Child was taken from her, it being suspected she had infected the Child. She also had a mole upon her right hand and some breakings out upon her Limbs. I inquiring into the cause saw this poor Womans Child which was born within the year very well complexioned and sound.[38]

Here, although he does not make these points explicitly, his impression that the infant infected its nurse seems based on (a) signs of pox in the infant, (b) dermatological signs of pox on the nurse's breasts rather than her genitals, suggesting that the point of contagion was the infant's mouth to her breast, and further nondermatological signs confirming the pox diagnosis for the nurse, and (c) the absence of any signs of pox in a child born to the nurse herself before she nursed the infant in question. In the eighteenth century, most medical men who commented on the phenomenon seem to have agreed with Wiseman that some infants were born afflicted with venereal disease, that those infants were capable of infecting healthy wet-nurses, and that infected wet-nurses could infect healthy infants.

HUNTER AND BELL DEBATE INFANT–NURSE CROSS-TRANSMISSION

Despite the degree of consensus with Wiseman's views about infant–nurse cross-transmission, one of the most important medical writers of the late eighteenth century, John Hunter, challenged this consensus in his 1786 *Treatise on the Venereal Disease*. Hunter was a member of the Company of Surgeons, a teacher and staff surgeon at St. George's Hospital in London, a fellow of the Royal Society, and a serious student of comparative anatomy and physiology. His prestige and influence were sufficient to provoke another major medical writer, Benjamin Bell, surgeon at the Royal Infirmary of Edinburgh and fellow of the Edinburgh Royal College of Surgeons, to publish a counterargument in his *Treatise on Gonorrhoea Virulenta and Lues Venera* (1793). Bell, who respected Hunter, nevertheless disputed Hunter's argument in the *Treatise on the Venereal Disease* that gonorrhea and pox were a single disease, offered a counterargument that they were separate diseases, and strenuously disputed Hunter's contention that pocky infants could not infect wet-nurses. This debate has a different intellectual style from the earlier observations of physicians like Harris.

Hunter viewed a number of the received ideas about venereal disease skeptically and tried to test them by observation, experiment, and reason. He was particularly interested to understand what he referred to the "matter of disease"—matter that had the capacity to infect—and to understand where and how that matter could act. Thus, he did a famous self-experiment, injecting himself with the discharge from the penis of a man with gonorrhea, carefully observing over three years his developing symptoms (which included signs of pox), and chronicling them in his *Treatise*.[39] He concluded that pox was a later stage of clap. Hunter doubted that semen could "become venereal" and, after conception, "grow into a pocky child." He further doubted that a fetus could be infected:

> It is also supposed, that a foetus in the womb of a pocky mother may be infected and have the disease from her, as it were naturally interwoven with it. This I should doubt very much, both from what may be observed from the secretions, and of finding that even the matter from such constitutional inflammation is not capable of communicating the disease as before mentioned. However, one can conceive the bare possibility of a child being affected in the womb of a pocky mother, not indeed from the disease of the mother, but from a part of the same matter which contaminated the mother and was absorbed by her, and whether irritating her solids to action or not may possibly be conveyed to the child, pure as

> absorbed; and if so it may affect the child exactly the same way it did or might have done the mother.[40]

He did, however, think infants might be born infected as a consequence of contact with matter of disease in their mother's birth canal or genitals during the birth process. He also doubted that a child born poxed could infect a wet-nurse or that the milk of a wet-nurse could infect a child.

Although Hunter's writing is not always perspicuous, he tried to get away from earlier, more essentialist conceptions of disease to develop a model involving the action of specific matter of the particular disease on specific body surfaces. He points out that it does not follow from the fact that a person is infected with a contagious disease that all secretions from that person's body have the power to infect. Thus, he notes, saliva from a mad dog can cause rabies, but other secretions from the mad dog—its urine, for instance—do not.[41] Hunter reports experiments he designed and conducted on hospital patients to test the common assumption that all matter produced by venereal inflammations is "also venereal and poisonous." Before he began treating a new venereal hospital patient, he took matter from a pocky sore or chancre on the tip of a lancet, and then injected it in a place on the patient's body free from any lesions. Matter from the chancres produced new chancres, but matter from the pocky sores produced no new local lesions.

Hunter was acutely aware that the differential diagnosis of venereal disease was problematic. In the *Treatise*, he argues that cases *wrongly* diagnosed as venereal led to what he considered the erroneous idea that a child could be born with venereal disease and communicate the disease to a wet-nurse. He offers an unusual case to illustrate this point. On September 30 1776 a lady gave birth to a "weakly" child. The child had difficulty nursing, though the mother had "abundant" milk. Thus, a neighbor child was procured "to assist in keeping the breasts in a proper state." The lady nursed her child at her right breast, the neighbor child at her left. In weeks, the lady's left breast developed ulcers around the nipple and swollen glands in the axilla. The neighbor child was found "short breathed, had the thrush, and died tabid [emaciated]," with many sores on different parts of its body. The lady developed more ulcers and pains, was treated with mercury, but in 1779

> was delivered of another child in a diseased state. This child was committed to the care of a wetnurse, and lived about nine weeks; the cuticle peeling off in various parts, and a scabby eruption covering the whole body; the child died.

Then the wet-nurse developed breast ulcers and was treated with mercury, but "the bones" of her "nose and palate exfoliated, and in a few months she also died tabid."[42] Hunter here produces a narrative that carefully describes observable symptoms in a set of related patients, apparent pathways of contagion, and responses to treatment, each of which he understands as a criterion on which to base a diagnostic conclusion. He knows that many contemporary surgeons considered that a pocky infant—with or without detectable symptoms—could infect a wet-nurse and that development of pocky symptoms in a wet-nurse was a diagnostic sign that the infant was infected. Yet Hunter concluded—perhaps in part because the lady for a time responded to mercurial treatment but also because he did not believe in congenital pox or child–nurse transmission—that "new poisons" had arisen that were "very similar to the venereal in many respects, although not in all."

Hunter was more a close observer and analyst of phenomena and was more of an experimenter than a grand theorist. Insofar as he invokes higher-order principles, they tend to be disabling, making him shy away from more obvious conclusions to be drawn from his own observations—as Bell realized. Denying that contagion could be "infinite," he was bothered by the model of transmission from parent to infant to nurse, the nurse getting venereal disease "as it were third hand." He commented: "If however it were possible to contaminate once in this way it would be possible to contaminate forever."[43]

Bell, however, expressed "astonishment" that Hunter could deny what seemed to Bell facts firmly established in a surgeon's experience:

> that children may receive, and frequently do receive the venereal disease from their parents . . . that ulcers in the mouths of these children will, and frequently do produce the same disease upon the nipples of women whom they suck. Nay, that these will give it again to other children, and these to other nurses.[44]

One reason Bell argued that gonorrhea and syphilis were different diseases was that he observed that patients with gonorrhea did not transmit pox to their infants. He agreed it was possible that some infants were infected with pox "during labour from venereal sores in the genitals of the mother" but pointed out that this could only be an explanation in cases where the pocky signs did not appear until two or three weeks after birth: "But where a child is instantly upon delivery found to be covered with a venereal eruption, . . . which I have met with in various instances, the infection must necessarily have been communicated a considerable time before birth."[45] Bell gives elaborate clinical descriptions of the rash and other signs on the newborn pocky infant and observes that what he calls the syphilis virus in the newborn is exceptionally infectious. He warns that risk of infecting wet-nurses is so high that they should not be used in such cases.

In Bell's case reports, a newborn's capacity to communicate syphilis to its wet-nurse is a diagnostic sign that the infant is infected. Thus, in one case, a father admitting to having been poxed before marriage believes himself adequately treated and cured. His wife gives birth to a child the husband considers not pocky, but Bell thinks it is. Bell treats the child and adds that an "unfortunate proof" that the child was infected subsequently appeared when first one wet-nurse and then her replacement developed pocky lesions on their breasts, one developed throat ulcers, and the other's own infant, after being nursed by her, died.[46]

TEXTS AND BODIES

There is a kind of self-inflicted, self-denying ordinance frequently articulated and often observed by modern sophisticated medical historians. This self-denying ordinance requires the historian to limit his inquiry to the knowledge of the period he or she is considering and to abstain from consideration of what—from the perspective of later science—might emerge as an alternative narrative or explanation of the phenomena being studied. In considering venereal disease, for example, this method would require restricting oneself to terms current in the period being studied (like *pox* or *lues venerea*), not second-guessing eighteenth-century authors' diagnoses, not making judgments about the therapeutic efficacy of mercury or sarsaparilla, and so on. Thus, Arrizabalaga, Henderson, and French, in their study of "the French Disease" in Renaissance Europe, eschew efforts to think back and forth between Renaissance medicine and modern science and decline to consider whether the little organisms we now know as *Treponema pallidum* were at work in Renaissance bodies. They are content to investigate how the "quiddity" the Renaissance called "the French disease" "arose from the society and culture of the time" and to explain the "success" of Renaissance physicians in treating the disease in terms of how they managed their "relationship with other social and professional groups" and satisfied the desires of patients.[47] Certainly in the case of venereal diseases, there are good reasons to support the skepticism that lies behind this self-denying ordinance of much sophisticated modern medical historiography. Indeed, some of these reasons were developed by early modern medical writers, including the important idea that a disease might not be transhistorical. Enlightenment writers interested themselves in the history of diseases, particularly lues venerea, and debated not only the sources of its apparently rather sudden appearance in Renaissance Europe but the questions of whether it was becoming less virulent or showing different symptomatology in the seventeenth and eighteenth centuries. And, as we have seen, most were aware that the differential diagnosis of pox was a problem, try as some did to disentangle from pox—a disease that seemed to attack so many different organs over such long periods—other disease entities like rickets or scrofula in infants, or gonorrhea or cancers in adults.

Nevertheless, it seems to me that modern skepticism has often gone too far and that it bars some reasonable argument and lines of inquiry. I worry that the current historiographical concentration on social historical subjects in eighteenth-century medical history, worthy as they are, may blind us to the serious engagement of medical men like Hunter and Bell with the important project of creating a dialogue between the realm of the text and the realm of extratextual phenomena of natural history, including bodies and disease-causing organisms. Hunter reached some conclusions that were almost certainly wrong, but his cultivation of a detailed and relatively exact method of reporting his observations of bodies made it possible for Bell and others to suggest alternative, more persuasive, readings of his reports; the precision of his reports made second-guessing his diagnoses a useful activity. Syphilis has reasonably been called "the great imitator" for its capacity to mimic so many other diseases. But, I venture to say, the rashes of other diseases in newborn infants do not simultaneously follow a history of pox in parents, frequently lead to infant death, produce characteristic chancres on wet-nurses' breasts, and eventuate in the destruction of those nurses' noses and the birth of pocky infants to the infected nurses. Both Hunter and Bell demonstrated their commitment to the proposition that the authority of texts could be challenged by nontextual experience.

I would hope that historians of eighteenth-century medicine and of the body, who must study texts and are rightly skeptical about their relationship to historical bodies, nevertheless—recognizing that such discussion, like most historical discussion, will lack certainty—do not entirely foreclose discussion that appeals beyond historical texts to nontextual phenomena like *Treponema pallidum*, mercury, and bodies.

CHAPTER NINE

Cultural Representations

Rogue Literature and the Reality of the Begging Body

TIM HITCHCOCK

It was a commonplace of eighteenth-century literature that the beggars who sat huddled on cold street corners, who crowded round the coaches as they disembarked their wealthy passengers, whose whines and cries for relief assaulted the ears of polite pedestrians were, quite simply, charlatans. With make-up and a complex superstructure of straps and crutches, with rented babies pinched to cry and bawl, with vermin-ridden hair, and with theatrical rags they played the ultimate urban confidence trick on the better off. Ned Ward, in his 1709 *History of London Clubs* depicts the scene at the beggars' club, after its members had retired from their curbside posts for the night:

> where by the vertue of sound tipple, the pretenders to the dark are restor'd instantly to their sight, those afflicted with feign'd sickness, recover perfect health, and others that halt before they are lame, stretch their legs without their crutches. . . . their dirty handkerchiefs and night caps are slipt into their pockets, their crippled legs and arms taken out of their slings and return'd from their cramping postures to their ease and liberty.[1]

George Parker similarly entertained his readers with an account of a false cripple who nearly lost a leg through the kindness of strangers:

> An old soldier had gagg'd [begged] about London many years. His mode for provoking compassion was to get some sheep's blood and a handful of flour which he put so artfully upon his knee, as to make the passengers who saw it believe it to be a mortification in his leg and thigh. He had taken his stand one morning in a part of the Borough where a young surgeon who walked to St Thomas Hospital happened to pass. . . . The young surgeon gave him sixpence, and promised to get him into the hospital. . . . The next thing was to get two men with a hand-barrow and some straw, . . . to fetch the old Gagger. The young surgeon arrived first and desired him to be of good cheer, . . . and he doubted not but that they should make a cure of him without cutting his leg off. When the gagger saw the two men, the hand-barrow, the straw, and the surgeons with their sleeves and aprons on, he jumps up and scampers thro' the crowd as if the devil was in him, to the admiration of the mob, who huzza' the surgeons back to the hospital by way of applauding their skill and perfect acquaintance in surgery.[2]

In part, this perception of beggars as conniving charlatans can be traced to a three-century-long tradition of so-called rogue literature, which suggested that a complex and well-ordered world of beggars, with a clear hierarchy, secret language, and malevolent intent, existed among the excluded and vagrant of sixteenth- and seventeenth-century European society. In recent years this literature has been reanalyzed as an essentially free-floating series of generic conventions that reflected the fears and assumptions of the middle classes and the elite, rather than the behavior of the beggarly poor.[3] Perhaps surprisingly, however, there is some evidence to support the rather jaundiced assessment retailed by Ward and Parker, and apparently rooted in genre writing. Every now and then a false beggar was discovered.

REAL, FAKED, OR BORROWED

On April 8, 1742, Henry Goodwin found himself before the Court of Bridewell, charged "for being a loose disorderly person wandring abrod & begging disguised by a long beard & pretending to be lame"; William Harding and Ann Middleton were taken up together in May 1747, "for being a . . . imposture and begging in the street yesterday with his hand tyed up and pretending to be lame and otherwise misbehaving."[4] Goodwin in particular was a professional, who was familiar with the ways of London. He had been apprenticed to a wood turner in Distaff Lane in the city, and supported a wife and child with his begging.[5] In 1702 the President and Governors of the Poor of London gleefully recorded their discovery of Richard Alegil:

> a boy of eleven years of age, who pretended himself lame of both his legs, so that he used to go shoving himself along on his breeches; they ordered him to be taken into their workhouse, intending to make him a tailor, upon which he confessed that his brother, a boy seven years of age, about four years ago, by the advice of other beggars, contracted his legs and turned them backwards, so that he never used them from that time to this, but followed the trade of begging; that he usually got five shillings a day, sometimes ten shillings; that he hath been all over the West of England, where his brother carried him on a horse, and pretended that he was born so, and cut out of his mother's womb. He hath also given an account that he knows of other beggars that pretend to be dumb and lame, and some that tie their arms in their breeches, and wear a wooden stump in their sleeve.[6]

Women participated in the same charade, most commonly by pretending to be pregnant or by passing off parish children as their own. In early March 1698 or 1699 Ann Wheatley was taken up for begging: "now pretends to be wth child . . . skulking about the streets"; a couple of decades later, Elizabeth Arnold was punished for "pretending to be in labor, which she has done for some months past."[7]

It is clear from these records that the judges and clerks who recorded the behavior and punished the bodies of the poor had a series of stereotypes derived from rogue literature at the backs of their minds. But we should not assume that beggars themselves were unaware of this literature, and, more importantly, we must acknowledge that although some entrepreneurial beggars faked their physical ailments, or borrowed parish children to give verisimilitude to their cries, a much larger number of beggars on the streets of London were in fact disabled in some way or were burdened with small children. They were missing limbs, they were blind and ill, they were mentally disturbed, or they were simply insane. They had been deserted by husbands and lovers or left destitute through widowhood. Poverty, disability, and parenthood went hand in hand. For the vast majority of people, both men and women, their livelihood depended on their ability to do hard physical work. When that ability was removed, poverty haunted their lives and destroyed their hopes. Most workhouse inmates, for instance, found themselves constrained to apply for relief because they were incapable of labor through disability or old age. Of the 841 paupers supported in the workhouse belonging to St. Giles in the Fields in 1730, 370 were described as ill or disabled, 126 were "overburdened with children," and a further 162 were above the age of seventy. With the exception of orphaned and abandoned children, most paupers were incapable of hard physical work because of a disability or the presence of small children.[8] And although the comprehensive system of poor relief was intended to provide a safety net for those unable to labor, many disabled men and pregnant women fell through this net entirely, or they combined parish relief with street begging.[9]

Some beggars refused parish relief because it undermined their sense of self-reliance. Samuel Badham, a professional beggar who walked painfully on crutches and who was active throughout the 1730s, came home one day to find his common-law wife wearing a parish gown. He said, "Sukey, there's no body that ever belonged to me ever wore a parish gown." He immediately "went out and asked charity, and with what I got I bought her another gown, and got the other made in to a petticoat for her."[10] But even the parish poor went out begging. In the late seventeenth century the parish vestry of St. Martin's in the Fields passed an order restricting its pensioners from begging door to door, and in the hours of the afternoon.[11] In the early nineteenth century, John Thomas Smith, in his unique account of London beggars *Vagabondiana*, ends his volume with an image of three disabled men, limping away from the artist, above a caption reading, "Beggars leaving town for their Workhouse."[12]

Contemporaries also believed that London itself attracted disabled beggars and unmarried mothers from around the country. By the end of the eighteenth century, London possessed an impressive array of lying-in hospitals and specialist workhouse wards that promised desperate women the possibility of a successful delivery. Complaints recorded at the lying-in hospital at Brownlow Street, that women who "have customarily received alms of parishes, others coming very dirty, ragged and others of bad behaviour" had obtained admission, suggests that the hopes of poverty-ridden pregnant women were not always illusory.[13] The disabled found London equally attractive. Joshua Gee, in his 1729 *Trade and Navigation of Great-Britain Considered*, suggested that the beggars of London were more likely to be disabled than their country cousins:

> If any Person is born with an defect or deformity, or maimed by fire, or any other casualty, or by any inveterate distemper, which renders them miserable Objects, their way is open to London; where they have free Liberty of shewing their nauseous sights to terrify People, and force them to give money to get rid of them; and those vagrants have for many years past, removed out of several parts of the three kingdoms, and taken their stations in the Metropolis, to the interruption of Conversation and Business.[14]

Nor do we need to make too much of the distinction between real and faked wounds, or between the parish poor and street mendicants. Certainly, for many eighteenth-century commentators, the boundary was much more subtle and immaterial than a modern perception might suggest:

> We say that when nature is deficient in one part, she makes amends in another, and the Observation is no where so true as in the Beggars in this parish; for when she sends a creature from the womb legg-less, and of consequence a dependent upon the next Turner's shop for deputy supporters, she ever supplies him with much brown for a natural cushion, as knowing him more inclinable to the sedentary than the peripatetick philosophy. If she puts out his eyes, she enlarges the sense of feeling, and makes him an acute distinguisher between brass and silver. If she chops off his arms, she in return stretches the wind-pipe, dilates the thorax and makes him capable of talking longer, and more to the purpose than a female scold: For what says the honest beggar upon his truss of straw in Moorfields of the passengers, I will have some of their money without stirring one foot from this spot of earth. Accordingly he tunes his voice raises his pipe to a pity-drawing pitch, and a shower of copper falls into his lap, which he converts at the next brandy-shop into true sterling.[15]

But whether real, faked, borrowed, or somewhere in between, wounds, disabilities, and babies were self-consciously deployed by the poor in order to excite compassion. Faced with the necessity of making a living from charity excited by sympathy, the disabled were forced to expose their broken bodies to the jaundiced gaze of pedestrian London. The physically whole were obliged to use what bodily strategies they could to give the impression of both disability and poverty, while parents were forced to drag their children about the streets as evidence of their need. All beggars, however, needed an equally instrumentalist approach to their own physical presence. In his *Essay on Deformity,* published in 1754, William Hay explicitly addressed the difficult choices available to the disabled in particular:

> A deformed Person will naturally consider, where his Strength and his Foible lie; and as he is well acquainted with the last, he will easily find out the first; and must know that (if it is any where) it is not, like *Sampson*'s, in the Hair; but must be in the Lining of the Head.[16]

In Hay's view, disability forced individuals back on their intellectual resources in new ways, and while for Hay himself this implied a life of writing and campaigning, for the poor, it implied trickery and careful self-presentation.[17]

The motive force driving the behavior of beggars was the effectiveness of physical infirmity and crying children in wringing charity from the tightly closed purses of the well-to-do. Exposed wounds and missing limbs, the physical signs of pregnancy, and the bawling demands of infants *worked*:

> When sores are very bad, or seem otherwise afflicting in an extraordinary manner, and the Beggar can bear to have them exposed to the cold air, it

> is very shocking to some People; 'tis a shame they cry such Sights shou'd be suffer'd: the main reason is, it touches their Pity *feelingly*, at the same time they are resolv'd, either because they are covetous, or count it an idle Expense, to give nothing, which makes them more uneasy. They turn their Eyes, and where the Cries are dismal, some would willingly stop their Ears, mend their Pace, and be very angry in their Hearts, that Beggars shou'd be about Streets.[18]

At the end of the eighteenth century, Francis Grose grumbled that exposing physical deformities was such an effective means of encouraging charity that beggars had come to view their disabilities as so many business opportunities:

> In this community, natural defects or bodily misfortunes are reckoned advantages and pre-eminences. A man who has lost one leg yields the pass to him who wants both; and he, who has neither legs nor arms, is nearly at the head of his profession, very extraordinary deficiencies excepted;—an instance of which was given in a sailor, who had but one eye, one leg and no arms. This man asking in marriage the daughter of a celebrated blind man, was answered by her father—that he thanked him for the honour intended, which he should have accepted, had not his daughter received some overtures from a man who crawled with his hinder parts in a porridge-pot.[19]

Begging has always required its practitioners to deploy a complex language and visual rhetoric. You need to cajole without threatening and draw attention to yourself without attracting the wrong sort of attention. You need to present a compelling case for charity in the split second during which a walker's eyes slide over the one aspect of your being that cannot be reduced or denied, your physical presence. This chapter is a preliminary attempt to explore the ways in which beggars used and presented their bodies and their babies, often crippled and damaged, or squalling and crying, as a constituent part of a broader rhetorical strategy. It is also an attempt to understand how beggars regarded and used their own physicality, how they self-consciously deployed aspects of their own bodies to discomfit and discompose the middle classes and the elite. In the process it will ask what beggars' willingness to expose their private weaknesses to the full glare of publicity says about how eighteenth-century people inhabited that transitory sack of pus and humors, that prison of pain, brief pleasure, and expectation, that is the human body?

THE INFIRM BODY AS A RESOURCE

That the London poor used and thought about their bodies differently is beyond doubt. Mary Cut and Come Again, the "queen of the blackguards, pilferers, and ballad singers," went to her death at Tyburn on June 7, 1745, refusing to give her correct name, Mary White, until it was almost too late. "On the credit of her voice, [she] commenced ballad [singing], which employed her when she had nothing worse to do," until on the March 27, 1745, she was arrested for assaulting Elizabeth Turner near Leicester Fields and for stealing a bundle of clothes worth three shillings and fifteen pence.[20] What is remarkable is not her crime, or even her refusal to give her name until after her fate was sealed, but rather her reaction to her arrest. It was around eight o'clock at night, and according to Mary McPherson:

> she would not be taken away till a proper officer came: she pulled her breasts out, and spurted the milk in the fellows faces, and said, d—n your eyes, what do you want to take my life away? She licked the prosecutrix before the Justice, and said, she longed for it; and she said, she would spit upon the Justice's seat, and she did so. The Justice said, he would send her to Newgate, she said, d—n my eyes then I shall have a ride for the money. Then the Justice ordered her to be fettered and handcuffed; and she said, if he would take the handcuffs off, she would tell him her right name, otherwise she would not.[21]

Mary clearly inhabited a visceral world in which the body's products, its spit and milk, shit and piss, had a meaning and power that we have since lost.[22] But her remarkable behavior and bravado, in the face of what she correctly assumed to be certain death, reflects a substantial emotional distance between her sense of self and her physical body. At a basic level she seems to claim that whatever the authorities do to her body, she would insist on gaining a benefit, of having "a ride for the money." Indeed, the equation between the destruction of her body and money reflects the everyday reality of women in her position, who were forced, day in and day out, to sell their bodies for labor or sex or as extras in the theater of urban life. The world Mary inhabited was one in which paupers needed to distance themselves from their physical existence for the very purpose of survival.

Mary's poverty was absolute. In response to her claim at her trial at the Old Bailey that the goods she was accused of stealing were her own, Henry Jurratt said, "I have seen many a ballad singer, and I never saw one with two aprons." But more than this, her willingness to gamble a painful death against a prideful anger at her treatment by the justice system suggests a full experience of using her body, of separating her sense of self, from the form it took. And this is the point. Long before Adam Smith used the subtle tricks of political economy to turn human beings—or at least poor human beings—into mere ciphers for physical labor, the beggars and paupers of London had been obliged to learn the trick of thinking of their own bodies as a resource to be used and expended for profit. Most would have agreed with John Bowes who, when caught sodomizing Hugh Ryley up against the rails of the churchyard at Covent Garden one early December evening in 1718, declared "Sirrah what's that to you, can't I make use of my own body?"[23]

The poor both sold their bodies for whatever price they could get, and at the same time maintained a strong sense of emotional and personal selfhood. A clear mind–body duality was a resource the poor could not afford to live without. This was not a duality based on some high-flown Cartesian model; this pauper identity was the price they paid every single day in their efforts to wrest a living from the streets.

Although there is some excellent recent work on the experience of living in an early modern body—particularly a woman's body[24]—most of the literature on poor bodies in the eighteenth and nineteenth centuries has emphasized the ways in which they were perceived and regulated. The bodies of the poor have been associated with the grotesque and the carnivalesque, described as porous and uncontrolled, and, in general, shoe-horned to fit a broad narrative of the evolution of prisons, workhouses, and hospitals, and by extension of an ever more intrusive state. Or, alternatively, their largely unexamined bodies have been counted and collated to create a history of movement and migration. In contrast, the bodies of the middle and upper classes have been described as increasingly contained and disciplined—modeled on the smooth outline of classical statuary.[25] But for beggars, what was important was the extent to which physical infirmity could be used for financial gain.[26] Their bodies were not so much grotesque as simply useful.

The accident, the tragedy that marked the beginnings of a life on streets, effectively forced beggars to look to the one resource they had left, their bodies. In around 1732, Samuel Badham, a shoemaker, fell ill. His feet became infected, and he could no longer wear shoes. In response to these events, to the decay of his feet, he went begging, "a thick bundle of rags tyed under the soles of his feet, and with a stick in each hand." For the next eight years his life fell into a regular pattern, with an established route through London, allowing him to show his ragged appendages and wave his crutches at a new audience each day.[27] That journey, from gainful employment to beggary, was also made by Henry Buxton around the same time. He was arrested for begging in the City of London on June 12, 1738, and carefully explained to the lord mayor, John Barnard, that

> He is about twenty six years of age, that about two years ago he received a violent fall which has occasioned his lameness & great pains in his back & bowels. That he is unable to provide for himself . . . About four months ago . . . he came to St Thomas's Hospital in this city to be cured of his illness, from whence he was discharged incurable.[28]

To be literally lame, to have a withered arm or leg, or better still a missing one, was a comprehensive and undeniable claim to the charity of the physically whole. In an image that captures one of the central ironies of body history, the poor waved their missing limbs for attention. In doing so they also scared the wits out of their fellow Londoners. The assumed danger posed to pregnant women provided an excuse for commentators' opprobrium. Civicus, writing in the *London Journal* 1731, thought crippled beggars "ought not to be suffer'd to wander the streets, exposing their distorted limbs, and filthy sores, such sights being frequently attended with the worst consequences to women with child."[29] Joshua Gee agreed:

> As to those creatures that go about the streets to show their maim'd limbs, nauseous sores, stump hands, or feet, or any other deformity, I am of opinion, that they are by no means objects fit to go abroad; . . . considering the frights and pernicious impressions which such horrid sights have given to pregnant women (and sometimes even to the disfiguring of infants in the womb).[30]

Even the Grand Jury for the City of London justified its attempts to clear the streets of vagrants by claiming the presence of beggars was "productive of many bad consequences . . . in a particular manner to child-bearing women."[31]

Despite these concerns, the number of people with a missing limb begging on the streets of the capital simply increased. In part this was the result of the changing nature of warfare, along with the adventurism of the new fiscal-military state. From its inception, the Old Poor Law had made a special case for shipwrecked sailors and returning soldiers, excluding them from the punishments meted out to vagrants. But as the scale and scope of military activity increased over the course of the eighteenth century, so too did the number of individuals who fell in to these categories. At the same time, the establishment of the Royal Hospitals at Greenwich and Chelsea at the end of the seventeenth century ensured that a high proportion of Britain's disabled ex-servicemen could be found in London, either among the full-time residents of the hospitals or else drawn to the capital four times a year to collect their quarterly pensions.[32] Faced with the limited possibilities open to a damaged person in this most physical of worlds, more and more ex-sailors and -soldiers found themselves on the streets of London. A measure of the impact of the military on the beggars of the capital can be found in Matthew Martin's late eighteenth-century survey of street beggars. Among the 2,000 adult beggars he interviewed, over 18 percent of the men and 13 percent of the women had a direct military connection.[33]

For well-to-do victims of military disasters and triumphs—such as Admiral Horatio Nelson, who lost an arm in battle in 1797—there was at least the

FIGURE 9.1: Ex-soldier with no arms and peg-leg sailor drinking in a tavern. Etching by Isaac Cruikshank, ca. 1791. Iconographic Collections, Wellcome Library, London.

chimera of beautifully crafted false arms and limbs. For them, the services of Gavin Wilson of Canongate were readily available. In the words of George Galloway:

> Thank heav'n! I'm safely landed frae Ostend,
> My broken ribs and shattered arms to mend
> By famous Wilson of Canongate;
> These wings of my poor trunk he'll reinstate.
> . . .
> Legs, thighs an' arms, to equip our battered hulls
> Toes, fingers, noses he must bring bushels.[34]

For most people, this reconstructive fantasy was not possible. The poor were faced with the stark choice of begging or starving. This was certainly the choice that Thomas Dargaval faced in 1780 when he lost his arm to canister shot and the surgeon's knife, while serving before the mast on the ship *Vigilant*. His response was to craft the story of his lost arm into a compelling military narrative, and to ask the ship's carpenter to help create a prop—a small wooden model of the ship. Re-armed with the story, his empty sleeve, and the ship, he made a living on the streets of London.[35] When in 1763 John Collier wanted to depict a deserving beggar who could give voice to his criticism of a wealthy

FIGURE 9.2: A "deserving beggar" approaches a pluralist for alms. Etching with engraving by John Collier, 1770. Iconographic Collections, Wellcome Library, London.

pluralist, he chose "A soldier maimed and in the beggars' list," who had lost his "leg and thigh" at the battle of Guadeloupe.[36]

But, if the empty sleeves and missing limbs of London's beggars formed a self-evident case for charity, and a clear resource from the beggars' perspective, these most obvious wounds were just one fragment of a larger repertory of physical representation. Blindness was perhaps the next most-used disability. The *Blind Beggar of Bethnal Green* was one of those irritatingly saccharine plays that was reprinted regularly throughout the century, but, more importantly, it reflects the equation in popular imagination between begging—and tolerated and decriminalized begging at that—and blindness.[37] Workhouses and hospitals, even the Foundling Hospital, regularly trained blind children in music, frequently in how to play the violin.[38] The expectation was that they would support themselves through public begging, justified with a tune. The largest single figure included in Paul Sandby's painting *Asylum for the Deaf* (ca. 1760) is a blind cellist accompanied by a boy.[39]

The extent to which blindness both justified begging and disarmed criticism can be measured by the case of Thomas Cooper. Charged with theft in May 1755, he claimed:

> The prosecutor wanted me to go into divers parts of the kingdom, to shew how the roads lay to the French; I was to lead about a blind man, but to enter into the French service at the same time; and he wanted me to go to the priest about it, but I would not.[40]

That a blind beggar could be conceived as a cover for espionage is a measure of the power of blindness in constructing a beggar persona that discouraged further inquiry. A more distressing case is that of Anne Martin, alias Chapney, who, in 1761, was imprisoned in Newgate for two years for "putting out the eyes of children, with whom she went a begging about the country."[41]

Of course, the other thing about blindness was that it was one disability—along with deafness and the inability to speak—that could be readily faked. In 1705 Thomas Newby was arrested "for binding up his hand pretending to be lame & his eyes blind."[42] And in 1730, Samuel Gold was prosecuted, "for being an imposture & pretending to be blind when he hath [been seen] . . . to play at cards."[43]

Disability, blindness, and the more general category of lameness were gendered in a range of specific and complex ways. Partly because of the clothes men and women wore, a missing leg or arm was more obvious when it had become detached from a man. The dresses and cloaks worn by women tended to hide a wooden appendage or empty sleeve. More than this, and in whatever way the limb had been lost, men could claim a military explanation for its absence that induced guilt and admiration in equal measure—a narrative strategy that was unavailable to women.

STATEMENTS OF VICTIMHOOD

Women were not entirely excluded from begging on the basis of a disability, but physical impairment was less central to their rhetorical strategies. Indeed, the substantial number of male beggars caught cross-dressing in pursuit of alms suggests that female attire was in itself a compelling case for charity. In the bowdlerized second edition of the biography of Bampfylde-Moore Carew,

FIGURE 9.3: "Blind" beggar checks the contents of his collecting tin. Etching by John Thomas Smith, 1816. Iconographic Collections, Wellcome Library, London.

"king of the beggars," Bampfylde borrows a friend's "child, a gown and one of her petticoats; and being thus accoutred with the child in his arms . . . [he pretends] to be an unfortunate woman, whose house had been burnt . . . coughing very violently and making the child cry."[44] Isaac Fairbrother appeared before the Court of Bridewell in early April 1742, accused of "pretending to be an old woman & beggin in the parish of St Buttolph Bishopsgate."[45] In the minds of most eighteenth-century commentators, male beggars were disabled while female beggars were simply poor. In 1749 the author of *Low Life* bracketed the two in his description of beggarly crossing sweepers as "Poor women and lame men whose bread depends chiefly on sweeping the passages of churches, chapels and meetings."[46]

For women, the most important narrative and physical strategy open to them was an appeal to seduction and desertion, evidenced by pregnancy or the presence of children. Ann Fretwell, alias Davis, was arrested in Newgate Street, "pretending to be with child & imposing upon people thereby with intent to move compassion."[47] Many more women begged the streets with an all-too-real child on the way. The significant thing is that drawing attention to their condition formed a powerful begging strategy that underpinned a clear statement of victimhood. Pregnancy formed compelling evidence for seduction, desertion, or widowhood.

Babies, once born, continued to have a valuable role in begging. Elizabeth Evan's defense, when she was tried for infanticide in the Spring of 1740, was that "she found . . . [the baby] dead, and therefore laid it away from her; and before she would have killed it, she would have gone a-begging with it."[48] There are very few images of ballad sellers, or female hawkers, or indeed female beggars, that do not include a baby, swaddled and slung across its mother's back. The most striking figure in William Hogarth's 1741 image *The Enraged Musician* is a ballad singer, babe in arms, bawling out a song of female ruin, "The Ladies Fall."[49] And the virtuous female beggar Hogarth depicts collecting the scraps from the wedding feast in his *Industrious 'Prentice Out of His Time & Married to His Master's Daughter* (1747) has a small child strapped to her back. There are even real examples of women borrowing children for the purpose of begging. In 1760 Julian Harrington was arrested, "wandering about the streets & receiving alms under colour of a pass pretending to be the mother of a boy she had with her who did not belong to her."[50] The power of an infant child to wrest reluctant alms from the well-to-do, however, can be more effectively measured in the frequent claims that beggars rented out small children from parish nurses and, more fancifully, from fully organized baby rental agencies. The anonymous author of the 1735 *Trip through Town* recorded:

> Having one day rambled into the heart of the good parish of St Giles in the Fields . . . I came to a place call'd the Infant Office, where young children stand at livery and are let out by the day to the town mendicants. . . . A woman above 70 would needs hire a baby that was sucking at the breast, and another who had a complexion as sallow as a Portuguese sailor, must forsooth be accommodated with a child as fair as a smock fac'd parson. . . . A beggar woman, who was vastly in arrears for the let of children was refus'd. . . . [and she] threw an old ragged riding hood over her shoulders, cursing them for a parcel of unchristian old

FIGURE 9.4: Ballad singer, with babe in arms and small girl, bawls out a song of female ruin. Detail of engraving by William Hogarth, 1741. Iconographic Collections, Wellcome Library, London.

FIGURE 9.5: Female beggar with baby collecting scraps from the Industrious 'Prentice wedding feast. By William Hogarth, 1747. Iconographic Collections, Wellcome Library, London.

> B—tches in forcing her to tell the town ten thousand lyes by saying she had three poor infants sick at home.[51]

Some forty years later the *Annual Register* reported as credible fact that

> Among the vagrants found begging in the streets of London, and carried before the Lord Mayor to be passed to their respective parishes; was a woman with a child in her arms, which, upon her examination, appeared to be hired at the rate of eight-pence a day of its mother in Petticoat-lane. She was committed to Bridewell to hard labour, and the child returned to its parent.[52]

The power of women to wrest reluctant charity from pedestrians rested on a broader concern with children and pregnancy, and a recognition of the profoundly disadvantaged position of women under patriarchy. But we should not lose sight of women's ability to manipulate the realities of pregnancy and motherhood to construct a powerful case for relief. In the jaundiced assessment of one author, female beggars frequently pinched the child, "in her Arms, like a tuneable Instrument to be *set to Musick*, when she came in the view of any seemingly well-disposed People."[53] In other words, the extreme physical signs of disability and motherhood conveyed powerful claims to legitimate relief and were used self-consciously by the poor to justify their requests. This was by no means the limit of the strategies and stratagems forced on the poor. Instead these are just two aspects of a complex visual rhetoric shared by all beggars. Nakedness, bare feet, and filth, and most importantly rags, which both pretended to hide nakedness and drew attention to that nakedness, were also a part of a claim to charity.[54] And if disability was used more frequently by men than by women, and if pregnancy and the presence of children was the basis of many female claims to alms, raggedness, filth, nakedness, and vermin were something shared by both sexes. When John Webb took up his place at Grosvenor Gate in June 1762, he was "very offensive and stunck very much having no clothes or linnen to shift himself."[55] Whatever else, you can be certain that his poverty was not questioned.

The authority and power of these strategies is recorded by the erstwhile contributors of alms. Jonathan Swift evokes the distress and sense of revulsion beggars engendered with their bodies in his description of Gulliver's reaction to the public beggars of Brobdingnag:

> the beggars watching their opportunity, crouded to the sides of the coach and gave me the most horrible spectacles that ever an . . . eye beheld. There was a woman with a cancer in her breast . . . full of holes. . . . There

> was a fellow with a wen in his neck. . . . And another with a couple of wooden legs. . . . But the most hateful sight of all was the lice crawling upon their cloaths. . . . It perfectly turned my stomach.[56]

In the process of exposing these physical signs of distress, beggars made a refusal to give alms all the more difficult. Swift's physical revulsion is a measure of the authority of the beggars. Indeed, the insistence on the part of most commentators that wounds were faked and children were rented reflected the success of these strategies rather than the commonality of the practice. In 1709, in *The Tatler*, Richard Steele claimed rather unconvincingly, that

> Going through an alley the other day, I observed a noisy impudent beggar bawl out, that he was wounded in a merchantman, that he had lost his poor limbs, and showed a leg clouted up. All that passed by, made what haste they could out of sight and hearing. But a poor fellow at the end of the passage, with a rusty coat, a melancholy air, and a soft voice, desired them to look upon a man not used to beg. The latter received the charity of almost every one that went by.

His observation that soft-voiced poverty was more effective than a brash appeal to suffering is open to real question, but his subsequent observation, that "The strings of the heart, which are to be touched to give us compassion, are not so played on but by the finest hand" is surely correct.[57]

For the poor, for beggars on the street, they had no choice but to use their bodies in a self-conscious and deliberate way, to weave a story as they waved for attention. As a result they necessarily developed a distanced relationship with their own physical presence and a self-conscious personal narrative, in which body and biography were interwoven. More than anyone, beggars were forced to examine and use their bodies in imaginative and well-thought-out ways, to rely, in William Hay's words, on the "Lining of the Head" to distance themselves from their own physicality and to exploit and manipulate the assumptions and sympathies of a broader population.[58]

To end with the other side of this equation, losing a leg, being blind, suffering the uncertainty of pregnancy alone on the streets, struggling to support your children in the face of real poverty, dying slowly of gangrene, going naked and unclean, or allowing oneself to be eaten by lice are awful and painful things. Their usefulness, and the necessity of making use of them to formulate an appeal for money, does not make them less tragic. James Dawson Burn was a beggar boy who could not afford shoes even if his begging life would allow him to wear them:

> During the winter my feet were hacked in innumerable fissures from which blood was continually starting, when I washed them at night before going to bed (which was as seldom as possible) my sufferings were intense, added to this, my heels were . . . elongated . . . with the action of the frost, which caused me either continual pain or an itching, which was nearly as bad to bear.[59]

This reflects the tiniest fragment of the real human suffering, of the pain, death, and distress experienced daily by the beggars of London.

CHAPTER TEN

Self and Society

Attitudes toward Incest in Popular Ballads

RUTH PERRY

Ballads are a great unsung body of texts that hover on the margins of eighteenth-century literary history without quite being acknowledged by modern scholars of the period. But ballads were a crucial cultural phenomenon in eighteenth-century society, a common experience of rich and poor, so embedded in the soundscape as not to be remarked, any more than the air people breathed. Joseph Addison famously wrote about ballads in three *Spectator* papers of 1711,[1] praising the beauties of "The Ballad of Chevy Chase" and "Babes in the Wood." He said: "an ordinary Song or Ballad that is the Delight of the common People, cannot fail to please all such Readers as are not unqualified for the Entertainment by their Affectation or Ignorance."[2] He cautioned his audience not to let the ballad's simplicity of style prejudice them against its "true poetical Spirit," or its passionate and beautiful sentiments. He cites the example of the "late Lord Dorset," a man with "the greatest Wit tempered with the greatest Candour . . . one of the finest Criticks as well as the best Poets of his Age" who "had a numerous Collection of old *English* Ballads, and took a particular Pleasure in Reading of them."[3]

Ballads were many people's first literary experience in eighteenth-century England and Scotland, whether these were simple broadsides from which they learned to read or the earliest sung stories that moved them to tears or ignited their imaginations. Oliver Goldsmith loved ballads, and several find their way into "The Vicar of Wakefield" (1766). He wrote in his essay "Happiness" (1759), "The music of Mattei[4] is dissonance to what I felt when our old dairymaid sung me into tears with 'Johnny Armstrong's Last Good Night,' or the cruelty of 'Barbara Allen.' "[5]

Ballads were more present in eighteenth-century England than literary critics and historians tend to remember. We are so used to making our way in the world with our eyes that we forget what Walter Ong calls the "lifeworld of the oral/aural past,"[6] an environment, in the city at least, of street cries and rhymes, bells ringing and chants, work songs and lullabies, an environment in which rags were sought and strawberries sold to the accompaniment of words and musical phrases so familiar that the hearer did not have to be able to understand the words to recognize which peddler was abroad exercising his or her lungs. Ballads were sung in taverns and camps, in dimly lit laborer's cottages, as well as in the blazing halls of the wealthy; they were memorized and transmitted orally by ordinary people as well as by professional singers and actors. They were printed on broadsides and in chapbooks and garlands, and they often circulated from print to oral transmission and back again. A ballad might be learned by ear and then written down to save or to remember, or it might be learned from a broadside, remembered, and then passed along orally.

From the earliest dawn of printing, broadsides were the comic books and the poster art of the poor: single sheets decorated with woodcuts on which were printed the texts of ballads with the title of the familiar tune to which those words could be sung. Between the middle of the sixteenth century and the beginning of the seventeenth century, more than 3,000 broadsides were officially entered in the Stationer's Register, although it is likely that five times as many were printed and sold in that period.[7] The walls of taverns and cottages were pasted up with as many as twenty, thirty, or more of these productions for decoration and quick reference as *aides memoire*. However ballads were composed, however far back they go, historically speaking, our earlier printed record of them has often been in the form of a broadside. They were bought and sold at fairs, and peddlers carried them the length and breadth of England and Scotland along with cloth and thread and needles and thimbles and other household goods. Ballads were the literary art of the common people and, more than novels, the popular culture of all classes. To examine a ballad that was circulating in eighteenth-century Scotland therefore as cultural evidence of attitudes toward sex, self, and society is not an empty exercise. What complicates the use of ballads as cultural evidence is that there is never one single definitive text, but many variants, so that one has to refer to themes, episodes, formulas, and storylines rather than a precise sequence of words.

"SHEATH AND KNIFE"

The ballad I want to examine here is "Sheath and Knife," known to Robert Burns (at least a few verses of it) and the melody notated by him in the *Scots*

Musical Museum (1787–1803), which he compiled with James Johnson. It was subsequently collected by William Motherwell from the recitation of Mrs. King of Kilbarchan parish on February 9, 1825, and appears in his *Minstrelsy: Ancient and Modern* (1827). Sir Walter Scott also remembered this ballad from his childhood: "I have heard the 'Broom blooms bonnie' sung by our poor old nursery-maid as often as I have teeth in my head, but after cudgeling my memory I can make no more than the following stanzas."[8] Scott puts a ballad in the mouth of his character, Effie Deans, in *The Heart of Midlothian* (1818) when she sings a stanza that appears to be a composite of several songs, with the two refrain lines: "The broom grows bonny, the broom grows fair" and "And we daurna gang down to the broom nae mair."[9] She is just coming back from seeing her lover when she sings it, defying her sister's inquiries. For her, it marks an end of innocence and the beginning of suffering. Ironically, the illegitimate child that Effie bears from this affair returns at the end of the novel—like the repressed impulse he represents—to kill his father, imbricating sex and blood and murder like the ballad that this refrain alludes to.

Here is the text as Motherwell collected it.

Sheath and Knife

It is talked the warld all over
The brume blooms bonnie and says it is fair
That the king's dochter gaes wi child to her brither.
And we'll never gang doun to the brume onie mair.

He's taen his sister doun to her father's deer park,
The brume blooms bonnie, etc.
Wi his yew-tree bow and arrow fast slung to his back
And we'll never, etc.

'Now when that ye hear me gie a loud cry
The brume blooms bonnie, etc.
Shoot frae thy bow an arrow and there let me lye.
And we'll never, etc.

'And when that ye see that I am lying dead
The brume blooms bonnie, etc.
Then ye'll put me in a grave, wi a turf at my head
And we'll never, etc.

Now when he heard her gie a loud cry
The brume blooms bonnie, etc.

His silver arrow frae his bow he suddenly let fly
Now they'll never, etc.

He has made a grave that was lang and was deep,
The brume blooms bonnie, etc.
And he has buried his sister, wi her babe at her feet.
And they'll never, etc.

And when he came to his father's court hall,
And the brume blooms bonnie, etc.
There was music and minstrels and dancing and all
But they'll never, etc.

'O Willie, O Willie, what makes thee in pain,
The brume blooms bonnie, etc.
'I have lost a sheath and knife that I'll never see again.'
For we'll never, etc.

'There is ships o your father's sailing on the seas
The brume blooms bonnie, etc.
That will bring as good a sheath and knife unto thee.'
And we'll never, etc.

'There is ships o my father's sailing on the sea
The brume blooms bonnie, etc.
But sic a sheath and a knife they can never bring to me.'
Now we'll never, etc.[10]

Like most other Scottish or English ballads, this one has all the hallmarks of oral composition: repetitive textual, metrical, and melodic structures; patterned arrangements of narrative; recurrent formulaic phrases and epithets (a grave that is "long and deep"; a "yew-tree bow"); conceptual parallelism and incremental repetition (in one verse she tells him to put her in a grave; in the next verse he digs a grave and buries her in it); and formalized refrain lines: "the brume blooms bonnie" and "we'll never go doun to the brume onie mair."[11] The repetitions serve not only the memory of the singer but permit the listener to think about what has gone before, to let his or her attention lapse for a moment, to sink into fantasy, to paint some pictures in the mind. They pace the story.

The action usually unfolds in event and speech—sometimes dialogue—without narrative comment and begins somewhere in the middle. Historical time is irrelevant; what happens in ballads is timeless or out of time: "It is talked the warld all over." Ballad tone is neither sentimental nor moralizing but stark and ritualistic; there is little or no subjective consciousness behind it. This creates the necessary distance so that the powerful and tragic events of ballads can be safely shared by narrator and audience. The rhythm of the episodic structure of ballads has been described as leaping and lingering—that is, as leaping instantaneously, with no attempt at realism, over great swathes of time and space and then lingering in a single moment, letting its complications sink in. In this episodic structure, focusing on vivid scenes that appear and disappear according to some familiar logic of the mind, ballads are like nothing so much as dreams.

In "Sheath and Knife" we move effortlessly from one scene to another—the deer park, the grave, walking in the broom, the court hall filled with merriment; even the enigmatic but suggestive reference to the sheath and knife springs forth as from intuition rather than reasoned thought. The action feels symbolic rather than literal, represented impersonally from the outside rather than narrated subjectively; the starkness of the narration together with the horror of the situation is what moves us. We do not know how the sister feels, but she seems to be cooperating in her own death when she tells her brother to shoot an arrow from his bow when he hears her give a loud cry, which is apparently the cry of childbirth. On the other hand, the brother has staged this tragedy. "He has taen his sister down to her father's deer park . . . With his yew-tree bow and arrow slung fast to his back"; yew trees are associated with death and sorrow and with graveyards, where they are often planted. Moreover, he has brought with him a silver arrow—a special, precious arrow—the instrument of execution for a woman who is a king's daughter and his only sister. When he returns to his father's court hall, he finds himself in the midst of festivities. The music and dancing are evocative of sexuality and bodily pleasure—music always signifies passion and the lapsing of reason—which sets off his sadness in contrastive relief. He seems all the more alone because everyone around him is gay. And when he invokes the phallic imagery of a sheath and knife, we know that he is speaking of his sister both as his lost sexual partner, unique among women and specially suited to him (nowhere in the whole world where those ships of his father's are sailing will there be another like her) but also as the carrier, the sheath, of his child.

This brother and sister are two parts of one whole; they could not be closer. Their bodies belong together as intimately as a fetal child within its mother's womb. They are made for each other in the way they are presented in the ballad, siblings with the same father, interchangeable and identical so far as the description goes, except that one is male and one is female. Their sexual interests are mirrored in each other; they begin and end in one another. Male and female, they are not opposed or different but positioned as two interlocking parts. The problem is that as a woman, the sister's body shows the evidence of their congress, and they both seem to understand that she must be made to disappear. The reasons are not specified but one knows that the world, which is already gossiping about them, will not tolerate this incestuous, too-close union any more. The offspring of their illicit union cannot be allowed to see the light of day.

There is no reference to consequences—no sense of what will happen when the "king's dochter" is found missing—only the brother's deep sense of bereavement after the killing, as if half of him has been shot away, removed, irretrievably lost, and he will never be the same. The brother expresses no guilt, which is, after all, an emotion about the self. The pain of loss he feels and expresses with this haunting metaphor is all the more terrible and unrelieved for his lack of guilt. The chorus lines, too, underscore the sense of ritual and irretrievable loss. An eighteenth-century audience in a social situation, having heard this song many times, might have joined in on these lines, participating in the story and giving vent to their feelings. In a village this song might have been the property of a local singer, and when people gathered to work or to talk in the evening, the person who knew "Sheath and Knife" might have been prodded to sing it and these chorus lines would have been caroled back by those listening.

"The broom blooms bonnie and they say it is fair" obviously puns on the word "fair"—the blooming broom is so beautiful and life is so unjust and so tragic. The broom, a yellow-flowering shrub weed that grows in the margins of cultivation, like the "greenwood," marks a liminal place outside the established limits of society, outside the law, in the bushes, so to speak, where illicit sex always takes place.[12] In Scottish ballads, people always lie down in the broom, away from prying eyes.[13] The poignant "And we'll never go down to the broom any more" signals the end of innocence for us as well as for the principals in the story, and also the end of sexual pleasure and of companionable strolling for the brother and sister. This is how the ballad was known at the end of the eighteenth century, and how Francis James Child copied it from Motherwell and Scott, the great collectors of the Romantic period.

"THE SHEATH AND THE KNIFE OR LEESOME BRAND"

More recently, evidence has surfaced of a much earlier version of this ballad that poses some puzzling questions. This much earlier version of "Sheath and Knife" was written down in the commonplace book of Robert Edwards, along with other poems and musical examples. Robert Edwards, born in 1617, was a minister of Murroes parish near Dundee, in eastern Scotland. His commonplace book dates to about 1630. He titles this entry "The Sheath and the Knife or Leesome Brand" (Leesome Brand being someone's name), and his text includes some of the features of Child no. 15, "Leesome Brand," which is an entirely separate ballad from "Sheath and Knife" in the Child canon.[14] In the version of "Leesome Brand," collected by Peter Buchan in his *Ancient Ballads and Songs of North of Scotland* (1828) as it was included in Child's collection, the speaker's son, a ten-year-old boy named Leesome Brand, comes to an "unco" land and falls in love with a gay lady at court who is roughly his age.

> This ladye was scarce eleven years old,
> When on her love she was right bauld;
> She was scarce up to my right knee,
> When oft in bed wi men I'm tauld.

When "nine months have come and gone" this lady turns "pale and wane" and says to Leesome Brand, "In this place I can nae mair stay." She sends him to her father's stable for horses for them both and to her mother's coffers for her tocher or dowry. It is she who has parents at court, not he, and she who provides the means for their escape. He follows her instructions and they mount and ride, but after a while his "true love then began to fail" and she tells him that she feels as if her back is breaking. She wishes she had a midwife but he tells her that they are far from any midwife. He offers to do for her whatever a man can do, but she rejects his offers of assistance and sends him away to hunt while she gives birth. She dies in childbirth and so does her baby. After he finds her lying there dead, he returns to his own home to his mother (his father in other versions), who questions him about his sadness. He replies that he has lost a golden knife and a gilded sheath. When his mother asks if there are not goldsmiths "here in Fife" who can make another knife and another sheath, he replies "I've lost my ladye I lov'd sae dear, / Likeways the son she did me bear."

According to Child, versions of "Leesome Brand," have been found throughout Scandinavia (there are versions in Norwegian, Swedish, Danish, Finnish, and Icelandic), and there are also variants in German, French, and Dutch. The Danish ballad, "Redselille og Medelvold" (Roselilly and Ole), is a variant of "Leesome Brand" about unlucky lovers who are not siblings. It begins when a mother discovers that her daughter's breasts run with milk and she extorts from the girl a confession that she has been beguiled by a man. The mother then threatens various punishments for both of them. The girl tells her lover, and they take two horses and ride away together. After they have ridden awhile, the girl finds that her time has come and her lover lifts her off her horse and spreads his cloak for her. He offers to bandage his eyes and render such service as a man may, but she does not want him to know the pangs a woman suffers and sends him off to hunt. He comes back to find her dead with two sons dead too; he buries them all and then falls on his sword.

FIGURE 10.1: A knight and his lover flee from death. Engraving by Edward Harding after Diana Beauclerk, 1796. Miss Beauclerk was alleged to have had an incestuous relationship with her brother, Lord Bolingbroke. Iconographic Collections, Wellcome Library, London.

What is interesting about the ballad "Sheath and Knife or Leesome Brand," found in Robert Edwards's 1630 commonplace book—a variant not known by Child when he compiled *The English and Scottish Popular Ballads*—is that it combines elements from both "Sheath and Knife" and "Leesome Brand" to create an incest ballad with very different affective qualities than either of the two contributory ballads as they were collected separately two centuries later. What follows is a transcription of the variant in Robert Edwards's 1630 manuscript book:

The Sheath and the Knife or Leesome Brand

Ther was a sister and a brother
the sun gois to under the wood
who most intirelie lovid other
god give we had nevir beine sib.

Sayes "sister I wald lay the by
the sun gois to etc.
and thou wald not my deuds cry"
god give we, etc.

"Alas brother wald ye doe so
the sun gois to etc.
I rather nou death undergoe
alas give we had newer bein sib

the morrne is my fathir's feast
the sun etc.
Weil in my clothis I most be least
god give we had newir bein sib.

When they conwining al at ons
to royal feasting in the hal
it me behovith them amongs
ge dekit in a goun of pa;

and when I lout me to my to
the sun etc.
my lesse wil brak and go in tuo
god give etc.

and when I lout me toi my kni
the sun etc.
my lesse will brak and go in thrie
god give etc.

and it wil go from on to uthir
the sun etc.
until it come to Jhon by brother
lord give etc.

and Jhon my brothir is most il
the sun etc.
he wil hus both burne on a hil.
lord god we had etc.

I sal go to my fathirs stable
the sun etc.
and tak a stid both wight and able
lord give etc.

and we sal ryd til tym we spend
the sun etc.
until we see our trystis end."
lord give etc.

She had not riden a myle but ane
the sun etc.
when she gan quaking gran and gran
lord give etc.

"Is ther water in your shoes
or comes the wind into your glowes

Or think ye me to simple a knight
to ryd or go with you alnyght?"

"and when ye heire me loud loud cry
ye bend your bow and ran tharby

and when ye se my ly ful stil
so souing your horne come me til.

I wald give al my fathirs land
for on woman at my command."

when that he cam soon hir besead
[the bab was borne the lady dead.]

Ther he has tain his yong yong sonne
and borne to a milk womane.

he dreu his suord him wonding sore
from this tyme to wrid newir more.

"mother" quoth he "can so mak my bed
can se mak it long and nothing bread.

mother alas I tint my knife
I lovid better then my lyffe.

mother I have als tint my shead
I lovid better then them bead.

ther is no cutlar in this land
can make a kniffe so at my command."

he turned his faced to the wa
gave up the goast and gaid his way.

the on was layid in Marie Kirk
othir in Marie Queire

out throch the on there greu a birke
and out throch hir a breir.

ye may knou surlie by their signes
They wer tuo lowirs neire.[15]

The most stunning difference between the Robert Edwards's 1630 "Sheath and Knife" and the one sung by Mrs. King of Kilbarchan in 1825 is that the infant lives and is given up to "a milk woman"—a wet-nurse. Thus the incestuous relationship is not simply damned in this variant: something living and hopeful comes from it. Furthermore, the refrain lines of the Edwards variant, while very different from the refrain lines Mrs. King sang in 1825, are also meaningful. The repeated second line, "The sun gois to under the wood," evokes the hour of sunset, the hour that daylight ceases and darkness covers all, like the hour before death. This refrain line is very old; it appears in a four-line, anonymous, medieval lyric of the thirteenth century: "Nou goth sonne under wode," the "sonne" being Christ, the son of God, as well as the sun, and "wode" being the wood of the cross as well as a stand of trees.[16] It evokes both a simple sunset and the crucifixion, the loss of light from the world. The fourth line of every stanza of Robert Edwards's "Sheath and Knife," also repeated throughout the ballad, is "God give we had nevir bein sib." "God give" means "God grant," an invocation unusual in ballads that rarely have Christian sentiments. "Sib" is a capacious word in Middle English, denoting many different kinds of blood kinship as well as nonbiological relatedness. It connotes very special closeness and intimacy. Best friends might be "sib." But siblings are the most "sib" of all because of their special relationship in the English kinship system, a point to which I will return.

Another major difference between the Motherwell variant collected from Mrs. King in 1825 and this older version of the ballad found in Robert Edwards's commonplace book is that the brother–sister relationship represented is much more equalitarian in the older 1630 version: "Ther was a sister and a brother / the sun gois to under the wood / who most intirelie lovid [each] other / god give we had nevir beine sib" rather than "He takes his sister down to her father's deer park," as in the eighteenth-century variant. In the 1630 version we are told that "they most intirely" loved one another. He then says, "sister I wald lay the by [sister I would lay thee by] . . . and thou wold not my deuds cry"—that is, sister, I would hide thee so that you (your pregnancy) would not proclaim my deeds to the world. And she replies: "Alas, brother wald ye doe so / The sun gois to under the wood / I rathir nou death undergoe / God give we had nevir been sib."

FIGURE 10.2: A wet-nurse receives a high-born child. Etching by Robert de Launay after Etienne Aubry, ca. 1740. Iconographic Collections, Wellcome Library, London.

Then she says it is her father's feast day and she must be laced in her clothes and join the guests. When she bends over, her laces will break, presumably because of her pregnant girth. She says that *she* will go to her father's stable and get a horse for them to ride away on.

They had not ridden but a mile when her waters break and he asks if he is too simple, or lowly, to ride or stay with her all night. She tells him that when he hears her loud, loud cry to bend his bow and run to the sound, and when he sees her lying still to sound his horn—or rather, to make his horn sigh—and to come to her. She says, "I wald give al my fathirs land / for on woman at my command," a poignant reminder that their sexual relations have dis-severed her from her natural gender group. One assumes that he does as she tells him, but:

When that he cam soon hir besead,
The sun gois to under the wood
The bab was borne the lady dead
God give we had nevir been sib.

So in this earlier version he does not kill her, does not use his yew-tree bow to shoot a silver arrow at her, but he is perhaps hunting as in "Leesome Brand" when she delivers her baby and dies in the attempt. Then comes the verse about the living offspring:

Ther he has tain his yong yong sonne
The sun gois to under the wood
And borne to a milk womane
God give we had nevir beine sib.

After that he draws his sword and mortally wounds himself and asks his mother to make his bed for him:

Mother alas I tint my knife
The sun gois to under the wood
I lovid better than my lyffe.
God give we had nevir beine sib.

Mothir I have als tint my shead
The sun goes to under the wood
I lovid better than them bead.
God give we had nevir beine sib.

Then he turns his face to the wall and gives up the ghost. The ballad ends with the familiar formula about true lovers: one was laid in the church and the other in the choir, and out of one grows a birch and out of the other a briar.

Ye may knou surlie by thir signes
The sun goes to under the wood
They wer tuo lowirs neire.
God give we have nevir beine sib.

This is an interesting early variant for the ballad "Sheath and Knife" circulating at the end of the eighteenth century. Although the biological weight of the liaison falls on the woman, and although the brother speaks of wanting to keep *his* deeds from being known in the world, we are told in the first verse that these two love each other entirely and equally. Rather than the brother simply taking his sister to the woods and killing and burying her, these two try to run away together. Moreover, when she is gone, he wounds himself with his sword and dies, rather than just standing around in his father's hall with music and minstrels, muttering enigmatically about losing a sheath and knife. Most significantly, in this version their child is alive with a wet-nurse at the end, not dead and buried with the mother. It is an altogether more positive vision of their connection both in terms of their love, their death, and the lament in which they wish that they were not sib, which means more than wishing they were not siblings. It also means "I wish we were not related" or "I wish we were not so close or intimate."

This earlier version has a more fluid continuous narrative line rather than being made up of disjointed, symbolic episodes. From the moment the brother worries about the world finding out about them until the rose and the briar grow from their graves, the sequence is clear. In the eighteenth-century/early nineteenth-century variant of "Sheath and Knife," the story ends elliptically with the brother's sadness in the midst of court revelry, and the sexual connection of the brother and sister is foregrounded with the enigmatic phrase, "sheath and knife," repeated three times. This figure for a set of genitals that go together but are irreplaceably lost ends the ballad; this haunting image still resonates when the song is over. The 1630 variant, by contrast, ends with the familiar trope of the rose and briar, telling us that these familiar symbolic plants marked two "lovers neire." That appears to be more significant than the brother's lament that "no cutlar in this land/can make a kniffe so at my command," a complaint that also confounds the sexual figuration since it is his sister who has died and their child who is living. In the 1630 variant, the incestuous sex matters less than their tragic relationship; they are star-crossed lovers who happen to be brother and sister rather than a brother and his sister, the king's daughter, who has been having illicit sex with her brother and must pay for it with her life.

In the eighteenth-century version collected in the nineteenth century, the woman is much more passive and she alone must die—the weight of the sexual transgression falls on her. The sister in the seventeenth-century version has a great deal more agency; she is the one who goes to the stable and gets a horse for them to ride; she tells her lover what to do; and when she dies with tragic inevitability, her brother no longer wants to live but takes his own life. The drama is told entirely from the outside: what happens, what he says, what she says. That impassivity still marks the later eighteenth-century version, but by then all sense of the particularity of the woman has disappeared, for she says almost nothing. Insofar as anyone is affected by the drama, the only self that matters in the late eighteenth-century version is the brother's self.

INTERPRETING ATTITUDES TO INCEST

What are we to make of the change in the ballad "Sheath and Knife" over the course of two centuries? Does it mean anything at all? Can we interpret the changes in it as signifying changes in society? These are only variants, after all, and there is no way of knowing if there were others circulating that make this way of reading them moot. Extrapolating from a single variant found in a

1630 commonplace book must be tentative and provisional. Nevertheless, the cultural critic cannot resist asking whether these differences signify changes in attitudes toward incest, or sex, or brother–sister relations? In interpreting texts from an oral tradition, it is worth remembering Walter Ong's dictum that "oral traditions reflect a society's present cultural values" rather than preserving its past. "The integrity of the past [is] subordinate to the integrity of the present" in any oral tradition, he writes.[17] In other words, oral transmission tends to update texts; what is no longer relevant drops out of an oral repertoire. So the change in this ballad, the breaking apart of "Sheath and Knife" and "Leesome Brand" into two separate ballads, probably does tell us something about a shift in cultural attitudes between 1630, when Robert Edwards compiled his manuscript book, and the later eighteenth century when Walter Scott's nursemaid sang "Sheath and Knife" to him, and when Mrs. King learned it, while Agnes Lyle learned "Leesome Brand" as a separate ballad.[18]

There are quite a few other ballads about incest collected in eighteenth-century Scotland, and all of them (there are six in Child's canon) are about brother–sister incest.[19] Not a one of them is about intergenerational incest between fathers and daughters or uncles and nieces, nor is that because brother–sister incest was the only kind of incest that occurred in the British Isles. Court records, in England at least, provide plenty of evidence of intergenerational father–daughter and uncle–niece incest, although none of it seems to have been memorialized in ballads. The exclusive reference to brother–sister incest in Scottish and Irish ballads may reflect, instead, the lines of transmission for this form of literature and the fact that women tended to be the carriers of the old ballads; women may have found it too upsetting to sing about intergenerational incest. Indeed, when one of these incest ballads was collected from Irish traveling people in the twentieth century, the collector noted that it was considered an "unlucky" or "forbidden" song. And Sara Cleveland, an American singer, who learned one of these brother–sister incest ballads from her mother, reported that her mother rarely sang it and then only when alone or working around the house.[20] So it is possible that the exclusive focus on brother–sister incest in orally transmitted ballads is an artifact of the gendered patterns of transmission of these ballads.

But it is also possible that these incest ballads all focus on a similar form of sibling incest because they all come out of one cultural moment, traceable to roughly the same period and location—northern Scotland in the late eighteenth/early nineteenth centuries. One can only speculate about what they might reveal about kinship, siblings, self, or society at this particular historical juncture. In four of the seven brother–sister incest ballads listed in note 19, the incest is committed knowingly;[21] in "Sheath and Knife," "Lizie Wan," and "The Rich Man's Daughter," the brother kills the sister when he discovers that she is pregnant. In "Brown Robyn's Confession," the erring brother goes to sea when suddenly neither the sun nor moon nor stars is visible. The sailors are sure that someone on board among them is evil and they decide to cast lots to determine who is bringing them bad luck and needs to be cast out. When the lot falls to Brown Robyn, he confesses to incest with his mother and his sister and he bids his fellow sailors bind him to a plank and throw him overboard, which they do. In the other three incest ballads, "The Bonny Hind," "The King's Dochter Jean,"[22] and "Babylon," the brother, who has been abroad for many years and does not recognize his sister and is himself unrecognizable, assaults and rapes her not knowing that he is her brother. When he learns what he has done, he kills himself or she kills herself. It is thought that incest is also at the root of the murderous rage between brothers in "Two Brothers" and in "Edward." The incestuous desire of one brother for their sister—"a little bush that might have been a tree"—and the protective resistance of the other brother is what incites the struggle between them that leads to the fratricidal murder of one brother by the other.

Nor is outright incest the only form of fraternal conflict related to sexual conduct in ballads collected in the eighteenth-century British Isles. Brothers in ballads often seek to control their sisters' sexuality by determining who they can and cannot marry. And when a sister is not obedient to their wishes, they kill her or they kill her lover. In "Bonnie Susie Cleveland," the brother lights the fire under the pyre to which his sister is tied. In "Andrew Lammie," the girl's mother and father beat her badly, but it is her brother who breaks her back for loving Andrew Lammie. In the "Cruel Brother," the brother kills his sister with a little penknife because although her lover has asked permission of all her kin to marry her—her mother, her father, and her sisters—he neglects to ask her brother's permission. In "Dowie Dens of Yarrow" and in "Clerk Saunders," the brothers kill their sister's lover because he comes from the wrong class.

CHANGING BROTHER–SISTER RELATIONS

Sibling solidarity is a hallmark of the British kinship system in which lineage is traced through both the mother and the father. Siblings are the closest of all kin in a bilateral, cognatic system because only siblings have identical kin. Through their common mother, they are uniquely related to a set of maternal relatives as well as paternal relatives. Furthermore, siblings are the only link between these sets of maternal and paternal relatives because although marriage binds their parents together, it creates no recognized connection between the blood relations of the two spouses: the mother's relatives are not considered kin to the father's relatives. Siblings are thus a significant link in the chain of kinship. One linguistic trace of the significance of siblings in the British kinship system is the fact that the terms for nieces and nephews—the children of one's siblings—often meant "grandchild" as well. In other words, the children of one's siblings were as significant intergenerationally as the direct line through one's own progeny.[23] Then, too, brothers and sisters were often each others' earliest sustained cross-sex relationship and childhood companions.

In the course of the eighteenth century, the primacy of this brother–sister relationship changed for a number of reasons, among them the development of capitalism, the drive to accumulate property, and changes in inheritance law. Less and less were sisters considered equal members with their brothers of their clans or kin groups, and less and less were they thought to be female versions of their brothers but increasingly instead "chickens raised for other men's tables," as James Harlowe so eloquently puts it in Richardson's novel *Clarissa*. The social expectation that sisters would be protected and cherished by their brothers, and the expectation that brothers would be responsible for their sisters, diminished in the course of the eighteenth century. These ballads about brothers who kill their sisters, who rape them violently, or who kill their lovers, that is, who appropriate, direct, or terminate their sexuality, all register the sense that brothers own their sisters. Their cruelty is magnified by the betrayal of the closeness expected between a brother and a sister, a betrayal of the protection a brother once owed his sister and a betrayal of the sibling love that once distinguished the British kinship system. By the eighteenth century, brothers are represented in ballads as dangerous to the interests of their sisters, whether because of biology, character, or evolving male privilege, or some combination of all three.

When "Sheath and Knife or Leesome Brand" was divided into two separate ballads—one in which two lovers come to grief and one in which a brother kills his sister because she carries his child and he wants to hide the evidence of his incest—it signaled a new economy in the relations between brothers and sisters. Brothers seem to have significantly more power than their sisters; they keep a tight rein on their sisters' sexual favors, whether for themselves or by choosing their sisters' lovers. Incest has become dangerous for a woman in this later world, because her pregnancy reveals it, rather than being tragic for both lovers. The story has changed from being a story of star-crossed lovers to a story in which the woman must be sacrificed. Incestuous sex has become so shameful that it must be hidden at all costs. The ballad focuses on the brother's acts and feelings, and we hardly hear from the sister in the later variant "Sheath and Knife"; she has a lot more to say in the earlier version. The Romantic period—the early nineteenth century—is of course the period associated with brother–sister incest, the refusal to accommodate to an other but the insistent narcissistic mirroring of the self instead, in a loved second self. The sheath-and-knife trope for brother–sister incest in ballads collected at the turn of the century might also be seen as a figure for the intense, solipsistic enthusiasm for the self in its context, an emphasis that privileges the alter ego of the sibling rather than looking further afield for a loved object.

Juliet Mitchell, analyzing brother–sister incest in our own society, notes that the mother is usually absent in the clinical configuration of sibling incest.[24] Writing psychoanalytically about sibling incest, she describes the "narcissistic love-of-the-other-as-the-self" that can explode into murderous hatred because of the threat of self-annihilation if there is no mother present to supervise her children and to keep familial relationships in order. According to Mitchell, brothers and sisters drawn to one another both love one another as versions of themselves, as alter egos, and at the same time hate each other for supplanting each other in their families. When the mother is absent, there is no one to reinforce the differentiation between mothers and sisters, no one to make sure that mothers and sisters do not become confused.[25] The confusion that results when sibling incest ends in pregnancy is that a man becomes a father/brother and a woman becomes a sister/mother. According to Mitchell, love and hate, sexual attraction, and murderous rivalry coexist in modern-day sibling relationships. "Siblings provide a way of learning to love and hate the same person," she says,[26]a truth that still resonates for contemporary audiences of this ballad. "Sheath and Knife" is too abbreviated to identify any elements of hate or rivalry in it but there is also no mention of the true love that so dominates the 1630 version of the ballad. However, the mother is certainly missing (as she is from much of the fiction of the end of the eighteenth century).[27] Although there is a father with a court and ships sailing on the seas, there is no mother in the ballad. The brother–sister mirroring is figured in the repeated image of the sheath and knife as well as the dangerousness of their closeness. Sex and death are inextricably entwined in this ballad in the killing of the sister the moment after she gives birth, and in the burial of the infant with the sister/mother.

I have been arguing that "Sheath and Knife," a ballad that Sir Walter Scott heard as a young boy and that William Motherwell collected from an "old singing woman" in 1825, carries cultural attitudes of the eighteenth century toward the self, sexuality, and the family, and that these cultural attitudes come into sharper focus when compared with a variant of the ballad in circulation two centuries earlier. The way the brother controls his sister's fate, her relative silence, his isolation and self-regarding misery, the way the death punishment is dealt to the woman of the couple that engages in transgressive sex, the stillborn infant, the absence of the mother in this tragedy—all of these elements bespeak a configuration of kinship and sexuality familiar to us from other sources.[28] The ballad illustrates women's sexual victimization, a focus on male subjectivity—even a kind of pre-Romantic narcissism—the absence of adult women, and bodies that are not polymorphously perverse but that are formally male and female, designed functionally to fit together like a sheath and knife, by a kind of divine workmanship. The loss of one half renders the other incomplete. But the perfect fit, and indeed sexuality itself, is unholy in the ballad

FIGURE 10.3: A brother and sister play with a baby. The relationship between the trio is left ambiguous. Engraving by R. C. Bell after W. Mulready. Iconographic Collections, Wellcome Library, London.

because the incest tinges it with horror and it is not romanticized with love. That these meanings were available in ballads sung by ordinary people in the period suggests that they were not confined to the novel-reading literate classes but that they permeated all levels of society. The ballad form itself also testifies to the wellsprings of human creativity that wrest beauty out of tragedy and that can make art with just the human voice and mind.

NOTES

Introduction

1. Roy Porter, *Bodies Politic: Disease, Death and Doctors in Britain, 1650–1900* (London: Reaktion Books, 2001), 28; Porter, *Enlightenment: Britain and the Creation of the Modern World* (London: Penguin Books, 2001), 283.
2. Thomas Laqueur, "Bodies, Details, and the Humanitarian Narrative," in *The New Cultural History: Essays,* ed. Lynn Hunt (Berkeley: University of California Press, 1989), 176–204.
3. Ros Ballaster, "The Body as Fiction: Eighteenth-Century Tales of the Orient," in *The Eighteenth-Century Body: Art, History, Literature, Medicine,* ed. Angelica Goodden (Oxford, UK: Peter Lang, 2002), 17–32.
4. Fernando Vidal, "Extraordinary Bodies and the Physicotheological Imagination," in *The Faces of Nature in Enlightenment Europe,* ed. Lorraine Daston and Gianna Pomata (Berlin: BWV, 2003), 61–96.
5. For Feijóo and the Enlightenment see Rebecca Haidt, *Embodying Enlightenment: Knowing the Body in Eighteenth-Century Spanish Literature and Culture* (New York: St. Martin's Press, 1998); Vidal, "Extraordinary Bodies."
6. Thomas Paine, *The Age of Reason: Being an Investigation of True and Fabulous Theology,* part 1 (Paris, 1795), chapter 3.
7. P. Linebaugh, "The Tyburn Riot against the Surgeons," in *Albion's Fatal Tree,* Douglas Hay et al. (Harmondsworth, UK: Penguin Books, 1977), 65–117.
8. Ibid.
9. Ruth Richardson, *Death, Dissection and the Destitute* (London: Penguin, 1988), 63.
10. Most bodies in a Christian cemetery were laid on their backs, aligned west to east, with the head at the west end so that at the Resurrection the dead would sit up to face east. Suicides and unbaptized children were among those who would not be resurrected; Chris Thomas, *London's Archaeological Secrets: A World City Revealed* (New Haven, CT: Yale University Press, 2003), 130.
11. George Minois, *History of Suicide: Voluntary Death in Western Culture* (Baltimore, MD: Johns Hopkins University Press, 1999), chapters 8 and 9.

12. Mary Douglas, "The Construction of the Physician: A Cultural Approach to Medical Fashions," in *The Healing Bond. The Patient-Practitioner Relationship and Therapeutic Responsibility,* ed. Susan Budd and Ursula Sharma (London: Routledge, 1994), 23–41.
13. Marc Bloch, *The Royal Touch: Sacred Monarchy and Scrofula in England and France*, trans. J. E. Anderson (London: Routledge & Kegan Paul, 1973).
14. Roy Porter, *Mind Forg'd Manacles: Madness and Psychiatry in England from Restoration to Regency* (London: Penguin Books, 1987), 63–64.
15. John Donne, *Devotions upon Emergent Occasions*, ed. Anthony Raspa (Montreal: McGill-Queen's University Press, 1975), 51, 117.
16. Porter, *Bodies Politic,* 41.
17. Sander L. Gilman, *Sexuality: An Illustrated History* (New York: Wiley, 1989), chapter 6.
18. Letter to James Watt, January 6, 1781, quoted in Desmond King-Hele, ed., *The Letters of Erasmus Darwin* (Cambridge, UK: Cambridge University Press, 1981), 104.
19. James Sharp, *Instruments of Darkness. Witchcraft in England, 1550–1750* (London: Hamish Hamilton, 1996); Keith Thomas, *Religion and the Decline of Magic: Studies in Popular Beliefs in Sixteenth and Seventeenth-Century England* (London: Weidenfeld & Nicolson, 1971).
20. Daniel Schäffer, " 'That Senescence Itself Is an Illness': A Transitional Medical Concept of Age and Ageing in the Eighteenth Century," *Medical History* 46 (2002): 525–48.
21. Porter, *Bodies Politic,* 43.
22. Jonathan Andrews et al., *The History of Bethlem* (London: Routledge, 1997). Later, in the nineteenth century, physiognomy was extended to the new science of criminology incorporating theories of degeneration.
23. Porter, *Mind Forg'd Manacles.*
24. Roger Shattuck, *Forbidden Knowledge. From Prometheus to Pornography* (New York: St. Martin's Press, 1996); Lorraine Daston and Katherine Park, *Wonders and the Order of Nature, 1150–1750* (New York: Zone Books, 1998); Porter, *Enlightenment,* 228.
25. Richard D. Altick, *The Shows of London: A Panoramic History of Exhibitions, 1600–1862* (Cambridge, MA: Belknap Press, 1978); John Brewer and Roy Porter, eds., *Consumption and the World of Goods* (London: Routledge, 1993).
26. Byrne's skeleton is now in the Hunterian Museum, Royal College of Surgeons of England, London. Roy Porter, "John Hunter: A Showman in Society," *Transactions of the Hunterian Society* 52 (1993–1994): 19–24; Wendy Moore, *The Knife Man* (London: Bantam, 2005), chapter 12.
27. Jeremy Bentham, "On the Use of the Dead to the Living," reprinted in *Bentham's Auto-Icon and Related Writings*, James E. Crimmins (Bristol, UK: Thoemmes Press, [1824] 2002).
28. Andrew Cunningham, "Auto-Icon: Jeremy Bentham's Three Bodies, the Moral Laws of Nature, and the Ideology of Industrial Capitalism," in *The Faces of Nature in Enlightenment Europe,* ed. Lorraine Daston and Gianna Pomata (Berlin: BWV, 2003), 181–210.
29. Norbert Elias, *The Civilizing Process*, vol. 1 (New York: Urizen Books, 1978), chapter 2, V–VII.
30. Porter, *Bodies Politic,* chapter 9.

31. Roy Porter, *Flesh in the Age of Reason* (London: Allen Lane, 2003), chapter 7.
32. N.D. Jewson, "The Disappearance of the Sick-Man from Medical Cosmology, 1770–1870," *Sociology* 10 (1976): 225–44; Mary Fissell, "The Disappearance of the Patient's Narrative and the Invention of Hospital Medicine," in *British Medicine in an Age of Reform,* ed. Roger French and Andrew Wear (London: Routledge, 1991), 92–109.
33. Saul Jarcho, ed. and trans., *The Clinical Consultations of Giambattista Morgagni* (Boston: Francis A. Countway Library of Medicine, 1984); Jarcho, ed. and trans., *Clinical Consultations and Letters by Ippolito Francesco Albertini, Francesco Torti, and other Physicians: University of Bologna MS 2089–1* (Boston: Francis A. Countway Library of Medicine, 1989).
34. Séverine Pilloud and Micheline Louis-Courvoisier, "The Intimate Experience of the Body in the Eighteenth Century: Between Interiority and Exteriority," *Medical History* 47 (2003): 451–72.
35. Thomas Dixon, "Patients and Passions: Languages of Medicine and Emotion, 1789–1850," in *Medicine, Emotion and Disease, 1700–1950,* ed. Fay Bound Alberti (Basingstoke, UK: Palgrave Macmillan, 2006), 22–52.
36. Fay Bound Alberti, "Introduction: Medical History and Emotion Theory," in *Medicine, Emotion and Disease, 1700–1950,* ed. Fay Bound Alberti (Basingstoke, UK: Palgrave Macmillan, 2006), xiv–xxix.
37. Christopher Lawrence, "The Nervous System and Society in the Scottish Enlightenment," in *Natural Order: Historical Studies of Scientific Culture,* ed. Barry Barnes and Steven Shapin (Beverly Hills, CA: Sage, 1979), 19–40.
38. Roy Porter, "Bodies of Thought: Thoughts about the Body in Eighteenth-Century England," in *Interpretation and Cultural History,* ed. Joan H. Pittock and Andrew Wear (Basingstoke, UK: Macmillan, 1991), 82–108; Porter, "'Expressing Yourself Ill': The Language of Sickness in Georgian England," in *Language, Self and Society: The Social History of Language,* ed. Peter Burke and Roy Porter (Cambridge, UK: Polity Press, 1991), 276–99.
39. Pilloud and Louis-Courvoisier, "The Intimate Experience of the Body."
40. Porter, *Flesh in the Age of Reason,* 403.
41. Roy Porter and G.S. Rousseau, *Gout: The Patrician Malady* (New Haven, CT: Yale University Press, 1998); Clark Lawlor and Akhihito Suzuki, "The Diseases of the Self: Representing Consumption, 1700–1830," *Bulletin of the History of Medicine* 74 (2000): 458–94.
42. Emma C. Spary, "The Performance of Surgery in Enlightenment France," *Endeavour* 23 (1999): 180–83.
43. Susan C. Lawrence, *Charitable Knowledge: Hospital Pupils and Practitioners in Eighteenth-Century London* (Cambridge, UK: Cambridge University Press, 1996), 242.
44. W. F. Bynum, *Science and the Practice of Medicine in the Nineteenth Century* (Cambridge, UK: Cambridge University Press, 1994), 42–43.
45. Anne Borsay, "An Example of Political Arithmetic: The Evaluation of Spa Therapy at the Georgian Bath Infirmary, 1742–1830," *Medical History* 45 (2000): 149–72.
46. John Blunt [Samuel William Fores], *Man-Midwifery Dissected . . .* (London: Published for the author, 1793), frontispiece.
47. Jason S. Zielonka, "'A Man-Midwife.' Etching, Hand Coloured, by S. W. Fores, London, 1793. New Haven, Yale Medical Library, Clements C. Fry Collection." *Journal of the History of Medicine* 30 (1975): 259; Porter, *Mind Forg'd Manacles,* 206–32.

48. George S. Rousseau, "The Pursuit of Homosexuality in the Eighteenth Century: 'Utterly Confused Category' and/or Rich Repository?" in *Eighteenth-Century Life*, vol. 9., ed. Robert P. Maccubbin (Williamsburg, VA: College of William and Mary, 1985).
49. Lawrence Stone, *The Family, Sex and Marriage in England, 1500–1800* (London: Weidenfeld & Nicolson, 1977); Tim Hitchcock, *English Sexualities, 1700–1800* (Basingstoke, UK: Macmillan, 1997).
50. John Cleland, *Fanny Hill or Memoirs of a Woman of Pleasure* (Harmondsworth, UK: Penguin, [1748–1749] 1998).
51. Cunningham, "Auto-Icon."
52. Jeremy Bentham, "Offenses against One's Self," *Journal of Homosexuality* 3 (1978): 389–405; 4 (1979): 91–109.
53. Roy Porter and Lesley Hall, *The Facts of Life: The History of Sexuality and Knowledge from the Seventeenth Century* (New Haven, CT: Yale University Press, 1994).
54. James Graham, *A Lecture on the Generation, Increase, and Improvement of the Human Species* (London: 1782), 27.
55. Stone, *The Family, Sex and Marriage in England.*
56. Porter, *Bodies Politic,* 75.
57. Andrew Wear, "The History of Personal Hygiene," in *Companion Encyclopedia of the History of Medicine,* vol. 2, ed. W. F. Bynum and Roy Porter (London: Routledge, 1993), 1283–1308.
58. Morag Martin, "Doctoring Beauty: The Medical Control of Women's *Toilettes* in France, 1750–1820," *Medical History* 49 (2005): 351–68.
59. William Buchan, *Domestic Medicine; or, the Family Physician . . .* (Edinburgh, Scotland: Balfour, Auld and Smellie, 1769), 89.
60. Lucia Dacome, "Useless and Pernicious Matter: Corpulence in Eighteenth-Century England," in *Cultures of the Abdomen: Diet, Digestion and Fat in the Modern World,* ed. Christopher E. Forth and Ana Carden-Coyne (Basingstoke, UK: Palgrave Macmillan, 2005).
61. Ibid.
62. Janet Semple, *Bentham's Prison: A Study of the Panopticon Penitentiary*. Oxford: Clarendon Press, 1993); Cunningham, "Auto-Icon."

Chapter 1

1. David Cressy, *Birth, Marriage, and Death: Ritual, Religion, and the Life-Cycle in Tudor and Stuart England* (Oxford: Oxford University Press, 1997), 384.
2. Charles Dickens, *Martin Chuzzlewit* (Harmondsworth, UK: Penguin, [1844] 1999), 303.
3. Doreen Evenden, *The Midwives of Seventeenth-Century London* (Cambridge, UK: Cambridge University Press, 2000).
4. Jeremy Black and Roy Porter, *A Dictionary of Eighteenth-Century World History* (Oxford: Blackwell, 1994), 587, 710, 257; Massimo Livi-Bacci, *A Concise History of World Population*, 3rd ed. (Malden, MA: Blackwell, 2001), 27.
5. John Graunt, *Natural and Political Observations Mentioned in a Following Index, and Made upon the Bills of Mortality* (London: Royal Society, 1676), 68–70.
6. Charles-Louis de Secondat Montesquieu, *The Persian Letters*. Translated by C. J. Betts. Harmondsworth: Penguin, [1721] 1973), 202–4.

7. A. Walzer, "Logic and Rhetoric in Malthus's *Essay on the Principle of Population, 1798*," *Quarterly Journal of Speech* 73 (1987): 1–17.
8. Lisa Cody, *Birthing the Nation: Sex, Science, and the Conception of Eighteenth-Century Britons* (Oxford: Oxford University Press, 2005), 285–92.
9. Stephen Jay Gould, *The Mismeasure of Man*, revised ed. (New York: Norton, 1996).
10. Kenneth Pomeranz, *The Great Divergence: China, Europe, and the Making of the Modern World Economy* (Princeton, NJ: Princeton University Press, 2000).
11. Black and Porter, *A Dictionary,* 146; Livi-Bacci, *A Concise History,* 42–48.
12. Roger Schofield, David Sven Reher, and Alain Bideau, eds., *The Decline of Mortality in Europe* (Oxford: Clarendon Press, 1991).
13. Edward A. Wrigley and Roger Schofield, *The Population History of England 1541–1871* (Cambridge, UK: Cambridge University Press, 1989).
14. Tim Hitchcock, *English Sexualities, 1700–1800* (New York: St. Martin's Press, 1997), 25–26.
15. Thomas Laqueur, *Making Sex: Body and Gender from the Greeks to Freud* (Cambridge, MA: Harvard University Press, 1990).
16. Katherine Park and Robert Nye, "Destiny Is Anatomy," *The New Republic,* February 18, 1991, 53–57; Mary Fissell, "Gender and Generation: Representing Reproduction in Early Modern England," *Gender and History* 7 (1995): 433–56; Michael Stolberg, "A Woman Down to Her Bones: The Anatomy of Sexual Difference in the Sixteenth and Early Seventeenth Centuries," *Isis* 94 (2003): 274–99; Laura Gowing, *Common Bodies: Women, Touch and Power in Seventeenth Century England* (New Haven, CT: Yale University Press, 2003).
17. Elizabeth Nihell, *A Treatise on the Art of Midwifery* (London), 101–2.
18. Barbara Duden, *The Woman Beneath the Skin: A Doctor's Patients in Eighteenth-Century Germany* (Cambridge, MA: Harvard University Press, 1991); Jacques Gélis, *History of Childbirth: Fertility, Pregnancy, and Birth in Early Modern Europe*, trans. Rosemary Morris (Boston, MA: Northeastern University Press, 1991); Fissell, "Gender and Generation."
19. Roy Porter and Lesley Hall, *The Facts of Life: The Creation of Sexual Knowledge in Britain, 1650–1950* (New Haven, CT: Yale University Press, 1995).
20. Gélis, *History of Childbirth,* 10–11, 58–65.
21. Fissell, "Gender and Generation," 447.
22. Duden, *The Woman Beneath the Skin.*
23. Angus McLaren, *Reproductive Rituals: The Perception of Fertility in England from the Sixteenth Century to the Nineteenth Century* (London: Methuen, 1984), 107–12.
24. Jane Sharp, *The Midwives Book . . .* (London: S. Miller, 1671), 3, 4.
25. Ruth Ginzberg, "Uncovering Gynocentric Science," in *Feminism and Science,* ed. Nancy Tuana (Bloomington: Indiana University Press, 1989).
26. Nihell, *A Treatise on the Art of Midwifery,* 90.
27. Cody, *Birthing the Nation.*
28. William Smellie, *A Treatise on the Theory and Practice of Midwifery*, vol. 1 (London: D. Wilson, 1752), 121–23.
29. In fact, the only known surgical delivery with both mother and child surviving, before 1800 was performed in 1739 by Mary Donally, an Irish country midwife who knew to cover the incision with egg whites, a natural antibiotic.
30. Adrian Wilson, *The Making of Man-Midwifery* (Cambridge, MA: Harvard University Press, 1995).

31. Evenden, *The Midwives of Seventeenth-Century London.*
32. Wilson, *The Making of Man-Midwifery.*
33. Nina Gelbart, *The King's Midwife: A History and Mystery of Madame du Coudray* (Berkeley: University of California Press, 1998); Lianne McTavish, *Childbirth and the Display of Obstetrical Authority in Early Modern France* (Burlington, VT: Ashgate Publishing, 2005).
34. Cody, *Birthing the Nation,* 145–51.
35. Clara Pinto-Correia, *The Ovary of Eve: Egg and Sperm and Preformation* (Chicago: University of Chicago Press, 1997); Cody, *Birthing the Nation.*
36. This widespread belief had the unfortunate corollary that no true rape could result in pregnancy if pleasure was required for reproduction. Sharp, *The Midwives Book,* 43–45.
37. McLaren, *Reproductive Rituals,* 13–29.
38. Cody, *Birthing the Nation,* 120–51.
39. John McManners, *Death and the Enlightenment* (Oxford: Oxford University Press, 1981), 10.
40. Roy Porter, *The Greatest Benefit to Mankind: A Medical History of Humanity* (New York: Norton, 1997), 236–37.
41. Ulinka Rublack, "Pregnancy, Childbirth and the Female Body in Early Modern Germany" *Past and Present* 150 (1996): 84–110.
42. Ralph Houlbrooke, *Death, Religion, and the Family in England, 1480–1750* (Oxford: Clarendon Press, 1998), 200–203.
43. Cremation, especially such rituals as suttee or widow burning in India, seemed to confirm for Europeans the primitivism and cruelty of other peoples and thus seemed to help justify their colonial presence around the world. Pompa Banerjee, *Burning Women: Widows, Witches, and Early Modern European Travelers in India* (New York: Palgrave Macmillan, 2003).
44. Philippe Ariès, *The Hour of Our Death*, trans. Helen Weaver (New York: Random House, 1981): 353–61.
45. Houlbrooke, *Death, Religion, and the Family in England,* 286–87.
46. Embalming for display would also be practiced by affluent eccentrics in the eighteenth century and beyond, for instance, when the dentist Martin van Butchell hired man-midwife and anatomist William Hunter to preserve his wife's corpse, which was dressed in finery and displayed to guests in the van Butchell home, or when the utilitarian philosopher Jeremy Bentham gave to University College London his fortune and remains, which are on display even today. Roy Porter, *Bodies Politic: Disease, Death and Doctors in Britain, 1650–1900* (Ithaca, NY: Cornell University Press, 2001), 73–74.
47. Houlbrooke, *Death, Religion, and the Family in England,* 342.
48. Ariès, *The Hour of Our Death,* 362.
49. McManners, *Death and the Enlightenment,* 166–72.
50. Jane Shaw, *Miracles in Enlightenment England* (London: Yale University Press, 2006).

Chapter 2

1. Antoinette Emch-Dériaz, "The Non-Naturals Made Easy," in *The Popularization of Medicine, 1650–1850,* ed. Roy Porter (London & New York: Routledge, 1992), 134–59.

2. John Cleland, *Institutes of Health* (London: T. Becket & T. Davies, 1761); Francis de Valangin, *A Treatise on Diet, or the Management of Human Life; by Physicians Called the Six Non-Naturals* (London: Author, 1768); George Cheyne, *Rules and Observations for the Enjoyment of Health and Long Life* (Leeds, UK: G. Wright & Sons, 1770).
3. David Cantor, ed., *Reinventing Hippocrates* (Aldershot, UK: Ashgate, 2002).
4. Andrew Wear, *Knowledge and Practice in English Medicine, 1550–1680* (Cambridge, UK: Cambridge University Press, 2000).
5. Lester S. King, *The Philosophy of Medicine: The Early Eighteenth Century* (Cambridge, MA: Harvard University Press, 1978), 64–94; Wear, *Knowledge and Practice,* 369–98.
6. Harold Cook, *The Decline of the Old Medical Regime in Stuart London* (Ithaca, NY: Cornell University Press, 1986), 148–49, 158; Wear, *Knowledge and Practice,* 399–433.
7. However, the continued influence of chemistry in the early Enlightenment is frequently overlooked. Antonio Clericuzio, *Elements, Principles and Corpuscles: A Study of Atomism and Chemistry in the Seventeenth Century* (New York: Springer, 1999); Allen George Debus, *The French Paracelsians: The Chemical Challenge to Medical and Scientific Tradition in Early Modern France* (Cambridge, UK: Cambridge University Press, 2002), 156–82.
8. Theodore M. Brown, *The Mechanical Philosophy and the "Animal Oeconomy": A Study of the Development of English Physiology during the Seventeenth and Eighteenth Centuries* (New York: Arno Press, 1981); Brown, "Medicine in the Shadow of the *Principia*," *Journal of the History of Ideas* 48 (1987): 629–48.
9. Rina Knoeff, *Herman Boerhaave (1668–1738): Calvinist Chemist and Physician* (Amsterdam: Royal Netherlands Academy of Arts and Sciences, 2002).
10. Jeremiah Wainewright, *A Mechanical Account of the Non-Naturals* (London: J. Clarke, 1737), 70.
11. Cullen moved beyond a purely Cartesian mechanical body to include a "vital principle" in the animal economy. On vitalism in the French context see Elizabeth Williams, *The Physical and the Moral: Anthropology, Physiology and Philosophical Medicine in France, 1750–1850* (Cambridge, UK: Cambridge University Press, 1994) and Williams, *A Cultural History of Medical Vitalism in Enlightenment Montpellier* (Aldershot, UK: Ashgate, 2003). Christopher Lawrence, "The Nervous System and Society in the Scottish Enlightenment," in *Natural Order, Historical Studies of Scientific Culture,* ed. Barry Barnes and Steven Shapin (Beverly Hills, CA: Sage, 1979), 19–40; Andrew Doig et al., eds., *William Cullen and the Eighteenth-Century Medical World* (Edinburgh: Edinburgh University Press, 1990).
12. Ann C. Vila, *Enlightenment and Pathology: Sensibility in the Literature and Medicine of Eighteenth-Century France* (Baltimore, MD: Johns Hopkins University Press, 1998); Hubert Steinke, *Irritating Experiments: Haller's Concept and the European Controversy on Irritability and Sensibility, 1750–90* (Amsterdam: Rodopi, 2005).
13. The completion of this project is often dated to the turn of the nineteenth century, linked to the establishment of large-scale hospitals, the rise of surgical thought, and the pathology of Xavier Bichat. See Michel Foucault, *The Birth of the Clinic: An Archaeology of Medical Perception,* trans. Alan Sheridan (New York: Vintage, 1973); William F. Bynum, *Science and the Practice of Medicine in the Nineteenth*

Century (Cambridge, UK: Cambridge University Press, 1994). George S. Rousseau, *Nervous Acts: Essays on Literature, Culture and Sensibility* (Basingstoke, UK: Palgrave Macmillan, 2004).

14. Henry Burdon, *The Fountain of Health* (London: The Author, 1734), 8, 11, 16, 32; John Huxham, *An Essay on Fevers*, 3rd ed. (London: J. Hinton, 1757), 139; Johann Georg Zimmermann, *A Treatise on the Dysentery*, trans. C.R. Hopson (London: J. & F. Rivington, 1771), 49–50, 215–16; Andrew Wear, "Fear, Anxiety and the Plague in Early Modern England: 'Religious and Medical Responses,'" in *Religion, Health and Suffering*, ed. John R. Hinnells and Roy Porter (London: Kegan Paul International, 1999), 339–63.
15. Vivian Nutton, "The Seeds of Disease: An Explanation of Contagion and Infection from the Greeks to the Renaissance," *Medical History* 27 (1983): 1–34; Jonathan Gil Harris, *Foreign Bodies and the Body Politic: Discourses of Social Pathology in Early Modern England* (New York: Cambridge University Press, 1998).
16. Walter Lynn, *An Essay Towards a More Easie and Safe Method of Cure in the Small Pox. Founded upon experiments, and a review of Dr Sydenham's works* (London: R. Knaplock, 1714), 14–15.
17. Herman Boerhaave, *A Treatise on the Venereal Disease*, trans. John Barker (London: T. Cox et al., 1729), 10–12.
18. Jonathan Gil Harris, *Sick Economies: Drama, Mercantilism, and Disease in Shakespeare's England* (Philadelphia: University of Pennsylvania, 2004).
19. James C. Riley, *Population Thought in the Age of the Demographic Revolution* (Durham, NC: Carolina Academic Press, 1985); Eileen Magnello and Anne Hardy, eds., *The Road to Medical Statistics* (Amsterdam: Rodopi, 2002); Andrea A. Rusnock, *Vital Accounts: Quantifying Health and Population in Eighteenth-Century England and France* (Cambridge, UK: Cambridge University Press, 2002).
20. Thomas Short, *New Observations, Natural, Moral, Civil, Political, and Medical, on City, Town, and Country Bills of Mortality* (London: T. Longman & A. Millar, 1750); John Heysham, *Observations on the Bills of Mortality, in Carlisle, for the year 1779* (Carlisle, UK: J. Milliken, 1780).
21. Michel Foucault, *The Order of Things: An Archaeology of the Human Sciences* (New York: Vintage, 1970), 71–76.
22. Roger French, "Sickness and the Soul: Stahl, Hoffman and Sauvages on Pathology," in *The Medical Enlightenment of the Eighteenth Century*, ed. Andrew Cunningham and Roger French (Cambridge, UK: Cambridge University Press, 1990), 99–102, 105–8; Julian Martin, "Sauvages's Nosology: Medical Enlightenment in Montpellier," in *The Medical Enlightenment of the Eighteenth Century*, ed. Andrew Cunningham and Roger French (Cambridge, UK: Cambridge University Press, 1990), 111–37.
23. Thomas Sydenham, *The Whole Works of That Excellent Practical Physician, Dr Thomas Sydenham*, 9th ed., trans. John Pechey (London: J. Darby et al., 1729), vi.
24. See, for example, the distinction made by Clifton Wintringham, *An Essay on Contagious Diseases* (York, UK: Francis Hildyard, 1721), 54–57.
25. Herman Boerhaave, *Boerhaave's Aphorisms: Concerning the Knowledge and Cure of Diseases*, trans. J. Delacoste (London: B. Cowse & W. Innys, 1715), 1.
26. Thomas Garnett, *A Lecture on the Preservation of Health* (London: T. Cadell jnr & W. Davis, 1800), 3.
27. Richard Mead, *A Mechanical Account of Poisons* (London: R. Smith, 1708).
28. Roy Porter and William F. Bynum, eds., *Medical Fringe and Medical Orthodoxy, 1750–1850* (London: Croom Helm, 1987); Roy Porter, *Quacks: Fakers and Char-*

latans in English Medicine (Stroud, UK: Tempus, 2003); Harold Cook, *Trials of an Ordinary Doctor: Joannes Groenevelt in Seventeenth-Century London* (Baltimore, MD: Johns Hopkins University Press, 1994).

29. James Kilpatrick, *An Essay on Inoculation* (London: J. Huggonson, 1743), iv, 15; John Huxham, *An Essay on Fevers*, 3rd ed. (London: J. Hinton, 1757), 34; Cleland, *Institutes of Health,* 100.
30. Huxham, *An Essay on Fevers,* 107–8.
31. Burdon, *The Fountain of Health,* 19.
32. Zimmermann, *A Treatise on the Dysentery,* 23–24; Dale Ingram, *An Historical Account of the Several Plagues That Have Appeared in the World . . . Since the Year 1346* (London: R. Baldwin & J. Clark, 1755), 53.
33. Samuel Brown, *An Inaugural Dissertation on the Bilious Malignant Fever . . .* (Boston: Manning & Loring, 1797), 28–30; Robert Robertson, *Essay on Fevers* (London: The author, 1790), 88.
34. Leonore Davidoff and Catherine Hall, *Family Fortunes: Men and Women of the English Middle Class, 1780–1850* (Chicago: University of Chicago Press, 1987); Robert Shoemaker, *Gender in English Society, 1650–1850: The Emergence of Separate Spheres?* (London: Longman, 1998); Anthony Fletcher, *Gender, Sex, and Subordination in England, 1500–1800* (New Haven, CT: Yale University Press, 1996).
35. Thomas Laqueur, *Making Sex: Body and Gender from the Greeks to Freud* (Cambridge, MA: Harvard University Press, 1990). For recent debates on this model see the exchange between Laqueur and Michael Stolberg: Stolberg, "A Woman Down to Her Bones: The Anatomy of Sexual Difference in the Sixteenth and Early Seventeenth Centuries," *Isis* 94 (2003): 274–99; Laqueur, "Sex in the Flesh," *Isis* 94 (2003): 300–306.
36. Mary Fissell, *Vernacular Bodies: The Politics of Reproduction in Early Modern England* (Oxford: Oxford University Press, 2004), 53–89; Patricia Crawford, "Attitudes to Menstruation in Seventeenth-Century England," *Past and Present* 91 (1981): 47–73; Ottavia Niccoli, "Menstruum Quasi Monstruum: Monsterous Births and Menstrual Taboo in the Sixteenth Century," in *Sex and Gender in Historical Perspective,* ed. Edward Muir and Guido Ruggiero (Baltimore, MD: Johns Hopkins University Press, 1990), 1–25.
37. Elizabeth Williams, "Hysteria and the Court Physician in Enlightenment France," *Eighteenth-Century Studies* 35 (2002): 247–55.
38. Kevin Siena, "Pollution, Promiscuity, and the Pox: English Venereology and the Early Modern Discourse on Social and Sexual Danger," *Journal of the History of Sexuality* 8 (1998): 553–74; Marie E. McAllister, "Stories of the Origin of Syphilis in Eighteenth-Century England: Science, Myth, and Prejudice," *Eighteenth-Century Life* 24 (2000): 22–44.
39. Ruth Perry, "Colonizing the Breast: Sexuality and Maternity in Eighteenth-Century England," *Journal of the History of Sexuality* 2 (1991): 204–34.
40. Jean Astruc, *A Treatise on the Diseases of Women*, 2 vols. (London: J. Nourse, 1762).
41. Robertson, *Essay on Fevers,* 101.
42. Ibid., 87.
43. John Huxham, *An Essay on Fevers*, 3rd ed. (London: J. Hinton, 1757), 79–80.
44. The term comes from James C. Riley's authoritative account of disease and the environment. James C. Riley, *The Eighteenth-Century Campaign to Avoid Disease* (London: MacMillan, 1987); Carlo M. Cipolla, *Miasmas and Diseases: Pub-*

lic Health and the Environment in the Pre-Industrial Age*, trans. Elizabeth Porter (New Haven, CT: Yale University Press, 1992); Andrew Wear, "Making Sense of Health and the Environment in Early Modern England," in *Medicine in Society: Historical Essays,* ed. Andrew Wear (Cambridge, UK: Cambridge University Press, 1992), 119–47.

45. John Tylston, *The Country Physician* (Edinburgh: 1701), unnumbered preface.
46. John Arbuthnot, *Essay Concerning the Effects of Air on Human Bodies* (London: J. Tonson, 1733), 172.
47. Mark Harrison, "'The Tender Frame of Man': Disease, Climate and Racial Difference in India and the West Indies, 1760–1860," *Bulletin of the History of Medicine* 70 (1996): 68–93; Alan Bewell, *Romanticism and Colonial Disease* (Baltimore, MD: Johns Hopkins University Press, 1999); Harrison, *Climates and Constitutions: Health, Race, Environment and British Imperialism in India, 1600–1850* (New Delhi: Oxford University Press, 1999).
48. Bewell, *Romanticism and Colonial Disease,* 37.
49. Ingram, *An Historical Account,* 61.
50. Colin Chisholm, *An Essay on the Malignant Pestilential Fever Introduced into the West Indian Islands from Boullam on the Coast of Guinea, as It Appeared in 1793 and 1794* (London: C. Dilly, 1795), 94–97; Harrison, "The Tender Frame of Man," 75; Bewell, *Romanticism and Colonial Disease,* 29–46.
51. James Lind, *An Essay on Diseases Incidental to Europeans in Hot Climates* (London: T. Becket & P. A. de Hondt, 1768), 2–3.
52. Harrison, "The Tender Frame of Man," 73–79.
53. Susan Klepp, "Seasoning and Society: Racial Differences in Mortality in Eighteenth-Century Philadelphia," *William and Mary Quarterly* 51 (1994): 473–506; Trevor Burnard and Kenneth Morgan, "The Dynamics of the Slave Market and Slave Purchasing Patterns in Jamaica, 1655–1788," *Morgan William and Mary Quarterly* 58 (2001): 205–28.
54. Christopher Lawrence has demonstrated the point for scurvy. Lawrence, "Disciplining Disease: Scurvy, the Navy and Imperial Expansion, 1750–1825," in *Visions of Empire: Voyages, Botany, and Representations of Nature,* ed. David Philip Miller and Peter Hanns Reill (Cambridge, UK: Cambridge University Press, 1996), 80–106.
55. John Bell, *An Inquiry into the Causes Which Produce, and the Means of Preventing Diseases among British Officers, Soldiers, and Others in the West Indies* (London: J. Murray,1791), 35–37; Ingram, *An Historical Account,* 146.
56. Thomas Trotter, *Observations on the Scurvy* (Edinburgh: C. Elliott, 1786), 22.
57. John Atkins, *The Navy Surgeon, or, Practical System of Surgery* (London: J. Hodges, 1742), 356–57.
58. Chisholm, *An Essay on the Malignant Pestilential Fever,* 219.
59. Nicholas Hudson, "From 'Nation' to 'Race': The Origin of Racial Classification in 18th-Century Thought," *Eighteenth-Century Studies* 29 (1996): 247–64; Roxam Wheeler, *The Complexion of Race: Categories of Difference in Eighteenth-Century British Culture* (Philadelphia: University of Pennsylvania Press, 2000), 1–48.
60. Norris Saakwa-Mante, "Western Medicine and Racial Constitutions: Surgeon John Atkins' Theory of Polygenism and Sleepy Distemper in the 1730s," in *Race, Science and Medicine, 1700–1960,* ed. Bernard Harris and Waltraud Ernst (London: Routledge, 1999), 29–57.

61. Londa Schiebinger has found a similar presentation of Africans in anatomists' depictions of skeletons. Schiebinger, *The Mind Has No Sex? Women in the Origins of Modern Science* (Cambridge, MA: Harvard University Press, 1991), 211–13.
62. Saakwa-Mante, "Western Medicine and Racial Constitutions," 46.
63. Samuel Stanhope Smith, *An Essay on the Causes of the Variety of Complexions and Figure in the Human Species* (London: John Stockdale, 1789), 53, 77, 79–81, 99, 101, 105.
64. The term was also key to Buffon, whose *Histoire Naturelle* was foundational to debates on race. Phillip R. Sloan, "The Idea of Racial Degeneracy in Buffon's *Histoire Naturelle*," in *Racism in the Eighteenth Century*, ed. Harold E. Pagliaro (Cleveland, OH: Case Western Reserve University, 1973), 293–321; Smith, *An Essay*, 26, 27ff, 41, 42ff, 101ff, 118.
65. Smith, *An Essay*, 15–17.
66. Ibid., 91.
67. Ibid., 101ff.
68. Londa Schiebinger, "The Anatomy of Difference: Race and Sex in Eighteenth-Century Science," *Eighteenth-Century Studies* 23 (1990): 387–405; Schiebinger, *Nature's Body: Sexual Politics and the Making of Modern Science* (London, Pandora, 1994), chapters 4 and 5; Randi Davenport, "Thomas Malthus and Maternal Bodies Politic: Gender, Race, and Empire," *Women's History Review* 4 (1995): 415–40; Nancy Leys Stepan, "Race, Gender, Science and Citizenship," *Gender and History* 10 (1998): 26–52; Kathleen Wilson, *The Island Race: Englishness, Empire and Gender in the Eighteenth Century* (London: Routledge, 2002).
69. Smith, *An Essay, passim.*
70. Ibid., *passim*; Anonymous, *The Art of Preserving Beauty* (London: 1789), 166–80.
71. On beauty and health as cultural markers of the classed body, see Simon P. Newman, *Embodied History: The Lives of the Poor in Early Philadelphia* (Philadelphia: University of Pennsylvania Press, 2003), 1–10.
72. Smith, *An Essay*, 50–52.
73. Smith makes the same point later, substituting "race" for "species." Smith, *An Essay*, 57, 69.
74. Brown, *An Inaugural Dissertation*, 53.
75. Ann Carmichael, *Plague and the Poor in Renaissance Florence* (Cambridge, UK: Cambridge University Press, 1986); Paul Slack, *The Impact of Plague in Tudor and Stuart England* (London: Routledge, 1985), 55, 60, 252–53, 290–93, 304–6; Brian Pullan, "Plague and Perceptions of the Poor in Early Modern Italy," in *Epidemics and Ideas: Essays on the Historical Perception of Pestilence*, ed. Terence Ranger and Paul Slack (Cambridge, UK: Cambridge University Press, 1995), 101–24.
76. Ingram, *An Historical Account*, 26, 68–69, 172.
77. Daniel Peter Layard, *Directions to Prevent the Contagion of the Jail Distemper, commonly called the Jail Fever* (London: James Robson, 1772), 5–6.
78. John Mason Good, *Dissertation on the Diseases of Prisons and Poor Houses* (London: Medical Society of London / C. Dilly, 1795), 26.
79. Ibid., 33.
80. Ibid., 26–27.
81. Layard, *Directions to Prevent the Contagion*, 8.

82. John Pringle, *Observations on the Diseases of the Army* (London: A. Millar et al., 1765), xi; John Coakley Lettsom, *Medical Memoirs of the General Dispensary in London, for part of the years 1773 and 1774* (London: E. & C. Dilly, 1774), 23.
83. Pringle, *Observations on the Diseases,* 16, 22, 37–47; James Lind, *An Essay on the Most Effectual Means of Preserving the Health of Seamen in the Royal Navy . . . Together with an Essay on the Jail Distemper* (London: 1774), 133; John Alderson, *An Essay on the Nature and Origin of the Contagion of Fevers* (Hull, UK: G. Prince et al., 1788), 7; John Heysham, *An Account of the Jail Fever, or Typhus Carcerum* (London: T. Cadell et al., 1782), 28–37.
84. Heysham, *An Account of the Jail Fever,* 36.
85. Noah Webster, ed., *A Collection of Papers on the Subject of Bilious Fevers, Prevalent in the United States for a few years past* (New York: 1796), 79–95, 112, 155, 232, 240, 245.
86. Ibid., 95.
87. Riley, *The Eighteenth-Century Campaign to Avoid Disease,* 89–112.
88. Lind, *An Essay on the Most Effectual Means,* 344–45.
89. Ibid., 341–48; Layard, *Directions to Prevent the Contagion,* 27–46.
90. "Minutes of the Committee for Rebuilding the Workhouse," MS 3226 (1–3), 1796, 3, London Guildhall Library, St. Sepulchre (London Division).
91. Layard, *Directions to Prevent the Contagion,* 34–35.
92. Geoffrey L. Hudson, "Disabled Veterans and the State in Early Modern England," in *Disabled Veterans in History,* ed. David A. Gerber (Ann Arbor: University of Michigan Press, 2000).
93. Anne Borsay emphasized the "economic rationality" of policies aimed at the eighteenth-century disabled. Borsay, *Disability and Social Policy in Britain since 1750: A History of Exclusion* (Basingstoke, UK: Palgrave Macmillan, 2004). Marjorie Levine-Clark demonstrates the importance of this definition in the early nineteenth century, noting gendered elements of its construction. Marjorie Levine-Clark, "Engendering Relief: Women, Ablebodiedness, and the Poor Law in Early Victorian England," *Journal of Women's History* 11 (2000): 107–30; Lynn Hollen Lees, *The Solidarities of Strangers: The English Poor Laws and the People, 1700–1948* (Cambridge, UK: Cambridge University Press, 1998), 82–114.
94. John Clark, *Observations on the Diseases in Long Voyages to Hot Countries, and particularly on those which prevail in the East Indies* (London: D. Wilson & G. Nicol, 1773), 17, 22, 24; Rice Charlton, *Cases of Patients Admitted into the Hospital at Bath* (Bath, UK: 1776), 30–39; Bell, *An Inquiry into the Causes,* 121.
95. Thomas Prior, *An Authentick Narrative of the Success of Tar-water* (London: W. Innys et al., 1746), 69, 98–99, 117.
96. Too many examples exist for a comprehensive list but see John Aikin, *Thoughts on Hospitals* (London: Joseph Johnson, 1771), 93–94; Chisholm, *An Essay on the Malignant Pestilential Fever,* 224–26, 230, 260; John Brisbane, *Select Cases in the Practice of Medicine* (London: G. Scott for T. Cadell, 1772), 38–40; Daniel Lysons, *Practical Essays upon Continual and Intermitting Fevers* (Bath, UK: 1783), 121–22; Thomas Beddoes, *A Collection of Testimonies Respecting the Treatment of the Venereal Disease by Nitrous Acid* (London: J. Johnson, 1799), 26–27; Clark, *Observations on the Diseases, passim.*
97. Compare also the language in cases of gentlemen, merchants and laborers in William Blair, *Essays on the Venereal Disease and Its Concomitant Affections, Illus-*

trated by a Variety of Cases* (London: J. Johnson, 1798), 28, 53–56, 73–74, 92, 93–94; Clark, *Observations on the Diseases, passim.*

98. Atkins, *The Navy Surgeon,* 23; Webster, ed., *A Collection of Papers,* 163–64.
99. Gout is the obvious example here. Roy Porter and George S. Rousseau, *Gout: The Patrician Malady* (New Haven, CT: Yale University Press, 1998).

Chapter 3

This chapter is dedicated to the memory of the late Lynn Salkin Sbiroli, who at the time of her tragic, untimely death in Rome in 1991 was researching these matters. She lived long enough to publish her preliminary study, under my supervision, of the editions of Hill's *Lucina* and their European translations: Sbiroli, *Libertine O Madre Illibate* (Venice: Marcilio, 1989). Had she lived longer, she would have gone on to interpret their complex histories and also, I think, to have written a version of this chapter. The dedication is a small tribute to her aborted work. My practice in the chapter is to refer readers who wish to pursue these matters to the relevant place in Sbiroli's book rather than cite each edition and translation involved. The book itself is not widely available and has not yet been translated into English.

1. For jokes and the eighteenth century, see Paula Findlen, "Jokes of Nature and Jokes of Knowledge: The Playfulness of Scientific Discourse in Early Modern Europe," *Renaissance Quarterly* 43 (1990): 292–329; Simon Critchley, *On Humour: Thinking in Action* (London: Routledge, 2002); Simon Dickie, "Hilarity and Pitilessness in the Mid-Eighteenth Century: English Jestbook Humor," *Eighteenth-Century Studies* 37 (2003): 1–22. See also Sigmund Freud, *Jokes and Their Relation to the Unconscious* (London: Routledge & Kegan Paul, 1960).
2. Jacques Lacan, *The Seminar of Jacques Lacan* (Cambridge, UK: Cambridge University Press, 1988).
3. This is the line adopted by Judith Halberstam. Judith Halberstam and Ira Livingston, eds., *Posthuman Bodies, Unnatural Acts* (Bloomington: Indiana University Press, 1995).
4. J. A. Leo Lemay and George S. Rousseau, *The Renaissance Man in the Eighteenth Century,* William Andrews Clark Memorial Library Seminar Papers (Los Angeles: University of California Los Angeles Press, 1978); John Hill and George S. Rousseau, *The Letters and Papers of Sir John Hill, 1714–1775* (New York: AMS Press, 1982).
5. Charles, Kerby-Miller, ed., *Memoirs of the Extraordinary Life, Works, and Discoveries of Martinus Scriblerus* (New Haven, CT: Yale University Press, 1950).
6. Catherine Gallagher and Thomas Laqueur, *The Making of the Modern Body: Sexuality and Society in the Nineteenth Century* (Berkeley: University of California Press, 1987); Halberstam and Livingston, *Posthuman Bodies.*
7. Halberstam and Livingston, *Posthuman Bodies*; Neil Badmington, *Posthumanism* (Basingstoke, UK: Palgrave, 2000); Ann Weinstone, *Avatar Bodies* (Minneapolis: University of Minnesota Press, 2004).
8. Halberstam states as much, and this portion of my argument is indebted to her. Halberstam and Livingston, *Posthuman Bodies,* 17.
9. Kelly Hurley, "Postmodern Bodies: Reading like an Alien," in *Posthuman Bodies, Unnatural Acts,* ed. Judith Halberstam and Ira Livingstone (Bloomington: Indiana University Press, 1995), 211.

10. Thomas Laqueur, "Orgasm, Generation, and the Politics of Reproductive Biology," in *The Making of the Modern Body: Sexuality and Society in the Nineteenth Century,* Catherine Gallagher and Thomas Laqueur (Berkeley: University of California Press, 1987): 1–42; Laqueur, *Making Sex: Body and Gender from the Greeks to Freud* (Cambridge, MA: Harvard University Press, 1990).
11. Laqueur also demonstrated that "the reinterpretation of women's reproductive biology solved ideological problems inherent in eighteenth- and nineteenth-century social and political practices": Laqueur, "Orgasm, Generation, and the Politics of Reproductive Biology," viii. Also citing Lacan, *The Seminar of Jacques Lacan*, Laqueur showed what "devastating attack . . . [the] linking of female orgasm to conception had come under" (Laqueur, "Orgasm, Generation, and the Politics of Reproductive Biology," 2) in the late eighteenth century. Had Laqueur known about Hill's *Lucina* it would have proved to be manna from heaven for his model about the newly diminished female orgasm. Laqueur, "Orgasm, Generation, and the Politics of Reproductive Biology," viii.
12. These views would have surprised Nathaniel Highmore, whose view of generation followed the old model. Highmore, *The History of Generation* (London: Printed by R.N. for J. Martin, 1651). For competing theories of reproduction see Joseph Needham, *A History of Embryology*, 2nd ed. (Cambridge, UK: Cambridge University Press, 1959).
13. Edward Shorter, "The Management of Normal Deliveries and the Generation of William Hunter," in *William Hunter and the Eighteenth-Century Medical World*, ed. W. F. Bynum and Roy Porter (Cambridge, UK: Cambridge University Press, 1985), 371–84; Mary Terrall, "Salon, Academy, and Boudoir: Generation and Desire in Maupertuis' Science of Life," *Isis* 87 (1996): 217–29; Valeria Finucci and Kevin Brownlee, *Generation and Degeneration: Tropes of Reproduction in Literature and History from Antiquity through Early Modern Europe* (Durham, NC: Duke University Press, 2001).
14. Dorothy Stimson, *Scientists and Amateurs: A History of the Royal Society* (London: Sigma Books, 1949).
15. Gallagher and Laqueur, *The Making of the Modern Body.*
16. One wonders whether *Lucina* had been spurred on by the new view of the male sodomitical body described in 1749 in the anonymous *Satan's Harvest Home: Reasons for the Growth of Sodomy* and published just as Hill was imagining his joke.
17. Sibly printed the book privately and paid for the print run himself. Ebenezer Sibly, *The Medical Mirror. Or Treatise on the Impregnation of the Human Female: Shewing the Origin of Diseases, and the Principles of Life and Death* (London: Printed for the author and sold by Champante and Whitrow, 1796).
18. Ibid., 35.
19. Ibid., 67.
20. Ibid., 35ff.
21. Ibid., 60ff.
22. See p. 58 for the reasons and the chronology.
23. Sibly, *The Medical Mirror,* 1.
24. Every biographer of Hunter has recounted the story of Home's plagiarism from, and subsequent destruction of, Hunter's vast manuscripts over three decades after his death in 1793, for whose investigation a parliamentary committee was formed.
25. Home was also interested in human reproductive abnormality: Everard Home, "An Account of a Child with a Double Head. In a Letter from Everard Home to John

Hunter," *Philosophical Transactions of the Royal Society of London* 80 (1790): 296–305. See also Home, "Some Observations on the Mode of Generation of the Kangaroo, with a Particular Description of the Organs Themselves," *Philosophical Transactions of the Royal Society of London* 85 (1795): 221–38; Home, "An Account of the Dissection of an Hermaphrodite Dog. To Which Are Prefixed, Some Observations on Hermaphrodites in General," *Philosophical Transactions of the Royal Society of London* 89 (1799): 157–78.

26. Jesse Foot, *The Life of John Hunter* (London: T. Becket, 1794); John Hunter and Everard Home, eds., *The Works of John Hunter. With Notes* (London: Longman, 1835; Jessie Dobson, *John Hunter* (Edinburgh: E. & S. Livingstone, 1969).
27. Home, "An Account of the Dissection of an Hermaphrodite Dog."
28. One reconstruction of the historical events is found in F. N. L. Poynter, "Hunter, Spallanzani, and the History of Artificial Insemination," in *Medicine, Science, and Culture: Historical Essays in Honor of Owsei Tempkin,* ed. Lloyd G. Stevenson and Robert P. Multhauf (Baltimore, MD: Johns Hopkins Press, 1968), 97–113.
29. Laurence Sterne, *Adventures of Tristram Shandy* (New York: Odyssey, [1759–1767] 1950), 1.
30. For the different resonance of organs of generation versus technology of reproduction, see Jacques Roger, *The Life Sciences in Eighteenth-Century French Thought* (Stanford, CA: Stanford University Press, 1997).
31. It was printed in December 1749. The *Gentleman's Magazine* and *Monthly Review* reviewed it in their January issues: *Gentleman's Magazine* 20 (Jan. 1750): 48; *Monthly Review* 2 (Jan. 1750): 255–56. The anonymous reviewer in *Monthly Review* lists it as "said to be Dr Hill."
32. For Hill's cross-gendered authorial phase and its role in shaping his public profile, see George Rousseau, "Curiosity and the *Lusus Naturae*: Proteus Hill," in *Curiosity and Wonder from the Renaissance to the Enlightenment,* ed. R. J. W. Evans and Alex Marr (Aldershot, UK: Ashgate, 2006), 213–49.
33. Abraham Johnson [John Hill], *Lucina sine concubitu. A Letter Humbly Address'd to the Royal Society; in Which Is Proved by most Incontestible Evidence, Drawn from Reason and Practice, That a Woman May Conceive and Be Brought to Bed Without Any Commerce with Man* (London: Printed and sold by M. Copper, 1750), 49–50.
34. George Rousseau, "The London Earthquakes of 1750," *Cahiers d'Histoire Mondiale* 11 (1968): 436–54.
35. Johnson [Hill], *Lucina sine concubitu,* 50.
36. William Wollaston, *The Religion of Nature Delineated* (Dublin: 1726).
37. Johnson [Hill], *Lucina sine concubitu,* 9–10.
38. Virgil, *Georgics,* III, which Hill read in Dryden's translation and which he made his epigraph in Roe, together with following passages cited from Ovid and Milton. Johnson [Hill], *Lucina sine concubitu,* 12.
39. Johnson [Hill], *Lucina sine concubitu,* 13.
40. Ibid., 15.
41. From early days Hill gravitated to the microscope and was prominent in its eighteenth-century histories: John Hill, *Essays in Natural History and Philosophy. Containing a Series of Discoveries, by the Assistance of Microscopes* (London: J. Whiston et al., 1752).
42. Johnson [Hill], *Lucina sine concubitu,* 17.
43. Ibid., 20.
44. Ibid., 31–49.

45. Ibid., 29.
46. John Milton, *L'Allegro*, in *Milton, The Compete Shorter Poems,* edited by John Carey (London: Longman, 2006), 137–38.
47. For jokes about Réaumur's chicken-breeding machines see Roger, *The Life Sciences,* 483. Gena Corea has studied the technologies of reproductive women birthing without the assistance of male sperm: Corea, *Man-Made Women: How New Reproductive Technologies Affect Women* (London: Hutchinson, 1985); Corea, *The Mother Machine: Reproductive Technologies from Artificial Insemination to Artificial Wombs* (London: Women's Press, 1987). See also Antoine Réaumur, *The Art of Hatching and Bringing up Domestick Fowl of All Kinds, at any Time of the Year, Either by Means of the Heat of Hot-Beds, or That of Common Fire* (London: C. Davis, A. Millar & J. Nourse, 1750).
48. Johnson [Hill], *Lucina sine concubitu*, 41–45.
49. Ibid., 45.
50. Lemay and Rousseau, *The Renaissance Man*; Hill and Rousseau, *The Letters and Papers of Sir John Hill.*
51. Johnson [Hill], *Lucina sine concubitu*, 46.
52. Ibid.
53. Ibid., 47.
54. For Parsons's career in the Royal Society see George S. Rousseau and David Haycock, "'Voices Calling for Reform': The Royal Society in the Mid-Eighteenth Century—Martin Folkes, John Hill, and William Stukeley," *History of Science* 37 (1999): 377–406. Parsons had also assisted Hill in the 1740s with his translation of Theophrastus, and the biographic record suggests a degree of animosity within mentorship between the two men. James Parsons, *The Microscopical Theatre of Seeds: Being a Short View of the Particular Marks, Characters, Contents, and Natural Dimensions of All the Seeds* (London: M. Cooper, 1745).
55. James Parsons, *A Mechanical and Critical Enquiry into the Nature of Hermaphrodites* (London: J. Walthoe, 1741); Parsons, "A Letter to the President Concerning the Hermaphrodite Shewn in London," *Philosophical Transactions* 47 (1751–1752): 142–45.
56. Hill and Rousseau, *The Letters and Papers of Sir John Hill,* 18–49; George S. Rousseau and David Haycock, "The Jew of Crane Court: Emanuel Mendes Da Costa (1717–1791), Natural History and Natural Excess," *History of Science* 38 (2000): 127–70.
57. Martin Schurig, *Spermatologia Historico-Medica* (Wittenberg, Germany: 1780); Robin Baker, *Sperm Wars: The Science of Sex* (New York: Basic Books, 1996).
58. For a mini-epitome of the history of semen see Sibly, *The Medical Mirror,* 20–26.
59. Laqueur's model change is often overlooked when interpreted in light of Roger's panoptic view. Needham, *A History of Embryology*; Roger, *The Life Sciences.*
60. Hill was influenced by Needham's treatise, just as he was imagining the plot of *Lucina.* John Turberville Needham, *Observations upon the Generation, Composition, and Decomposition of Animal and Vegetable Substances* (London: 1749).
61. Poynter, "Hunter, Spallanzani, and the History of Artificial Insemination."
62. Ibid.
63. Ibid.
64. It is puzzling that Poynter, "Hunter, Spallanzani, and the History of Artificial Insemination," omitted Hill for reasons unknown.
65. Perhaps no one more assiduously than chemist-playwright Carl Djerassi in his recent play about the ethical dilemmas immaculate conception poses. Carl Djerassi,

An Immaculate Misconception: Sex in an Age of Mechanical Reproduction (London: Imperial College Press, 2000).

66. The matter was so sensitive that it was not until 1854 that Pope Pius IX pronounced the doctrine infallible.
67. Sbiroli, *Libertine O Madre Illibate,* 77, 91.
68. Fantastic births were widely discussed in Hill's time partly under the sway of the recent English version of *Aristotle's master-piece: or, The secrets of generation displayed* (1694). "Birthing miracles" differed from strange fantastic births. Marie-Hélène Huet, *Monstrous Imagination* (Cambridge MA: Harvard University Press, 1993); Dudley Butler Wilson, *Signs and Portents: Monstrous Births from the Middle Ages to the Enlightenment* (London: Routledge, 1993); Lorraine Daston and Katherine Park, *Wonders and the Order of Nature, 1150–1750* (New York: Zone Books, 1998).
69. Pierre Darmon, *Trial by Impotence: Virility and Marriage in Pre-Revolutionary France* (London: Chatto & Windus, 1985); Darmon, *Damning the Innocent: A History of the Persecution of the Impotent in Pre-Revolutionary France* (New York: Viking, 1986).
70. Lisa Forman Cody, *Birthing the Nation: Sex, Science, and the Conception of Eighteenth-Century Britons* (Oxford: Oxford University Press, 2005).
71. The work of Angus McLaren enriches these areas: McLaren, *Reproductive Rituals: The Perception of Fertility in England from the 16th to the 19th Century* (London: Methuen, 1984). Without including Hill, he has also explored the Freudian implications: McLaren, "Contraception and Its Discontents: Sigmund Freud and Birth Control," *Journal of Social History* 12 (1979): 513–29.
72. Lemay and Rousseau, *The Renaissance Man*; Hill and Rousseau, *The Letters and Papers of Sir John Hill.*
73. Angus McLaren, "The Pleasures of Procreation: Traditional and Biomedical Theories of Conception," in *William Hunter and the Eighteenth-Century Medical World,* ed. William F. Bynum and Roy Porter (Cambridge, UK: Cambridge University Press, 1985), 323–41.
74. A copy can be found in the Wellcome Library, London. Sbiroli, *Libertine O Madre Illibate,* 157.
75. Printer Robert Dodsley reissued Roe and Johnson bound together at his own expense in 1761, 1762, 1765, and 1771, all during Hill's lifetime.
76. Nothing is known about the German translations, and no translator's name appears on the title page. There were also Dutch and Danish translations.
77. Roger, *The Life Sciences.*
78. Sbiroli, *Libertine O Madre Illibate,* 148.
79. Laqueur, "Orgasm, Generation, and the Politics of Reproductive Biology," viii.
80. Lemay and Rousseau, *The Renaissance Man,* 63–64.
81. Sbiroli, *Libertine O Madre Illibate,* 125–58.
82. Jean-François Sacombe, *La Luciniade: Ou, L'art Des Accouchemens, Poème Didactique* (Paris: Garnéry, 1792); Sacombe, *La Luciniade, Poème en Dix Chants, Sur L'art Des Accouchemens* (Paris: Courcier, 1799); Sacombe, *Lucine Francéaise, ou Recueil d'observations Médicales, Chirurgicales, Pharmaceutiques, Historiques, Critiques et Littéraires, Relatives é La Science Des Accouchemens* (Paris: Bidault, 1803).
83. In Lynn Hunt the suggestion is made that *Lucina* can figure into the traditions of eighteenth-century pornography: Hunt, *The Invention of Pornography: Obscenity and the Origins of Modernity, 1500–1800* (New York: Zone, 1993). See also Hunt, *Politics, Culture, and Class in the French Revolution* (Berkeley: University of California Press, 1984).

84. For example, see the preface to the 1795 translation published in Paris by Favre.
85. Sbiroli, *Libertine O Madre Illibate,* 136–39.
86. Bebaimb was a seventeenth-century anatomist and colleague of Malpighi. Sbiroli, *Libertine O Madre Illibate,* 153.
87. Ibid., 149.
88. Ibid., 158.
89. Only modern editions have restored *Lucina* to its earlier authenticity, removing layers of allegorical and political significance. For example, the reader of the 1930 "limited edition" (Berkshire, UK: Golden Cockerel Press) never suspected that Hill's revenge joke had such a terrific political afterlife.

Chapter 4

1. Anonymous, "Machine surprenante de l'homme artificiel de Monsieur Reyselius," *Journal des Sçavans* 23 (1677): 252. The reference regarding the twelfth issue of the same year is to a description of a pump used to model the fabric of the heart, pp. 127–29. Wirtemberg later became Württemberg. On Reisel, see Ralf Bröer, *Salomon Reisel: (1625–1701); barocke Naturforschung eines Leibarztes im Banne der mechanistischen Philosophie* (Leipzig, Germany: Barth, 1996); Thomas L. Hankins and Robert J. Silverman, *Instruments and the Imagination* (Princeton, NJ: Princeton University Press, 1995), 192; André Doyon and Lucien Liaigre, *Jacques Vaucanson, mécanicien de genie* (Paris: PUF, 1966), 117–18, 162–63. Reisel also wrote the first systematic description of the Wirtemberg Siphon, invented by Jean Jordan of Stuttgart: a siphon with two equal, curved legs in which, when the legs were level, water climbed up one side and flowed down the other: Anonymous, "Machine surprenante"; Denis Diderot and Jean d'Alembert, eds., "Syphon," in *Encyclopédie, ou dictionnaire raisonné des sciences, des arts et des métiers* (Paris, 1765), vol. 15.
2. René Descartes, *Traité de l'homme*, written between 1630 and 1633 but first published in 1662 in Latin. Charles Adam and Paul Tannery, eds., *Oeuvres de Descartes,* vol. 11 (Paris: J. Vrin, 1975), 119ff.
3. Jessica Riskin, "Eighteenth-Century Wetware," *Representations* 83 (2003): 97–125; Riskin, "The Defecating Duck," *Critical Inquiry* 29 (2003): 599–633.
4. Leonora Cohen Rosenfield, *From Beast-Machine to Man-Machine: Animal Soul in French Letters from Descartes to La Mettrie* (Oxford: Oxford University Press, 1941).
5. "Soul of mud" was a paraphrase of the abbé Pluche's characterization of John Locke's view of the soul: Noël Antoine Pluche, *Le Spectacle de la nature, ou Entretiens sur les particularités de l'histoire naturelle*, vol. 5 (Paris: 1746), 176–77; Aram Vartanian, ed., *La Mettrie's L'Homme-machine, A Study in the Origins of an Idea* (Princeton, NJ: Princeton University Press, 1960), 150.
6. See for example Jonathan Israel, *Enlightenment Contested: Philosophy, Modernity and the Emancipation of Man, 1670–1752* (New York: Oxford University Press, 2006), 803, 809, 810.
7. Gertrude Carman Bussey, ed., *La Mettrie, Man a Machine* (La Salle, IL: Open Court, 1912), 6.
8. Ibid.
9. Thomas Carlyle, *History of Friedrich II of Prussia, Called Frederick the Great*, vol. 4, bk. 16, *The Ten Years of Peace. 1746–1756* (London: Chapman and Hall, 1864), 386.

10. Bussey, ed., *La Mettrie,* 6.
11. Julien Offray de La Mettrie, *L'Homme-machine*, in *La Mettrie's L'Homme-machine, A Study in the Origins of an Idea,* ed. Aram, Vartanian (Princeton, NJ: Princeton University Press, 1960), 152–56.
12. Ibid., 183.
13. Bussey, ed., *La Mettrie,* 8.
14. La Mettrie, *L'Homme-machine*, 180, 183.
15. On La Mettrie, Haller, and Boerhaave, see Vartanian, ed., *La Mettrie's L'Homme-machine,* 75–89. Despite his policy of ambiguity and overt criticisms of Spinoza, Boerhaave was reputed to be a Spinozist, a charge against which Samuel Johnson defended him. Jonathan Israel, *Radical Enlightenment: Philosophy and the Making of Modernity, 1650–1750* (Oxford: Oxford University Press, 2001), 705; Samuel Johnson, "Hermann Boerhaave," *The Gentleman's Magazine*, January–April 1739, reprinted in *The Works of Samuel Johnson*, vol. 14 (Troy, NY: Pafraets Company, 1903), 154–84.
16. Israel, *Radical Enlightenment,* 704.
17. On Haller's physiology, see François Duchesneau, *La Physiologie des Lumières: Empirisme, modèles, theories* (The Hague, The Netherlands: Nijhoff, 1982), chapters 5 and 6; Shirley Roe, "Anatomia Animata: The Newtonian Physiology of Albrecht von Haller," in *Transformation and Tradition in the Sciences: Essays in Honor of I. Bernard Cohen,* ed. Everett Mendelsohn (Cambridge, UK: Cambridge University Press, 1984), 273–300; Hubert Steinke, *Irritating Experiments: Haller's Concept and the European Controversy on Irritability and Sensibility, 1750–90* (Amsterdam: Rodopi, 2005).
18. Indeed the line between vitalism and materialism was often very blurry in this period, as historians of the eighteenth and early nineteenth centuries have noted. I have taken up this blurriness in Riskin, "Eighteenth-Century Wetware."
19. Shirley Roe, *The Natural Philosophy of Albrecht von Haller* (Manchester, NH: Ayer, 1981), introduction.
20. Aram Vartanian, "Trembley's Polyp, La Mettrie and Eighteenth-Century French Materialism," *Journal of the History of Ideas* 11 (1950): 271.
21. Carlyle, *History of Friedrich II of Prussia,* 387.
22. On La Mettrie's influence on his contemporaries, see Israel, *Radical Enlightenment,* chapter 37; Vartanian, "Trembley's Polyp."
23. Carlyle writes of Frederick: "It is certain he could, especially in his younger years, put up with a great deal of zanyism, ingenious foolery and rough tumbling, if it had any basis to tumble on . . . By far his chief Artist in this kind, indeed properly the only one, was La Mettrie." Carlyle, *History of Friedrich II of Prussia,* 385–86.
24. Dieudonné Thibault, *Mes souvenirs de vingt ans to séjour à Berlin: ou Frédéric le Grand, sa famille, sa cour, son gouvernement, son académie, ses écoles, et ses amis littérateurs et philosophes*, 2nd ed., vol. 5 (Paris: F. Buisson, 1805), 405.
25. "If La Mettrie was detested by the rest of the Berlin Academy, he was nevertheless, at the wish of the Prussian monarch, made a full member": Israel, *Radical Enlightenment,* 671. "Much to the chagrin of almost the entire Prussian Royal Academy, Frederick . . . required them in July 1748 to adopt La Mettrie as a full member of their body": Israel, *Radical Enlightenment,* 722. See also Israel, *Radical Enlightenment,* 574; Israel, *Enlightenment Contested,* 803; Vartanian, ed., *La Mettrie's L'Homme-machine,* 8–9.
26. Vartanian, ed., *La Mettrie's L'Homme-machine,* 101–2, nn.18,19.

27. Published in *Epître à Mlle. A.C.P. ou la 'Machine terrassée'* (1749). This *Epître* was one of three anonymous, self-satirizing pamphlets, all of which appeared in 1749. The others were *Epître à mon Esprit, ou l'Anonyme persiflé* and *Réponse à l'auteur de la Machine terrassée*. See Pierre Lemée, *Julien Offray de La Mettrie, médecin, philosophe, polémiste; sa vie, son oeuvre* (Mortain, France: Editions Mortainais, 1954), 205–19; Vartanian, ed., *La Mettrie's L'Homme-machine,* 102, n.20. Francine Markovits, ed., *La Mettrie: Oeuvres philosophiques*, vol. 2 (Rennes, France: Fayard, 1987), 215–16.
28. Voltaire–Mme. Denis, November 14, 1751: M. Beuchot, ed., *Oeuvres de Voltaire*, vol. 55 (Paris: Lefèvre; Firmin Didot frères, 1831), 688–89.Voltaire's letters from the Prussian court to his niece and lover, Mme. Denis (Marie Louise Mignot) are in fact rewritten, literary versions of an original, genuine correspondence. For the full story on this rewriting and its belated discovery, see Jonathan Mallinson, "What's in a Name? Reflections on Voltaire's *Paméla*," *Eighteenth-Century Fiction* 18 (2005–2006): 157–68. For the story of La Mettrie's death, see also Frederick–Wilhelmina, November 21, 1751 and Christoph Friedrich Nicolai, *Anekdoten*, vi. 197–227, in Carlyle, *History of Friedrich II of Prussia,* 399–400. See also Beuchot, ed., *Oeuvres de Voltaire*, no. 1785, 688–89.
29. Beuchot, ed., *Oeuvres de Voltaire*, no. 1785, 688–89.
30. The two meanings of "farce" may in fact be related, their common ancestry being the medieval comic performances that were stuffed in between religious presentations to entertain the hoi polloi. Beuchot, ed., *Oeuvres de Voltaire*, no. 1783, 684–85.
31. Ibid., no. 1785, 688–89.
32. See for example Thibault, *Mes souvenirs de vingt ans to séjour à Berlin,* 407. Thibault also reports that La Mettrie had an uncontrolled imagination and that as an "absolute and pronounced materialist" he was "afraid of almost everything": Ibid., 406.
33. Beuchot, ed., *Oeuvres de Voltaire*, no. 1789, 696–99.
34. Carlyle, *History of Friedrich II of Prussia,* 399–400.
35. La Mettrie, *L'Homme-machine*, 154, 182–83, 186, 190.
36. Ibid., 189–90.
37. Denis Diderot, *Eléments de physiologie* (Paris: Didier, [1784] 1964), 20–21.
38. La Mettrie, *L'Homme-machine*, 192.
39. Julien Offray de La Mettrie, *L'Homme-plante,* in *Oeuvres philosophiques,* vol. 1 (Paris: Fayard, [1748] 1984), 281–306.
40. La Mettrie, *L'Homme-machine*, 173–75, 196.
41. Diderot, *Eléments de physiologie,* 32.
42. La Mettrie, *L'Homme-machine*, 145, 152, 164–65, 175.
43. Ibid., 176, 191–92, 194.
44. Ibid., 189, 192, 194, 196.
45. Julien Offray de La Mettrie, *Système d'Epicure,* in *Oeuvres philosophiques,* vol. 1 (Paris: Fayard, [1750] 1984), 369–70.
46. Diderot, *Eléments de physiologie,* 308.
47. For discussion of La Mettrie's proto-evolutionism, see Robert J. Richards, "The Emergence of Evolutionary Biology of Behaviour in the Early Nineteenth Century," *British Journal for the History of Science* 15 (1982): 241–80; Richards, *Darwin and the Emergence of Evolutionary Theories of Mind and Behavior* (Chicago: University of Chicago Press, 1987), 25, 30, 32; Richards, *The Meaning of Evolution: The Morphological Construction and Ideological Reconstruction of Darwin's Theory* (Chicago: University of Chicago Press, 1993), 64.

48. For an in-depth discussion of what La Mettrie had in mind when he spoke of "le grand Singe" and where he got his information on these animals, see Vartanian, ed., *La Mettrie's L'Homme-machine,* nn.45–48, 213–16. La Mettrie, *L'Homme-machine*, 160.
49. On La Mettrie's influence on Diderot, see Vartanian, "Trembley's Polyp." Vartanian attributes Diderot's about-face between the deistic *Pensées philosophiques* (1746) and the materialistic *Pensées sur l'interprétation de la nature* (1754) importantly to La Mettrie's *L'Homme-machine,* 274.
50. Denis Diderot, *Suite de l'entretien,* in *Oeuvres Philosophiques*, ed. P. Verniere (Paris: Garnier, [1784] 1961), 385.
51. La Mettrie, *L'Homme-machine*, 359–60; La Mettrie, *Système d'Epicure,* 353–86.
52. I have examined the optical example in arguments from design in Jessica Riskin, "The Divine Optician," *American Historical Review*, forthcoming.
53. La Mettrie, *L'Homme-machine*, 177.
54. Ibid., 162, 167; La Mettrie, *Système d'Epicure,* 361–62.
55. La Mettrie, *L'Homme-machine*, 163; La Mettrie, *Système d'Epicure,* 357–58.
56. La Mettrie, *L'Homme-machine*, 178; Vartanian, ed., *La Mettrie's L'Homme-machine,* 25.
57. Aristotle, *Aristotle's Physics*, books 1 and 2, trans. W. Charlton (Oxford: Oxford University Press, 1983), 198b17–198b33, 199b10–199b13; Philoponus, *On Aristotle on the Soul, 2.1–6*, trans. William Charlton (London: Duckworth, 2005), 412a27–412b9.
58. "Organization" had also appeared occasionally in later seventeenth-century writing on life, matter, and spirit, but carrying different meanings. Authors of arguments from design occasionally used the word "organization" as synonymous with "design." See, for example John Ray, *The Wisdom of God Manifested in the Works of the Creation*, 1st ed. (London: Sam Smith, 1660), 48–49; Nehemiah Grew, *Cosmologia Sacra: Or a Discourse of the Universe as It Is the Creature and Kingdom of God* (London: 1701), chapter V; John Toland, *Letters to Serena* (London: Bernard Lintot, 1704), 235; Ralph Cudworth, *The True Intellectual System of the Universe: The First Part; wherein All the Reason and Philosophy of Atheism Is Confuted . . .*, vol. 1 (London: 1743), 149; Pierre de La Touche Boesnier, *Préservatif contre l'irréligion, ou démonstrations des véritez fondamentales de la religion chrétienne* (The Hague: 1707). A different, landmark usage of "organization" figured in John Locke's understanding of the identity of a living thing: that which made it the same over time despite the continual transformation of its material parts: John Locke, *An Essay Concerning Human Understanding*, ed. Peter H. Nidditch (Oxford: Clarendon Press, [1690] 1975), 330–32.
59. Hermann Boerhaave, *A new method of chemistry* (London: 1727), 142, 150.
60. Gottfried Wilhelm Leibniz, *Essais de théodicée: sur la bonté de Dieu, la liberté de l'homme et l'origine du mal* (Paris: Aubier, [1710] 1962), 161, 189, 247, 365.
61. In addition to the instances already listed, see Jacques Bénigne Bossuet, *De la connaissance de Dieu et de soi-même,* in *Oeuvres completes*, vol. 23 (Paris: L. Vives, [1704] 1864), 235, 242; Etienne Bonnot de Condillac, *Essai sur l'origine des connaissances humaines* (Amsterdam: P. Mortier, 1746), 77.
62. La Mettrie, *L'Homme-machine*, 166–67, 180.
63. Georges Buffon, *Histoire naturelle, générale et particulière*, vol. 2 (Paris: Imprimerie royale, 1749), 37.

64. Ibid., vol. 2, 44–46, 51–53, 486–87.
65. Ibid., vol. 1, 12.
66. On La Mettrie and Maupertuis, see Vartanian, ed., *La Mettrie's L'Homme-machine,* 5, 7–8. On Maupertuis's presidency of the Prussian Royal Academy of Sciences, see Mary Terrall, *The Man Who Flattened the Earth: Maupertuis and the Sciences in the Enlightenment* (Chicago: University of Chicago Press, 2002), chapter 8.
67. Pierre-Louis Moreau de Maupertuis, *Essai sur la formation des corps organizés* (Berlin: 1754), 14, 66–67.
68. James Parsons, *Philosophical Observations on the Analogy between the Propagation of Animals and That of Vegetables* (London: 1752), 9.
69. Charles Bonnet, *Contemplation de la nature,* in *Collection complete des oeuvres*, vol. 4 (Neuchatel, Switzerland: S. Fauche, [1764] 1781) part 2, 29; Bonnet, *La palingénésie philosophique,* part 9 (Geneva: Philibert et Chirol, 1769), 320.
70. Charles Bonnet, *Considérations sur les corps organisés* (Amsterdam: M. M. Rey, [1762] 1768), 1, 8, 13; Bonnet, *Contemplation de la nature,* part 3, chapter 14; Bonnet, *La palingénésie philosophique,* 321.
71. Charles Bonnet, *Principes philosophiques sur la cause première et sur son effet,* in *Essai de psychologie . . . Auxquelles on a ajouté des Principles philosophiques* (London: 1755), 344, 363, 380–81; Bonnet, *Considérations,* 13, 139; Bonnet, *Contemplation de la nature,* part 8, 368; Bonnet, *La palingénésie philosophique,* 323.
72. Bonnet, *Contemplation de la nature,* part 8, 359–61.
73. Ibid., part 8, 359.
74. Bonnet, *La palingénésie philosophique,* part 15, 112–13.
75. Charles Bonnet, *Essai de psychologie,* in *Collection complete des oeuvres*, vol. 8 (Neuchatel, France: S. Fauche, [1754] 1783), 107; Bonnet, *Principes philosophiques,* 298.
76. Bonnet, *Contemplation de la nature,* part 11, 293.
77. Diderot, *Eléments de physiologie,* 138.
78. Denis Diderot, *Le rêve de d'Alembert,* in *Oeuvres Philosophiques*, ed. P. Verniere (Paris: Garnier, [1769] 1961), 300; Diderot, *Suite de l'entretien,* 375–79.
79. Paul Broca, *Sur le Transformisme* (Paris: Reinwald, 1871).
80. Jean-Baptiste Lamarck, *Philosophie zoologique, ou, Exposition des considérations relative à l'histoire naturelle des animaux* (Paris: Dentu, 1809), 111, 445–46.
81. For recent scholarship correcting the misapprehension of Lamarck as first transformist, see Pietro Corsi, "Before Darwin: Transformist Concepts in European Natural History," *Journal of the History of Biology* 38 (2005): 67–83.
82. Denis Diderot, *Entretien entre d'Alembert et Diderot,* in *Oeuvres Philosophiques*, ed. P. Verniere (Paris: Garnier: [1769] 1961), 271–74.
83. Ibid., 267–68.
84. Diderot, *Le rêve de d'Alembert,* 299, 302.
85. Diderot, *Entretien entre d'Alembert et Diderot,* 267–68.
86. Ibid., 268.
87. Diderot, *Le rêve de d'Alembert,* 303.
88. Ibid., 298.
89. I treat this episode in Riskin, "Eighteenth-Century Wetware."
90. Erasmus Darwin, "The Loves of Plants," part II of *The Botanic Garden, A Poem in Two Parts* (Lichfield, UK: 1789).
91. On Darwin's transformism, see Bentley Glass, ed., *The Forerunners of Darwin* (Baltimore, MD: Johns Hopkins University Press, 1968), chapters 3–5; Robert J.

Richards, "Influence of Sensationalist Tradition on Early Theories of the Evolution of Behavior," *Journal of the History of Ideas* 40 (1979): 85–105; Richards, *The Meaning of Evolution,* 64–65. See also Erasmus Darwin, *Zoonomia; or, the laws of organic life* (Dublin: 1794–1796), 183, 499.

Chapter 5

My thanks especially to Carole Reeves, Tim Hitchcock, Brian Hurwitz, Clare Gittings, and Peter Razzell. Thanks also to staff at the British Library, Wellcome Library, and Iconographic Collection. The passages on burial and body snatching are based on my book, *Death Dissection and the Destitute.* I have worked extensively on the archaeologists' manuscript notes from the Spitalfields Project, for access to which I thank Theya Molleson. I am also grateful to the curator of jewelry, Victoria and Albert Museum, London, for allowing me to study its collection.

1. Paul Slack, *The Impact of Plague in Tudor and Stuart England* (London: Routledge, 1985), 326–37.
2. Ruth Richardson, *Death Dissection and the Destitute* (Chicago: Chicago University Press, 2000), 3–29.
3. Vic Gatrell, *The Hanging Tree* (Oxford: Oxford University Press, 1994), 172, 351.
4. Mary Chamberlain and Ruth Richardson, "Life and Death," *Oral History Journal* 11 (1983): 31–43; Ruth Pitter, "The Old Woman," in *Poems 1926–66* (London: Cresset Press, 1968).
5. Richardson, *Death Dissection,* 8–9, 22–23, 298–99.
6. Peter Linebaugh, "The Tyburn Riot against the Surgeons," in *Albion's Fatal Tree,* ed. Douglas Hay et al. (London: Allen Lane, 1975), 65–118.
7. Ibid., 106–7.
8. Douglas Hay et al., *Albion's Fatal Tree* (London: Allen Lane, 1975), plates 7–8; Linebaugh, "The Tyburn Riot against the Surgeons," 111–13; Gatrell, *The Hanging Tree,* 35–36.
9. Henri Misson, *Memoirs and Observations in His Travels over England. Translated by Mr. Ozell* (London: Browne and others, [1698] 1719), 89; Ruth Richardson, "The English Shroud" (paper presented at The Art of Death Conference, London, Victoria and Albert Museum).
10. John Morley, *Death, Heaven and the Victorians* (London: Studio Vista, 1971), plate 51.
11. Theya Molleson and Margaret Cox, *The Spitalfields Project 2: The Anthropology* (York, UK: Council for British Archaeology, 1993), 198–203.
12. Robert Janaway, "An Introductory Guide to Textiles from 18th and 19th Century Burials," in *Grave Concerns,* ed. Margaret Cox (York, UK: Council for British Archaeology, 1998), 17–32.
13. Chamberlain and Ruth Richardson, "Life and Death"; Richardson, *Death Dissection,* 20–21.
14. Richardson, *Death Dissection,* 18.
15. Anonymous, Report of an Inquest on an "Unfortunate Woman," Mary Ann Chapman, Drowned in the London Dock," *True Sun,* January 12, 1832; Richardson, *Death Dissection,* 234–35.
16. Reproduced in Nigel Llewellyn, *The Art of Death* (London: Reaktion, 1991), 84.
17. Misson, *Memoirs and Observations,* 90.

18. Richardson, *Death Dissection,* 3–29.
19. Montague Summers, *A Gothic Bibliography* (London: Fortune Press, 1941), *passim.*
20. Julian Litten, "The English Funeral 1700–1850," in *Grave Concerns,* ed. Margaret Cox (York, UK: Council for British Archaeology, 1998), 3–16; Gwynne Stock, "Quaker Burial," in *Grave Concerns,* ed. Margaret Cox (York, UK: Council for British Archaeology, 1998), 129–43; Clare Gittings, "Eccentric or Enlightened? Unusual Burial and Commemoration in England, 1689–1823," *Mortality* 12 (2007): 322–49.
21. Misson, *Memoirs and Observations,* 88.
22. Lester S. King, *The Medical World of the Eighteenth Century* (Chicago: University of Chicago Press, 1958), 264–70; Theophile Bonet, *Sepulchretum,* 3 vols. (Geneva: Cramer, 1679–1700); Matthew Baillie, *The Morbid Anatomy of Some of the Most Important Parts of the Human Body* (London: Johnson, 1793), 32; Ruth Richardson, "Pennant's Serpent," *The Lancet* 357 (2001): 966.
23. Hay et al., *Albion's Fatal Tree,* 189–253.
24. *An Act for Better Preventing the Horrid Crime of Murder* (25 Geo. II c. 37, 1752); Richardson, *Death Dissection,* 35.
25. Andreas Vesalius, *De Humanis Corporis Fabrica* (Basle, Switzerland: Printed for the author, 1543), frontispiece.
26. Anonymous, *The Harlot's Progress: Founded Upon Mr. Hogarth's Six Paintings* (Dublin: Exshaw, ca. 1730).
27. Keith F. Russell, *British Anatomy 1525–1800: A Bibliography* (Winchester, UK: St. Paul's, [1963] 1987).
28. Richardson, *Death Dissection,* 72.
29. Ibid., 52–99.
30. Ibid., 30–72.
31. S. Chaplin, "John Hunter and the Anatomy of a Museum," *History Today* 55 (2005): 19–25.
32. William Cheselden, *Osteographia* (London: Printed for the author, 1733); William Hunter, *Anatomia uteri humani gravidi tabulis illustrata . . . The Anatomy of the Human Gravid Uterus Exhibited in Figures* (Birmingham, UK: J. Baskerville, & S. Baker & G. Leigh, etc., 1774).
33. John L. Thornton, *Jan Van Rymsdyk* (Cambridge, UK: Oleander, 1982), *passim.*
34. See especially John Bell, *Anatomy of the Bones, Muscles and Joints. With Plates* (London: Longman, [1797] 1810), book 2, plates.
35. Ibid., v.
36. Anonymous, *A Dissertation on Mr Hogarth's Six Prints* (London: Dickinson, 1751), 58.
37. Richardson, *Death Dissection,* 31n.5; Lynda Payne, *With Words and Knives: Learning Medical Dispassion in Early Modern England* (Aldershot, UK: Ashgate, 2007), *passim.*
38. Richardson, *Death Dissection,* 93–94.
39. Ibid., 131–158, 193–97.
40. Jeremy Bentham, *Auto-Icon, or, Farther Uses of the Dead to the Living* (London: Printed from the original manuscript for an unknown friend of the author, ca. 1842); Ruth Richardson and Brian Hurwitz, "Jeremy Bentham's Self Image: An Exemplary Bequest for Dissection," *British Medical Journal* 295 1987): 195–98.

41. Richardson, *Death Dissection,* 259.
42. Summers, *A Gothic Bibliography, passim.*
43. Richardson, *Death Dissection,* 75–99.
44. Ibid., 87, 90–93, 104, 202, 223–26; Michael Sappol, *A Traffic of Dead Bodies* (Princeton, NJ: Princeton University Press, 2002), 428.
45. Richardson, *Death Dissection,* 72.
46. Ibid., 75–99; Ruth Richardson and James Stevens Curl, "George Frederick Carden and the Genesis of the General Cemetery Company," In *Kensal Green Cemetery,* ed. J. S. Curl (Chichester, UK: Phillimore, 2001).
47. Ruth Richardson, "Commerce, Delicacy and the Decline of Old Mortality" (Paper presented at Death and Dying Conference, London, Dulwich College, 1994).
48. Maurice Rickards and Michael Twyman, *Encyclopaedia of Ephemera* (London: British Library, 2000), 155–58; Llewellyn, *The Art of Death,* 76.
49. Bentham, *Auto-Icon*; Ruth Richardson and Brian Hurwitz, "Jeremy Bentham's Self Image."
50. David Irwin, "Sentiment and Antiquity: European Tombs 1750–1830," in *Mirrors of Mortality,* ed. Joachin Whaley (London: Europa, 1981), 131–53; Julie Rugg "From Reason to Regulation," in *Death in England,* ed. Peter Jupp and Clare Gittings (Manchester, UK: Manchester University Press), 202–29.
51. David E. Stannard, *The Puritan Way of Death* (Oxford: Oxford University Press, 1979), 156–63, 172–73.

Chapter 6

1. I am grateful to Carys Turner for her comments on an earlier version of this chapter.
2. Robert W. Jones, *Gender and the Formation of Taste in Eighteenth-Century Britain: The Analysis of Beauty* (Cambridge, UK: Cambridge University Press, 1998), 11.
3. François Voltaire, *The Philosophical Dictionary for the Pocket* (London: S. Bladon, 1765), 32.
4. Jones, *Gender and the Formation of Taste*; Laura L. Runge, "Beauty and Gallantry: A Model of Polite Conversation Revisited," *Eighteenth-Century Life* 25 (2001): 43–63.
5. For other discussions of beauty in this period see Arthur Marwick, *Beauty in History: Society, Politics and Personal Appearance, c. 1500 to the Present* (London: Thames & Hudson, 1988), chapters 3 and 4; Roy Porter, *Bodies Politic: Disease, Death and Doctors in Britain, 1650–1900* (London: Reaktion, 2001), chapter 3.
6. Jones, *Gender and the Formation of Taste,* 2.
7. N.H., *The Ladies Dictionary* (London: John Dunton, 1694), 49.
8. *The Athenian Mercury* 1, no. 18 (May 23, 1691), question 1.
9. Porter, *Bodies Politic,* 69.
10. Fenela Childs, "Prescriptions for Manners in English Courtesy Literature 1690–1760, and their Social Implications" (D.Phil. thesis, Oxford University, 1984), 161–75, 246, 279.
11. *The Athenian Mercury* 1, no. 18 (May 23, 1691), question 1.
12. Ibid.
13. Harry Beaumont [Joseph Spence], *Crito: Or, A Dialogue on Beauty* (London: R. Dodsley, 1752), 28.
14. Isaac Fuller, *Iconologia: Or, Moral Emblems* (London. B. Motte, 1709), no. 37.
15. Beaumont, *Crito,* 13.

16. *The Athenian Mercury* 1, no. 18 (May 23, 1691), question 1.
17. Jones, *Gender and the Formation of Taste,* 1 and *passim.*
18. Londa Schiebinger, *Nature's Body: Gender in the Making of Modern Science* (Boston: Beacon Press, 1993), 126.
19. N. H., *The Ladies Dictionary,* 59.
20. This was a reprint of a debate that had originally appeared in the *Universal Spectator.* "Mr Galliard's Harangue in Praise of Beauty," *Gentleman's Magazine,* February 1736, 69.
21. See, for example, "Beauty's Cruelty or the Passionate Lover" (no. 171) and "The Tyrannical Beauty" (no. 364), both undated, in John Holloway, ed., *The Euing Collection of English Broadside Ballads in the Library of the University of Glasgow* (Glasgow: University of Glasgow Press, 1971), 22, 605.
22. "Love's Conquest; or the Powerful Force of Beautyes Charmes," Pepys Ballads 3.105 (n.d.), http://emc.english.ucsb.edu/ballad_project/large_detail.asp?id=21109 (accessed June 15, 2006).
23. Anthony Synnott, *The Body Social: Symbolism, Self and Society* (London: Routledge, 1993), 78–79; Sander L. Gilman, *Health and Illness: Images of Difference* (London: Reaktion, 1995), 52; Gilman, *Making the Body Beautiful: A Cultural History of Aesthetic Surgery* (Princeton, NJ: Princeton University Press, 2001).
24. *The Merry Dutch Miller* (London: E. Crowch, 1672), title page.
25. *Onania: Or, the Heinous Sin of Self-Pollution,* 16th ed. (London: J. Isted, 1737), 14.
26. William Hay, *Deformity: An Essay* (London: R. and J. Dodsley, 1754), 65.
27. *Look e're you Leap: Or, A History of the Lives and Intrigues of Lewd Women,* 10th ed. (London: Edward Midwinter, ca. 1760), 37, 46–47.
28. Morris Palmer Tilley, *A Dictionary of the Proverbs in England in the Sixteenth and Seventeenth Centuries* (Ann Arbor: University of Michigan Press, 1950), no. B163, 34.
29. Francis Osborne, *Advice to a Son* (Oxford: Henry Hall, 1656), 40.
30. A. B., *Learn to Lye Warm . . . Containing Reasons Wherefore a Young Man Should Marry an Old Woman* (London: H. Burgis, 1672), 16.
31. J. H., *Essays of Love and Marriage* (London: H. Brome, 1673), 3.
32. Hay, *Deformity: An Essay,* 65..
33. "Handsome Wife," *The New Spectator; with the Sage Opinions of John Bull,* June 29, 1784, 7–8.
34. Fidelis Morgan, *The Female Tatler* (London: Everyman, 1992), no. 49, 114.
35. Donald F. Bond, ed., *The Spectator*, vol. 1 (Oxford: Oxford University Press, 1965), 137–38.
36. Jones, *Gender and the Formation of Taste,* 79.
37. Ibid., 9.
38. *Beauty: Or the Art of Charming* (London: Lawton Gulliver, 1725), 3.
39. Henry Stonecastle, *The Universal Spectator*, vol. 2 (London: J. Pemberton et al., 1736), 71.
40. *The Matrimonial Preceptor* (London: Thomas Hope, 1759), 1.
41. *Beauty,* 2.
42. Donald F. Bond, ed., *The Tatler*, vol. 1 (Oxford: Oxford University Press, 1987), 87.
43. George Wright, *The Lady's Miscellany; or Pleasing Essays, Poems, Stories and Examples for the Instruction and Entertainment of the Female Sex in General, in every Station of Life* (London: Chapman & Co, 1793), 89.

44. Joseph Browne, *The London Belles: Or, A Description of the Most Celebrated Beauties in the City of London* (London: Printed for the author, 1707), 2; Antoine Le Camus, *Abdeker: or, the Art of Preserving Beauty. Translated from an Arabic Manuscript* (London: A. Millar, 1754), 5.
45. Porter, *Bodies Politic,* 71.
46. Thomas Jeamson, *Artificiall Embellishments. Or Arts Best Directions How to Preserve Beauty or Procure It* (Oxford: William Hall, 1665), 89.
47. Ibid.
48. Published in Paris in 1741; translated into English in 1743.
49. Nicolas Andry, *Orthopaedia: Or, the Art of Correcting and Preventing Deformities in Children,* 2 vols. (London: A. Millar, 1743), ii, 19.
50. Jeamson, *Artificiall Embellishments,* 134.
51. Andry, *Orthopaedia,* ii, 28.
52. Browne, *The London Belles,* 3.
53. Andry, *Orthopaedia,* 28; Runge, "Beauty and Gallantry," 59.
54. Johann Jacob Wecker, *Arts Master-Piece: Or, the Beautifying Part of Physick* (London: Nathaniel Brook, 1660), sig. A3.
55. Jeamson, *Artificiall Embellishments,* 115, 102–17, *passim.*
56. Ibid., 121.
57. For example, Andry, *Orthopaedia,* 16.
58. Arthur Wellesley Secord, ed., *Defoe's Review*, vol. 2 (New York: Columbia University Press, 1938), no. 15, 59.
59. Andry, *Orthopaedia,* i, 63.
60. Ibid., i: 77.
61. Ibid., i, 79, 81, 86)
62. Wecker, *Arts Master-Piece,* 116–19.
63. N. H., *The Ladies Dictionary,* 151–53.
64. Roy Porter, *Flesh in the Age of Reason* (London: Allen Lane, 2003), 234.
65. Beaumont, *Crito,* 48–49.
66. J. C. Beaglehole, ed., "Journal of William Wales," in *The Journals of Captain James Cook on His Voyages of Discovery*, vol. 2 (Cambridge, UK: Cambridge University Press, 1961), 809.
67. Jeamson, *Artificiall Embellishments,* 63.
68. Camus, *Abdeker,* 24, 25.
69. Porter, *Flesh in the Age of Reason,* 238.
70. Ibid., 27.
71. Jeamson, *Artificiall Embellishments,* 64.
72. Porter, *Flesh in the Age of Reason,* 238.
73. Ibid., 240–43.
74. Jeamson, *Artificiall Embellishments,* 65, 66.
75. Porter, *Flesh in the Age of Reason,* 233–34.
76. *The Ladies Mercury,* February 28, 1693, question 1.
77. Camus, *Abdeker,* 40.
78. See also Porter, *Flesh in the Age of Reason,* 233.
79. Camus, *Abdeker,* 38.
80. Ibid., 39–40, 44–45.
81. Jeamson, *Artificiall Embellishments,* 66.
82. Ken Albala, *Eating Right in the Renaissance* (Berkeley: University of California Press, 2002), 103, 146.

83. Porter, *Bodies Politic,* 36, 72.
84. Claude Quillet, *Callipaedia: Or the Art of Getting Beautiful Children*, 3rd ed., trans. N. Rowe (London: 1733), 99.
85. Jeamson, *Artificiall Embellishments,* 46; Christoph Heyl, "Deformity's Filthy Fingers: Cosmetics and the Plague in *Artificiall Embellishments, or Arts Best Directions how to preserve Beauty or procure it* (Oxford 1665)," in *Didactic Literature in England 1500–1800: Expertise Constructed,* ed. Natasha Glaisyer and Sara Pennell (Aldershot, UK: Ashgate, 2003), 142.
86. Jeamson, *Artificiall Embellishments,* 160.
87. Frances Reynolds, *An Enquiry Concerning the Principles of Taste, and of the Origin of Our Ideas of Beauty etc.* (Los Angeles: Augustan Reprint Society Publications, [1785] 1951), 25.
88. Samuel Stanhope Smith, *An Essay on the Causes of the Variety of Complexion and Figure in the Human Species* (Philadelphia: John Stockdale, 1789), 68–69.
89. Ibid., 56–57, 72.
90. Schiebinger, *Nature's Body,* 1993, 126–27.
91. Andry, *Orthopaedia,* i, 74–76.
92. *The Art of Preserving Beauty: Containing Instructions to Adorn and Embellish the Ladies* (London: T. Axtell, 1789), 169.
93. Johann Reinhold Forster, *The Resolution Journal of Johann Reinhold Forster 1772–1775,* ed. Michael E. Hoare, vol. 3 (London: Hakluyt Society, [1772–1775] 1982), 390–91.
94. Jeamson, *Artificiall Embellishments,* 83, 160.
95. Beaumont, *Crito,* 8–9.
96. Porter, *Bodies Politic,* 71.
97. Schiebinger, *Nature's Body,* 1993, 127.
98. Ibid.; Felicity A. Nussbaum, *The Limits of the Human: Fictions of Anomaly, Race and Gender in the Long Eighteenth Century* (Cambridge, UK: Cambridge University Press, 2003), 10, *passim.*
99. Quillet, *Callipaedia,* 18.
100. Smith, *An Essay on the Causes of the Variety of Complexion,* 84–87.
101. Ibid., 15, 46–47, 51, 76; Reynolds, *An Enquiry,* 26–27.
102. For example, Jeamson, *Artificiall Embellishments,* chapter 1.
103. Katherine Park and Lorraine Daston, "Unnatural Conceptions: The Study of Monsters in Sixteenth- and Seventeenth-Century France and England," *Past and Present* 92 (1981): 20–55; Marie-Hélène Huet, *Monstrous Imagination* (Cambridge, MA: Harvard University Press., 1993); Dennis Todd, *Imagining Monsters: Miscreations of the Self in Eighteenth-Century England* (Chicago: University of Chicago Press, 1995); Mary E. Fissell, *Vernacular Bodies: The Politics of Reproduction in Early Modern England* (Oxford: Oxford University Press, 2004); Lisa Forman Cody, *Birthing the Nation: Sex, Science and the Conception of Eighteenth-Century Britons* (Oxford: Oxford University Press, 2005); Kevin Stagg, "Representing Physical Difference: The Materiality of the Monstrous," in *Social Histories of Disability and Deformity,* ed. David M. Turner and Kevin Stagg (London: Routledge, 2006).
104. Eighteenth-century debates on the role of the maternal imagination are examined in Paul-Gabriel Boucé, "Imagination, Pregnant Women, and Monsters, in Eighteenth-Century England and France," in *Sexual Underworlds of the Enlightenment,* ed. George S. Rousseau and Roy Porter (Manchester, UK: Manchester University Press, 1987), 86–100; Cody, *Birthing the Nation,* 120–51; Jeamson,

Artificiall Embellishments, 11, 15; *Aristoteles Master-Piece, Or the Secrets of Generation Displayed in all the Parts Thereof* (London: J. How, 1684), chapter 5; N. H., *The Ladies Dictionary,* 475, 478; Quillet, *Callipaedia,* 66.

105. *Procreation Refin'd. A New Method for the Begetting of Children with Handsome Faces* (London: Richard Griffin, ca. 1710), 14; Quillet, *Callipaedia,* 39, 45–51.
106. *Satan's Harvest Home: Or the Present State of Whorecraft, Adultery, Fornication, Procuring, Pimping, Sodomy, and the Game at Flatts* (London: Printed for the editor, 1749), 46, 49.
107. Quillet, *Callipaedia,* 21–22.
108. *Procreation Refin'd,* 6, 8.
109. Jeamson, *Artificiall Embellishments,* 16–28.
110. *The Art of Preserving Beauty,* 166, 175–76.
111. Thomas Tuke, *A Treatise Against Painting and Tincturing of Men and Women* (London: Edward Marchant, 1616); Thomas Hall, *The Loathsomenesse of Long Haire . . . With an Appendix against Painting, Spots, Naked Breasts etc.* (London: J.G., 1654), 99–106, 118–23.
112. Misospilus, *A Wonder of Wonders: Or, A Metamorphosis of Fair Faces Voluntarily Transformed into Foul Visages* (London: R. Smith, 1662), 4, 6, 24–27.
113. J[ohn] B[ulwer], *Anthropometamorphosis: Man Transform'd: Or, the Artificiall Changling* (London: William Hunt, 1653), 260–62.
114. Hall, *The Loathsomenesse of Long Haire,* 118.
115. Misospilus, *A Wonder of Wonders,* sig. A2.
116. Ibid., 13
117. Hannah Wolley, *The Gentlewoman's Companion* (London: A. Maxwell, 1675), 58.
118. "Account of English Books," *Gentleman's Magazine,* December 1754, 601.
119. John Gauden, *A Discourse of Auxiliary Beauty or Artificiall Hansomeness. In Point of Conscience between Two Ladies* (London: R. Royston, 1656), 37, *passim.*
120. N. H., *The Ladies Dictionary,* 55, 56.
121. Samuel Pepys, *The Diary of Samuel Pepys,* 11 vols., ed. Robert Latham and William Matthews (London: Bell, 1970–1983), October 20, 1660: i, 269.
122. Ibid., March 8, 1664: v, 78; May 28, 1667: viii, 240; May 10 and 11, 1669: ix, 549, 551.
123. Roy Porter, *Health for Sale: Quackery in England 1650–1850* (Manchester, UK: Manchester University Press, 1989), 147–48.
124. *A Guide to Health, Beauty, Riches and Honour* (London: S. Hooper, 1785), 4, 5, 25, 27; Porter, *Health for Sale,* 157–72.
125. John Bate, *The Art of Beauty: A Poem Humbly Addressed to the Oxford Toasts* (London: R. Francklin, 1719), advertisement appended.
126. Porter, *Health for Sale,* 147.
127. David E. Shuttleton, "A Culture of Disfigurement: Imagining Smallpox in the Long Eighteenth Century," in *Framing and Imaging Disease in Cultural History,* ed. George S. Rousseau et al. (Basingstoke, UK: Palgrave-Macmillan, 2003), 70.
128. The story originally appeared in *The Rambler.* "Letter from Victoria," *Gentleman's Magazine,* June 1751, 255–57, continued July 1751, 318–20.
129. *The Art of Preserving Beauty,* title page.
130. *The Athenian Mercury* 2, no. 5 (July 26, 1692), question 1.
131. "Letter to the Editor from Martha Plainface," Ladies Monthly Museum. Or, Polite Repository of Amusement and Instruction, December 1799, 430–31.

132. See Naomi Wolf, *The Beauty Myth: How Images of Beauty Are Used against Women* (London: Vintage, 1991), 14–16.

Chapter 7

1. *London Gazette*, March 1, 1683, 2.
2. *Post Man and the Historical Account*, August 7, 1701, 2.
3. *London Gazette*, June 2, 1730, 2.
4. See, most recently, Dror Wahrman, *The Making of the Modern Self: Identity and Culture in Eighteenth-Century England* (New Haven, CT: Yale University Press, 2004).
5. This point is helpfully made by, amongst others, Karen Harvey, "The Substance of Sexual Difference: Change and Persistence in Representations of the Body in Eighteenth-Century England," *Gender and History* 14 (2002): 202–23.
6. Roy Porter, *In Sickness and in Health: The British Experience 1650–1850* (London: Fourth Estate, 1998), chapter 3.
7. Anna Bryson, *From Courtesy to Civility: Changing Codes of Conduct in Early Modern England* (Oxford: Clarendon Press, 1998).
8. Mark S. R. Jenner, "Body, Image, Text in Early Modern Europe," *Social History of Medicine* 12 (1999): 143–54.
9. Dana Rabin, "Searching for the Self in Eighteenth-Century English Criminal Trials, 1730–1800," *Eighteenth-Century Life* 27 (2003): 85–106.
10. J. Paul Hunter, *Before Novels: The Cultural Contexts of Eighteenth-Century English Fiction* (New York: Norton, 1990).
11. Juliet McMaster, *Reading the Body in the Eighteenth-Century Novel* (London: Palgrave, 2004.
12. Aphra Behn, *Love Letters between a Nobleman and His Sister*, ed. Janet Todd (London: Penguin, [1684–1687] 1996), 58, 86, 359.
13. William Hogarth, *The Rake's Progress: 7. The Rake in Prison,* 1734, oil on canvas, 630 × 755 mm, Sir John Soane's Museum, London; William Hogarth, *Marriage à-la Mode: 6. The Lady's Death,* 1745, oil on canvas, 699 × 908 mm, The National Gallery, London.
14. Mark Hallett, *The Spectacle of Difference: Graphic Satire in the Age of Hogarth* (London: Yale University Press, 1999), 208.
15. Tom Brown, *Amusements Serious and Comical, Calculated to the Meridian of London* (Paris: Les Editions D'Aujourd'hui, 1700), 20–21.
16. Ned Ward, *The London Spy,* ed. Paul Hyland (East Lansing, MI: Colleagues Press, [1709] 1993), 40.
17. Ibid., 59–61.
18. Roy Porter, *Flesh in the Age of Reason* (London: Allen Lane, 2003), 247.
19. Alex Werner, ed., *London Bodies: The Changing Shape of Londoners from Prehistoric Times to the Present Day* (London: Museum of London, 1998), 84–85.
20. Ibid., 108.
21. *Old Bailey Proceedings Online* (hereafter cited as *OPB*), http://www.oldbaileyonline.org (accessed November 4, 2005), Trial of Michael Tooley, July 1695, ref. t16950703-24.
22. *OBP,* Trial of Corbet Vezey, January 1732: ref. t17320114-12.

23. *London Gazette*, June 28, 1711, 2.
24. John Cleland, *Fanny Hill or Memoirs of a Woman of Pleasure* (Harmondsworth, UK: Penguin, [1748–1749] 1998), 148.
25. This is evident, for example, in the Old Bailey rape cases, which are dominated by assaults and rape of children.
26. Kevin Siena, *Venereal Disease, Hospitals and the Urban Poor* (Rochester, NY: University of Rochester Press, 2004), 54.
27. Richelle Munkhoff, "Searchers of the Dead: Authority, Marginality, and the Interpretation of Plague in England, 1574–1665," *Gender and History* 11 (1999): 1–29.
28. *OBP,* Trial of unnamed woman, July 1678, ref. t16780703-2.
29. Lynn Botelho, "Old Age and Menopause in Rural Women of Early Modern Suffolk," in *Women and Ageing in British Society Since 1500,* ed. Lynn Botelho and Pat Thane (Harlow, UK: Pearson, 2001), 56.
30. Sean Shesgreen, *Images of the Outcast: The Urban Poor in the Cries of London* (Manchester, UK: Manchester University Press, 2002), 51.
31. Tim Hitchcock, *Down and Out in Eighteenth-Century London* (London: Hambledon, 2005), 210–13.
32. *OBP,* Trial of John Stevens, Richard Marshall, et al., October 1732, ref. t17321011-29.
33. *OBP,* Trial of Catherine Pember, otherwise Buck, December 1733, ref. t17331205-57.
34. *OBP,* Trial of John Theobalds, April 1732. ref. t17320419-43.
35. Patricia Crawford, "Printed Medical Advertisements for Women Medical Practitioners in London, 1670–1710," *Bulletin of the Society for the Social History of Medicine* 35 (1994): 66–70; Fenja Gunn, *The Artificial Face: A History of Cosmetics* (Newton Abbot, UK: David and Charles, 1973).
36. *OBP,* Old Bailey Proceedings Advertisements, January 1707, ref. a17070115-1.
37. Gretchen Gerzina, *Black England: Life before Emancipation* (London: John Murray, 1995), 22.
38. *OBP,* Trial of Joseph Smith, October 1726, ref. t17261012-6.
39. *OBP,* Trial of Mary Harris, January 1702, ref. t17020114-18702.
40. See, for example, *Aristoteles Master-Piece, or the Secrets of Generation Displayed in All the Parts Thereof* (London: J. How, 1684), chapter 20.
41. *OBP,* Trial of Mary Cut and Come Again, April 1745, ref. t17450424-43.
42. The phrase is used here to describe bruised genitals (*OBP,* Trial of Charles Troop, July 1751, ref. t17510703-39); in another case a prostitute who shows a "Negro" her genitals says it is "as black as his face," although it is not clear why (*OBP,* Trial of Loglin Rennells, December 1736, ref. t17360908-39).
43. *OBP,* Thomas Lee, May 1782, ref. t17820515-20.
44. *OBP,* Francis Woodmash, April 1731, ref. 17310428-72.
45. Sander Gilman, *The Jew's Body* (New York: Cornell University Press, 1991), 171.
46. Ibid., 39.
47. Todd Endelman, *The Jews of Georgian England, 1714–1830: Tradition and Change in a Liberal Society,* 2nd ed. (Ann Arbor: University of Michigan Press, 1999), chapter 1.
48. *OBP,* Trial of John Theobalds, April 1732, ref: t17320419-43.
49. *OBP,* Trial of Solomon Isaacs, July 1740, ref. t17400709-35.
50. *OBP,* Trial of Mary Picart, alias Gandon, June 1725, ref. t17250630-17.

51. *OBP,* Trial of John Stevens, Richard Marshall, et al., October 1725, ref. t17321011-42.
52. Wahrman, *The Making of the Modern Self,* 26–27.
53. *OBP,* Trial of Charles Franklin, September 1747, ref. t17470909-17.
54. *OBP,* Trial of Ann Fell, July 1790, ref. t17900707-46.
55. *OBP,* Trial of Jeremiah Beck, September 1799, ref. t17990911-18.
56. Ibid.
57. Male cross-dressing featured much less frequently in contemporary complaint, until the late seventeenth century, when it began to be seriously associated with sodomy.
58. *OBP,* Trial of John Stevens, February 1738, ref. t17380222-24.
59. Theo van der Meer, "The Persecutions of Sodomites in Eighteenth-Century Amsterdam: Changing Perceptions of Sodomy," in *The Pursuit of Sodomy: Male Homosexuality in Renaissance and Enlightenment Europe,* ed. Kent Gerard and Gert Hekma (New York: Haworth Press, 1989), 263–307.
60. *OBP,* Trial of Thomas Gordon, July 1732, ref. t17320705-30.
61. Ibid.
62. Rictor Norton, *Mother Clap's Molly House: The Gay Subculture in England, 1700–1830* (London: GMP, 1992); Randulph Trumbach, *Sex and the Gender Revolution: Heterosexuality and the Third Gender in Enlightenment London* (Chicago: University of Chicago Press, 1998).
63. *OBP,* Trial of Loglin Rennells, December 1739, ref. t17391205-51.
64. *OBP,* Judge's Summing Up (Nicholas Bradshaw), December 1678, ref. x16781211-1.
65. *OBP,* Trial of Ann Nichols, January 1712, ref. t17120111-9.
66. *OBP,* Trial of Sarah Harwood, alias Badger, alias Radford, April 1729, ref. t17290416-67.
67. National Archives, Kew, ASSI 45 18/5/60, 62.
68. Lisa Cody, *Birthing the Nation: Sex, Science, and the Conception of Eighteenth-Century Britons* (Oxford: Oxford University Press, 2005).
69. *OBP,* Trial of Mary Cut and Come Again, April 1745, ref. t17450424-31.
70. Ibid.
71. For example, Gail Kern Paster, *The Body Embarrassed: Drama and the Disciplines of Shame in Early Modern England* (Ithaca, NY: Cornell University Press, 2003).

Chapter 8

1. John Graunt, *Natural and Political Observations, Mentioned in a Following Index, and Made upon the Bills of Mortality*, 2nd ed. (London: John Martin et al., 1662), 24.
2. William Black, *An Arithmetical and Medical Analysis of the Diseases and Mortality of the Human Species* (London, 1789), 230.
3. Ronald Paulson, *Hogarth's Graphic Works*, 2 vols. (New Haven, CT: Yale University Press, 1965), vol. 1, no. 127–32, vol. 2, 141–49; N. F. Lowe, "The Meaning of Venereal Disease in Hogarth's Graphic Art," in *The Secret Malady: Venereal Disease in Eighteenth-Century Britain and France,* ed. Linda E. Merians (Lexington: University Press of Kentucky, 1996), 168–82; Paulson, *Hogarth's Harlot: Sacred Parody in Enlightenment England* (Baltimore, MD: Johns Hopkins University Press, 2003), 101.

4. While madness was sometimes associated with late-stage pox in Renaissance and early modern literature, and convulsions were sometimes remarked on as signs of congenital pox in infants, neurological signs were typically not associated with late-stage pox in most eighteenth-century accounts. John Andree explicitly denied that the pox could attack the brain, although, like many, he noticed that the mercury treatments for venereal disease could lead to tremors: Andree, *Observations on the Theory and Cure of the Venereal Disease* (London: W. Davis, 1779), 143. For a discussion of the nineteenth-century attentions to effects of late-stage syphilis on the brain, see Claude Quétel, *History of Syphilis*, trans. Judith Braddock and Brian Pike (Baltimore, MD: Johns Hopkins University Press, 1990), chapter 7.
5. Tobias Smollett, *The Adventures of Roderick Random*, ed. Paul-Gabriel Boucé (Oxford: Oxford University Press, [1748] 1979), 115–16.
6. For a useful brief account see Kenneth M. Flegel, "Changing Concepts of the Nosology of Gonorrhea and Syphilis," *Bulletin of the History of Medicine* 48 (1974): 571–88.
7. An important bibliography is Johann Karl Proksch, *Die Litteratur über de venerischen Krankheiten von den ersten Schriften über Syphilis aus dem Ende des Fünfzehnten Jarhhundrets bis zum jahre 1889*, 3 vols. (Bonn, Germany: P. Hanstein, 1889–1890). Additional titles appear from searching eighteenth-century reviews and "Eighteenth Century Collections Online," available at: http://gale.cengage.com/EighteenthCentury/index.htm. I have read about a hundred of these books.
8. For a consideration of differences and similarities between contemporary physicians, surgeons, and quacks who treated venereal disease, one focused on changing professional norms, see William F. Bynum, "Treating the Wages of Sin: Venereal Disease and Specialism in Eighteenth Century Britain," in *Medical Fringe and Medical Orthodoxy, 1750–1850,* ed. Roy Porter (London: Croom Helm, 1987), 5–28. I agree with Bynum that there was considerable overlap between the practices of these three kinds of medical men in their ideas about, and treatment of, venereal disease. Philip K. Wilson considers the practices of Daniel Turner, a surgeon who later became a licentiate of the College of Physicians, in relation to back-and-forth accusations of quackery by various practitioners. See Wilson, "Exposing the Secret Disease: Recognizing and Treating Syphilis in Daniel Turner's London," in *The Secret Malady: Venereal Disease in Eighteenth-Century Britain and France,* ed. Linda E. Merians (Lexington: University Press of Kentucky, 1996), 68–84.
9. This is an expanded version of the first edition, published in 1717. Turner aspired to a medical degree and eventually received an honorary one in 1723, in absentia, from Yale. Daniel Turner, *Siphylis: A Practical Dissertation on the Venereal Disease*, 4th ed. (London: J. Walthoe et al., 1732), 360–61.
10. Sintelaer was born in Holland of a Scotch-descended family, trained in Paris, and resident in London for twenty-five years. John Sintelaer, *The Scourge of Venus and Mercury, Represented in a Treatise of the Venereal Disease* (London: G. Harris et al., 1709), 294.
11. Turner, *Siphylis,* 124, 126, 128.
12. Charles Hales, *A Letter Addressed to Caesar Hawkins: Containing New Thoughts and Observations, in the Cure of the Venereal Disease . . . With a Few Extraordinary Cases in That Disease* (London: 1750), 55–56, 59.
13. William Buchan, *Observations Concerning the Prevention and Cure of the Venereal Disease. Intended to Guard the Ignorant and Unwary against the Baneful Effects of That Insidious Malady* (London: T. Chapman et al., 1796), 203.

14. In this important study, Siena makes the point that the poor were convinced of their entitlement to medical care under the Poor Law. Rather touchingly, he shows poor people—having been treated for the pox at parish expense—either after having been discharged cured or after having eloped from the rigors of salivation in a workhouse or hospital—returning later to demand more treatment. Kevin Siena, *Venereal Disease, Hospitals, and the Urban Poor: London's "Foul Wards"* (Rochester, NY: University of Rochester, 2004).
15. John Hunter, *A Treatise on the Venereal Disease* (London: The Author, 1786), 369.
16. See, for example, Jon Arrizabalaga, John Henderson, and Roger French, *The Great Pox: The French Disease in Renaissance Europe* (New Haven, CT: Yale University Press, 1997).
17. Hunter, *A Treatise,* 287.
18. For a major study dealing with the persistence of Renaissance treatments through the eighteenth century, see Erwin H. Ackerknecht, *Therapeutics: From the Primitives to the Twentieth Century* (New York: Hafner Press, 1973).
19. Jean Astruc, *A Treatise of Venereal Diseases, in Nine Books . . . ,* vol. 2, trans. W. Barrowby (London: Printed for W. Innys and J. Richardson, C. Davis, J. Clarke, R. Manby, and H. S. Cox, 1754), 232.
20. Richard Bunworth, *A New Discovery of the French Disease, and Running of the Reins: Their Signs, with Plain and Easy Directions of Perfect Curing the Same*, 2nd ed. (London: Henry Marsh, 1662), 11.
21. *A New Method of Curing the French-Pox . . . As Also Dr Sydenham's Judgment of the Same . . .* (London: John Taylor & Thomas Newborough, 1690), 41.
22. Andree, *Observations,* 6–7.
23. Jacques Gautier D'Agoty, *Exposition anatomique des maux* vénériens*, sur les parties de l'homme et de femme, et les remèdes les plus usités dans ces sortes de maladies* (Paris: J. B. Brunet & Demonville, 1773).
24. Turner, *Siphylis,* 357–58.
25. Flegel, "Changing Concepts"; Roy Porter, "The Eighteenth Century," in *The Western Medical Tradition: 800 BC to 1800 AD,* Lawrence I. Conrad et al. (Cambridge, UK: Cambridge University Press, 1995), 411.
26. Benjamin Bell, *A Treatise on Gonorrhœa Virulenta and Lues Venera*, vol. 2, (Edinburgh: James Watson & Co., 1793), 501–3.
27. Jean Astruc, *A Treatise of Venereal Diseases, in Nine Books . . . ,* vol. 2 trans. W. Barrowby (London: Printed for W. Innys and J. Richardson, C. Davis, J. Clarke, R. Manby, and H. S. Cox, 1754), 37.
28. Elizabeth Lane Furdell, "Harris, Walter (1647–1732)," in *Oxford Dictionary of National Biography* (Oxford: Oxford University Press, 2004).
29. Walter Harris, *A Treatise of the Acute Diseases of Infants. To Which Are Added, Medical Observations on Several Grievous Diseases,* trans. (from the Latin) John Martyn (London: T. Astley, 1742), 5, 24.
30. Ibid., 200, 213.
31. Ibid., 209, 225.
32. Bynum, "Treating the Wages of Sin," 5–28.
33. John Kirkup, "Wiseman, Richard (bap. 1620?, d. 1676)," in *Oxford Dictionary of National Biography* (Oxford: Oxford University Press, 2004).
34. His own application of this principle, however, seems to me sparing.
35. Here again we see the idea of the disease existing without observable symptoms, in a latent form. Richard Wiseman, *Eight Chirurgical Treatises*, vol. 2 (London: B. Tooke, 1719), 301–2.

36. Ibid., 347.
37. Ibid., 301.
38. Ibid., 346.
39. Flegel says that, because of the passive and neutral form of Hunter's language, it is impossible to be sure whether he is speaking of self-inoculation in his experiment, but most commentators—and I agree—have considered it a self-experiment (Flegel, "Changing Concepts"). See also Hunter, *A Treatise,* 324–27.
40. Hunter, *A Treatise,* 291–92.
41. Ibid., 290.
42. Ibid., 386–87.
43. Ibid., 296.
44. Bell, *A Treatise on Gonorrhœa Virulenta,* 177.
45. Ibid., 418.
46. Ibid., 422.
47. Arrizabalaga, Henderson, and French, *The Great Pox,* 282.

Chapter 9

1. Edward Ward, *The History of the London Clubs; or the Citizens Pastime, Particularly the Lying Club, the Beggars Club, the Yorkshire Club,* part 1 (London: J. Bagnall, 1709), 7. Literary critics have made the depiction of beggars in the eighteenth-century literature a topic of substantial analysis. Historians have been much less interested in the topic. For some recent work in a literary vein see Celeste Lanban, *Romantic Vagrancy: Wordsworth and the Simulation of Freedom* (Cambridge, UK: Cambridge University Press, 1995); Tim Fulford, *Landscape, Liberty, and Authority: Poetry, Criticism, and Politics from Thomson to Wordsworth* (Cambridge, UK: Cambridge University Press, 1996); Judith Frank, *Common Ground: Eighteenth-Century Satiric Fiction and the Poor* (Stanford, CA: Stanford University Press, 1997); Elizabeth Helsinger, *Rural Scenes and National Representation: Britain, 1815–1850* (Princeton: Princeton University Press, 1997).
2. George Parker, *A View of Society and Manners in High and Low Life; Being the Adventures in England, Scotland, Wales, France, etc., of G. Parker* (London: 1781), 155–59.
3. Linda Woodbridge, *Vagrancy, Homelessness and English Renaissance Literature* (Urbana: University of Illinois Press, 2001); Martine Van Elke, "The Counterfeit Vagrant: The Dynamic of Deviance in the Bridewell Court Records and the Literature of Roguery," in *Rogues and Early Modern English Culture,* ed. Craig Dionne and Steve Mentz (Ann Arbor: University of Michigan Press, 2004).
4. Court of Governors, Bridewell and Bethlam, Minutes: January 12, 1737–1738, to April 4, 1751 (MS 33011/21), Isaac Fairbrother: 143, Henry Goodwin: 143, William Harding and Ann Middleton: 321. Guildhall Library, London.
5. Vagrant Books (4), Together with Five Loose Pages and a Bill of Mortality, 1738–1742 (Misc. MS 322.5): Examinations of Henry Buxton, June 12, 1738; Henry Goodwin, March 20, 1741. Corporation of London Records Office, London. London Metropolitan Archives.
6. C.J. Ribton-Turner, *A History of Vagrants and Vagrancy and Beggars and Begging* (London: Chapman & Hall, 1887), 178–79.

7. Court of Governors, Bridewell and Bethlam, Minutes: August 23, 1695 to May 2, 1701 (MS 33011/17), Ann Wheatley: 249. Court of Govenors, Bridewell and Bathlam, Minutes: 1701 to June 1713 (MS 33011/18), Elizabeth Arnold: 644. Guildhall Library, London.
8. There is a small literature on disability in seventeenth- and eighteenth-century Britain that concentrates on the role of institutional care. See for example Peter Rushton, "Lunatics and Idiots: Mental Disability, the Community, and the Poor Law in North-East England, 1600–1800," *Medical History* 32 (1988): 34–50; Anne Borsay, "Returning Patients to the Community: Disability, Medicine and Economic Rationality before The Industrial Revolution," *Disability and Society* 13 (1998): 645–63; Margaret Pelling, "Old Age, Poverty, and Disability in Early Modern Norwich: Work, Remarriage, and other Expedients," in *Life, Death and the Elderly: Historical Perspectives,* ed. Margaret Pelling and R.H. Smith (London: Routledge, 1991), 74–101. For primarily literary analyses of disability in the eighteenth century see Helen Deutsch, "The Body's Moment: Visible Disability, the Essay and the Limits of Sympathy," *Prose Studies* 27 (2005): 11–26; Kathleen James-Cavan, "'[A]ll in me is Nature': The values of deformity in William Hay's *Deformity: An Essay,*" *Prose Studies* 27 (2005): 27–38; David Turner and Kevin Stagg, eds., *Social Histories of Disability and Deformity* (London: Routledge, 2006). *The Case of the Inhabitants of the Parish of St Giles in the Fields, as to Their Poor, and a Work-House Designed to be Built for Employing Them* (London: 1730).
9. For three excellent recent accounts of the development of the Old Poor Law, none of which deal specifically with the treatment of disability, see Lynn Hollen Lees, *The Solidarities of Strangers: The English Poor Laws and the People, 1700–1948* (Cambridge, UK: Cambridge University Press, 1998); Paul Slack, *From Reformation to Improvement: Public Welfare in Early Modern England* (Oxford: Clarendon Press, 1999); Steven King, *Poverty and Welfare in England 1700–1850: A Regional Perspective* (Manchester, UK: Manchester University Press, 2000).
10. "Samuel Badham's Account of Himself," in *The Ordinary of Newgate, His Account of the Behaviour, Confession, and Dying Words, of the Malefactors, Who Were Executed at Tyburn, on Wednesday the 6th of August (1740). Being the Fourth Execution in the Mayoralty of the Rt. Hon. Sir John Salter, Knt. Number IV. For the Said Year* (London: Printed and Sold by John Applebee, in Bolt-Court, near the Leg-Tavern, Fleet-street, 1740).
11. Jeremy Boulton, "Going on the Parish: The Parish Pension and its Meaning in the London Suburbs, 1640–1724," in *Chronicling Poverty: The Voices and Strategies of the English Poor, 1640–1840,* ed. Tim Hitchcock, Peter King, and Pamela Sharpe (Basingstoke, UK: Macmillan, 1997), 33.
12. John Thomas Smith, *Vagabondiana: Or, Anecdotes of Mendicant Wanderers through the Streets of London* (London: 1817), 51.
13. I would like to thank Lisa Cody for this reference. For an excellent treatment of birth and lying-in during this period see Lisa Forman Cody, *Birthing the Nation: Sex, Science and the Conception of Eighteenth-Century Britons* (Oxford: Oxford University Press, 2005). For the experience of mothers of illegitimate children see Tim Hitchcock, "'Unlawfully Begotten on her Body': Illegitimacy and the Parish Poor in St Luke's, Chelsea," in *Chronicling Poverty: The Voices and Strategies of the English Poor, 1640–1840,* ed. Tim Hitchcock, Peter King, and Pamela Sharpe (Basingstoke, UK: Macmillan., 1997), 70–86. Brownlow Street Lying-in Hospital,

First Minute Book, 1750–1756, MS H10/CLM/A1/1/1: June 27, 1753. London Metropolitan Archives, London.

14. Ribton-Turner, *A History of Vagrants,* 187.
15. *A View of London and Westminster: Or the Town Spy*, part 1 (London: T. Warner, 1728), 42–43.
16. William Hay, *Deformity: An Essay* (London: R. and J. Dodsley, 1754), 70.
17. James-Cavan, "'[A]ll in me is Nature.'"
18. *A Trip from St James's to the Royal Exchange: The Manners, Customs and Amusements of the Inhabitants of London and Westminster* (London: Edward Withers, 1744), 12–13.
19. Francis Grose, *The Grumbler: Containing Sixteen Essays by the Late Francis Grose, Esq. FAS of London and Perth* (London: 1791), 35–36.
20. "Mary Cut and Come Again's Account of Herself," in The Ordinary of Newgate, Hiss Account of the Behaviour, Confession, and Dying Words of the Malefactors Who Were Executed at Tyburn, on Friday the 7th of June, 1745. Being the Third Execution in the Mayoralty of the Right Honble Henry Marshal, Esq., Lord Mayor of the City of london. Number III. For the said year (London: Printed, and sold by M. Cooper, at the Globe in Pater-noster Row, 1745).
21. Old Bailey Proceedings Online (hereafter *OBP*), http://www.oldbaileyonline.org, Trial of Mary Cut and Come Again, April 1745, ref. t17450424–31.
22. Blood had a profound religious meaning for eighteenth-century people. See Piero Camporesi, *Juice of Life: The Symbolic and Magic Significance of Blood*, trans. Robert R. Barr (New York: Continuum, 1995). For breast milk and breast-feeding in this period see Ruth Perry, "Colonizing the Breast: Sexuality and Maternity in Eighteenth-Century England," *Journal of the History of Sexuality* 2 (1991): 204–34. Spit has been less well served by historians, but it was nevertheless full of mystical and religious authority, which gave real power to Mary's curses. Just three years after Mary's death, Bridgett Bostock attracted national attention by the curative powers of her "fasting spittle." On one occasion 600 people were reported to have made their way to her hovel in Cheshire in hopes of being cured when "she rubs the parts affected with her fasting spittle," giving "the blessing of her gums": *Gentleman's Magazine,* September 1748, 413–14; *Gentleman's Magazine,* October 1748, 448–51.
23. *OBP,* Trial of John Bowes, Hugh Ryly, Sexual Offences, Sodomy, December 1718: ref. t17181205-24.
24. See in particular Laura Gowing, *Common Bodies: Women, Touch and Power in Seventeenth-Century England* (New Haven, CT: Yale University Press, 2003).
25. For example see Peter Stallybrass and Allon White, *The Politics and Poetics of Transgression* (Ithaca, NY: Methuen, 1986). For more recent work focused on the representation and depiction of bodies see Susanne Scholz, *Body Narratives: Writing the Nation and Fashioning the Subject in Early Modern England* (London: Macmillan, 2000); Karen Harvey, "The Substance of Sexual Difference: Change and Persistence in Representations of the Body in Eighteenth-Century England," *Gender and History* 14 (2000): 202–23; and others cited in note 1 of this chapter.
26. For a recent attempt to historicize wounds and missing limbs through an examination of their primary religious significance, see Stephen Greenblatt, "Mutilation and Meaning," in *The Body in Parts,* ed. David Hillman and Carla Mazzio (New York: Routledge, 1997): 221–41.
27. "Samuel Badham's Account of Himself"; Old Bailey, Trial of Samuel Badham, August 1740, ref. t17400709-2.

28. Vagrant Books (4), Together with Five Loose Pages and a Bill of Mortality, 1738–1742 (Misc. MS 322.5).
29. The impact of visual impressions on pregnancy was hotly disputed by elite medical practitioners in this period but was also widely accepted as fact by the broader population. See Paul-Gabriel Boucé, "Imagination, Pregnant Women and Monsters in Eighteenth-Century England and France," in *Sexual Underworlds of the Enlightenment,* ed. George Rousseau and Roy Porter (Manchester, UK: Manchester University Press, 1987), 86–100; Lisa Cody, " 'The Doctor's in Labour: Or a New Whim Wham from Guildford,' " *Gender and History* 4 (1992): 175–96. Ribton-Turner, *A History of Vagrants,* 191.
30. Ibid., 188–89.
31. London Sessions Papers: Grand Jury Presentment, October Sessions 1749; Thirty-one examinations of supposed vagrants, December Sessions 1782: Thomas Dargaval, November 16, 1782. Corporation of London Records Office, London. London Metropolitan Archives.
32. Throughout most of the eighteenth century the Royal Military Hospital at Chelsea housed 476 pensioners, and provided regular relief to a further thousand "out pensioners," who were required to make their way to Chelsea four times a year in order to collect their money. Of the sample of 469 settlement and bastardy examinations recorded in Chelsea (1733–1766), 84 concerned pensioners of the hospital. See Tim Hitchcock and John Black, *Chelsea Settlement and Bastardy Examinations, 1733–1766,* vol. 33 (London: London Record Society, 1999).
33. See the unpaginated folded table interleaved at the end of Matthew Martin, *Letter to the Right Hon. Lord Pelham on the State of Mendicity in the Metropolis* (London: 1803).
34. George Galloway, "To the Memory of Gavin Wilson (Boot, Leg and Arm maker)," in *The New Oxford Book of Eighteenth-Century Verse,* ed. Roger Lonsdale (Oxford: Oxford University Press, [1795] 1984), 805–6.
35. London Sessions Papers: Grand Jury Presentment, October Sessions 1749; Thirty-one examinations of supposed vagrants, December Sessions 1782: Thomas Dargaval, November 16, 1782. Corporation of London Records Office, London. London Metropolitan Archives.
36. John Collier, "The Pluralist and Old Soldier," in *The New Oxford Book of Eighteenth-Century Verse,* ed. Roger Lonsdale (Oxford: Oxford University Press, [1763] 1984), 511.
37. At least five versions of this ballad and one act play were published during the eighteenth century. Robert Dodsley, *The Blind Beggar of Bethnal Green* (London: 1741.
38. For a brief discussion of the Foundling Hospital's policy see Ruth McClure, *Coram's Children: The London Foundling Hospital in the Eighteenth Century* (New Haven, CT: Yale University Press, 1981), 229. For a much more detailed treatment of attitudes towards blindness in the eighteenth century see William R. Paulson, *Enlightenment, Romanticism and the Blind in France* (Princeton, NJ: Princeton University Press, 1987).
39. Reproduced as Plate VII in Celina Fox, *Londoners* (London: Thames & Hudson, 1987).
40. Old Bailey, Trial of Thomas Cooper, May 1755, ref. t17550515-21.
41. *Annual Register for the Year 1761* (London: J. Dodsley et al., 1762), 96.

42. Court of the President and Governors for the Poor of London, 1702–1705 (MS New 377C/1/22): Thomas Newby, fol. 200. Corporation of London Records Office, London. London Metropolitan Archives.
43. Mansion House Justice Room, Charge Book, 1728–1733 (MS New 392C): Samuel Gold, May 27, 1730. Corporation of London Records Office, London. London Metropolitan Archives.
44. For an authoritative edition of the biography that separates out the apocryphal from the legitimate see C.H. Wilkinson, ed., *The King of the Beggars: Bampfylde-Moor Carew* (Oxford: Clarendon Press, 1931). *An Apology for the Life of Bampfylde-Moore Carew* (London: Sold by W. Owen, 1749), 107.
45. Court of Governors, Bridewell and Bethlam, Minutes: January 12, 1737–1738, to April 4, 1751 (MS 33011/21), Isaac Fairbrother: 143, Henry Goodwin: 143, William Harding and Ann Middleton: 321. Guildhall Library, London.
46. *Low-Life: Or One Half of the World, Knows Not How the Other Half Live,* 2nd ed. (London: 1749), 30.
47. Court of Governors, Bridewell and Bethlam, Minutes: October 16, 1722 to December 15, 1737 (MS 3011.20), Ann Fretwell alias Davis: 81. Guildhall Library, London.
48. Old Bailey, Trial of Elizabeth Evans, April 1740, ref. t17400416-24.
49. For a detailed discussion of this image see Sean Shesgreen, *Images of the Outcast: The Urban Poor in the Cries of London* (Manchester, UK: Manchester University Press, 2002), 102–17.
50. Court of Governors, Bridewell and Bethlam, Minutes: May 22, 1751 to December 1, 1761 (MS 3011/22), Julian Harrington: 349. Guildhall Library, London..
51. *A Trip Through Town. Containing Observations on the Humours and Manners of the Age*, 4th ed. (London: 1735), 13–14.
52. *Annual Register for the Year 1772* (London: J. Dodsley et al., 1773), 138.
53. *A Trip from St James's to the Royal Exchange,* 26.
54. For a recent article on the meaning of rags as clothing see Christiana Payne, "'Murillo-like Rags or Clean Pinafores': Artistic and Social Preferences in the Representation of the Dress of the Rural Poor," *Textile History* 33 (2002): 48–62.
55. On smell see Mark Jenner, "Civilization and Deodorization? Smell in Early Modern English Culture," in *Civil Histories: Essays Presented to Sir Keith Thomas,* ed. Peter Burke, Brian Harrison, and Paul Slack (Oxford: Oxford University Press, 2000), 127–44. Coroner's Inquest: John Webb, July 6, 1762. The Library and Muniment Room, Westminster Abbey, London.
56. Jonathan Swift, *Gulliver's Travels*, ed. Harold Williams (Oxford: Blackwell, [1726] 1965), 112–13.
57. George A. Atkin, ed. *The Tatler*, vol. 2 (London: Duckworth,1899), 139–40.
58. The evolution of a humanitarian narrative has been substantially explored from an elite perspective, but the self-conscious manipulation and deployment of these narratives by the poor has not formed the subject of a similar amount of scholarship. See for example Thomas L. Haskell, "Capitalism and the Origins of the Humanitarian Sensibility" (part 1), *American Historical Review* 90 (1985): 339–61; Thomas Laqueur, "Bodies, Details, and the Humanitarian Narrative," in *The New Cultural History,* ed. Lynn Hunt (Berkeley: University of California Press, 1989), 176–204; Deutsch, "The Body's Moment."

59. James Dawson Burn, *The Autobiography of a Beggar Boy*, ed. David Vincent (London: Europa, [1855] 1978), 73.

Chapter 10

1. Joseph Addison, *The Spectator*, May 21, 1711, 1–2; Addison, *The Spectator*, June 7, 1711, 1–2.
2. Addison, *The Spectator*, May 21, 1711, 1–2.
3. Joseph Addison, *The Spectator*, May 25, 1711, 1–2.
4. Maria Colomba Mattei Trombetta di Roma, called le Romaninia or La Colonna or Signora Mattei. Her career began in Naples in 1743, and she sang in numerous Italian cities before reaching London in 1754–1755. She returned to London and finished her career there, in 1758–1762. See Philip H. Highfill, Jr., Kalman A. Burnim, and Edward A. Longhans, *A Biographical Dictionary of Actors, Actresses, Musicians, Dancers, Managers and Other Stage Personnel in London, 1660–1800*, vol. 10 (Carbondale: Southern Illinois University Press, 1973–1993), 137–38. Thanks to Lowell Lindgren for this reference.
5. Oliver Goldsmith, "Happiness," *The Bee,* October 13, 1759, 1.
6. Walter J. Ong, *Orality and Literacy* (London: Methuen, 1982), 43.
7. Claude M. Simpson, *The British Broadside Ballad and Its Music* (New Brunswick, NJ: Rutgers University Press, 1966), xi.
8. Francis James Child, *The English and Scottish Popular Ballads*, vol. 1 (Northfield, MN: Loomis House Press, [1882–1898] 2001), 259.
9. Walter Scott, *The Heart of Midlothian,* with an introductory essay and notes by Andrew Lang, vol. 1 (Boston: Estes and Lauriat, 1893), chapter 9.
10. Child, *Ballads*, no. 16.
11. These elements of oral composition are listed by (among others) David Atkinson, *The English Traditional Ballad* (Burlington, UK: Ashgate, 2002), 12.
12. Ibid., 148.
13. A further association between broom and sexuality is suggested by the folk marriage custom of jumping over a broom, an implement originally made of broom, whence its name.
14. A version of "Leesome Brand" was collected by Motherwell from Agnes Lyle of Kilbarchan parish at about the same time as he collected "Sheath and Knife" from Mrs. King of the same parish. It is very similar to "Sheath and Knife" except that the lovers are not said to be brother and sister, and they have with them a child, his "auld son," whom he carries in his "coat lap." This child, too, is killed when he shoots his arrow.
15. Helena Mennie Shire, ed., *Poems from Panmure House* (Cambridge, UK: Ninth of May Press, 1960), 13–19. Shire observes that at the time this ballad was circulating, Ford's play *Tis Pity She's a Whore* was on the stage. The similarities of incident and atmosphere between that play and the ballad lead her to conjecture that "Ford may have had a version of this ballad running in his head" when he wrote the play: Ibid., 22–23.
16. Stephen Greenblatt, ed., *The Norton Anthology of English Literature*, vol. 1 (New York: W. W. Norton, 2006). Thanks to Alan Levitan for this reference.
17. Ong, *Orality and Literacy,* 48.

18. Motherwell collected these ballads in 1825, but by then both these women were elderly—he called them "old singing women"—and had learned their repertoires many years earlier. See Mary Ellen Brown, *William Motherwell's Cultural Politics* (Lexington: University of Kentucky Press, 2001), 161–70.
19. "Babylon" (Child, *Ballads*, no. 14), "Sheath and Knife" (Child, *Ballads*, no. 16), "The Bonny Hind" (Child, *Ballads*, no. 50), "Lizie Wan" (Child, *Ballads*, no. 51), "The King's Dochter Lady Jean" (Child, *Ballads*, no. 52), "Brown Robyn's Confession" (Child, *Ballads*, no. 57). There is another brother–sister incest ballad from Ireland sung by Peta Webb called "The Rich Man's Daughter," which is not a Child ballad but follows this pattern. The brother kills the sister when she is about to give birth to their child.
20. Sara Cleveland sings "Queen Jane" on her CD *Folk Legacy*, the only published American version of "The King's Dochter Lady Jean." This feeling about the song is reported by Kenneth Goldstein in his notes to the CD and reported by Atkinson, *The English Traditional Ballad*, 113.
21. "Sheath and Knife" (Child, *Ballads*, no. 16), "Lizie Wan" (Child, *Ballads*, no. 51), and "Brown Robyn's Confession" (Child, *Ballads*, no. 57). For a more elaborate treatment of this theme in folksong see G. Brewster, "The Incest Theme in Folksong," *F F Communication* 80 (1972): 3–36.
22. Including the version sung by Sara Cleveland, "Queen Jane."
23. Naomi Tadmor, *Family and Friends in Eighteenth-Century England* (Cambridge, UK: Cambridge University Press, 2001), 121.
24. Juliet Mitchell, *Siblings* (Cambridge, UK: Polity Press, 2003.
25. Ibid., 29.
26. In a successful resolution of this tension, continues Mitchell, a person learns to transform narcissistic love into "object-love and murderousness into objective hatred for what is wrong or evil in the self and other." Mitchell, *Siblings,* 225.
27. The absence of the mother in the later version of this ballad is another sign of its eighteenth-century provenance, as those familiar with the literature of this period will recognize. Mothers begin to disappear from English novels by the second half of the eighteenth century. See Ruth Perry, *Novel Relations: The Transformation of Kinship in English Literature and Culture, 1748–1818* (Cambridge, UK: Cambridge University Press, 2004), especially chapter 8, "The Importance of Aunts."
28. Again, see Perry, *Novel Relations*, but there is plenty of other work on gender relations and sexuality in the eighteenth-century British Isles that could account for these particulars.

BIBLIOGRAPHY

A. B. *Learn to Lye Warm . . . Containing Reasons Wherefore a Young Man Should Marry an Old Woman*. London: H. Burgis, 1672.

"Account of English Books." *Gentleman's Magazine,* December 1754, 601.

Ackerknecht, Erwin H. *Therapeutics: From the Primitives to the Twentieth Century.* New York: Hafner Press, 1973.

An Act for Better Preventing the Horrid Crime of Murder (25 Geo. II c. 37), 1752.

Adam, Charles, and Paul Tannery, eds. *Oeuvres de Descartes.* Vol. 11. Paris: J. Vrin, 1975.

Addison, Joseph. *The Spectator*, May 21, 1711, 1–2.

Addison, Joseph. *The Spectator*, May 25, 1711, 1–2.

Addison, Joseph. *The Spectator*, June 7, 1711, 1–2.

Aikin, John. *Thoughts on Hospitals.* London: Joseph Johnson, 1771.

Albala, Ken. *Eating Right in the Renaissance.* Berkeley: University of California Press, 2002.

Alberti, Fay Bound. "Introduction: Medical History and Emotion Theory." In *Medicine, Emotion and Disease, 1700–1950,* edited by Fay Bound Alberti. Basingstoke, UK: Palgrave Macmillan, 2006.

Alderson, John. *An Essay on the Nature and Origin of the Contagion of Fevers.* Hull, UK: G. Prince et al., 1788.

Altick, Richard D. *The Shows of London: A Panoramic History of Exhibitions, 1600–1862.* Cambridge, MA: Belknap Press, 1978.

Andree, John. *Observations on the Theory and Cure of the Venereal Disease.* London: W. Davis, 1779.

Andrews, Jonathan, Asa Briggs, Roy Porter, Penny Tucker, and Keir Waddington. *The History of Bethlem.* London: Routledge, 1997.

Andry, Nicolas. *Orthopaedia: Or, the Art of Correcting and Preventing Deformities in Children.* 2 vols. London: A. Millar, 1743.

Annual Register for the Year 1761. London: J. Dodsley et al., 1762.

Annual Register for the Year 1772. London: J. Dodsley et al., 1773.

Anonymous. *The Art of Preserving Beauty.* London: 1789.

Anonymous. *A Dissertation on Mr Hogarth's Six Prints*. London: Dickinson, 1751.

Anonymous. *The Harlot's Progress: Founded upon Mr. Hogarth's Six Paintings*. Dublin: Exshaw, ca. 1732.

Anonymous. "Machine surprenante de l'homme artificiel de Monsieur Reyselius." *Journal des Sçavans* 23 (1677): 252.

Anonymous. "Report of an Inquest on an 'Unfortunate woman,' Mary Ann Chapman, Drowned in the London Dock." *True Sun*, January 12, 1832.

An Apology for the Life of Bampfylde-Moore Carew. London: W. Owen, 1749.

Arbuthnot, John. *Essay Concerning the Effects of Air on Human Bodies*. London: J. Tonson, 1733.

Ariès, Philippe. *The Hour of Our Death*, translated by Helen Weaver. New York: Random House, 1981.

Aristoteles Master-Piece, or the Secrets of Generation Displayed in All the Parts Thereof. London: J. How, 1684.

Aristotle. *Aristotle's Physics*. Books 1 and 2. Translated with introduction and notes by W. Charlton. Oxford: Clarendon Press, 1983.

Arrizabalaga, Jon, John Henderson, and Roger French. *The Great Pox: The French Disease in Renaissance Europe*. New Haven, CT: Yale University Press, 1997.

The Art of Preserving Beauty: Containing Instructions to Adorn and Embellish the Ladies. London: T. Axtell, 1789.

Astruc, Jean. *A Treatise on the Diseases of Women*. 2 vols. London: J. Nourse, 1762.

Astruc, Jean. *A Treatise of Venereal Diseases, in Nine Books; Containing an Account of the Origin, Propagation, and Contagion of This Distemper. As Also of the Nature, Cause, and Cure of All Venereal Disorders, Whether Local or Universal. Together with a Short Abstract of the Lives of the Authors Who Have Wrote on These Diseases and a List of Their Works*, Vol 2. Translated by W. Barrowby. London: Printed for W. Innys and J. Richardson, C. Davis, J. Clarke, R. Manby, and H. S. Cox, 1754.

The Athenian Mercury 1, no. 18 (May 23, 1691).

The Athenian Mercury 2, no. 5 (July 26, 1692).

Atkin, George A., ed. *The Tatler*. Vol. 2. London: Duckworth, 1899.

Atkins, John. *The Navy Surgeon, or, Practical System of Surgery*. London: J. Hodges, 1742.

Atkinson, David. *The English Traditional Ballad*. Burlington, VT: Ashgate, 2002.

Badmington, Neil. *Posthumanism*. Basingstoke, UK: Palgrave, 2000.

Baillie, Matthew. *The Morbid Anatomy of Some of the Most Important Parts of the Human Body*. London: Johnson, 1793.

Baker, Robin. *Sperm Wars: The Science of Sex*. New York: Basic Books, 1996.

Ballaster, Ros. "The Body as Fiction: Eighteenth-Century Tales of the Orient." In *The Eighteenth-Century Body: Art, History, Literature, Medicine,* edited by Angelica Goodden. Oxford: Peter Lang, 2002.

Banerjee, Pompa. *Burning Women: Widows, Witches, and Early Modern European Travelers in India*. New York: Palgrave Macmillan, 2003.

Bate, John. *The Art of Beauty: a Poem Humbly Addressed to the Oxford Toasts*. London: R. Francklin, 1719.

Beaglehole, J. C., ed. "Journal of William Wales." In *The Journals of Captain James Cook on His Voyages of Discovery*. Vol. 2. Cambridge: Cambridge University Press, 1961.

Beaumont, Sir Harry [Joseph Spence]. *Crito: Or, A Dialogue on Beauty*. London: R. Dodsley, 1752.

Beauty: Or the Art of Charming. London: Lawton Gulliver, 1735.

Beddoes, Thomas. *A Collection of Testimonies Respecting the Treatment of the Venereal Disease by Nitrous Acid*. London: J. Johnson, 1799.

Behn, Aphra. *Love Letters between a Nobleman and His Sister*, edited by Janet Todd. London: Penguin, [1684–1687] 1996.

Bell, Benjamin. *A Treatise on Gonorrhœa Virulenta and Lues Venera*. Vol. 2. Edinburgh: James Watson & Co.; G. Mudie & J. Murray, 1793.

Bell, John. *Anatomy of the Bones, Muscles and Joints. With Plates*. London: Longman, [1797] 1810.

Bell, John. *An Inquiry into the Causes Which Produce, and the Means of Preventing Diseases among British Officers, Soldiers, and Others in the West Indies*. London: J. Murray, 1791.

Bentham, Jeremy. *Auto-Icon, or, Farther Uses of the Dead to the Living*. London: Printed from the original manuscript for an unknown friend of the author, ca. 1842.

Bentham, Jeremy. "Offenses against One's Self." *Journal of Homosexuality* 3 ([ca. 1785] 1978): 389–405; 4 ([ca. 1785] 1979.): 91–109.

Bentham, Jeremy. "On the Use of the Dead to the Living." Reprinted in *Bentham's Auto-Icon and Related Writings*, James E. Crimmins. Bristol, UK: Thoemmes Press, [1824] 2002.

Beuchot, M., ed.. *Oeuvres de Voltaire*. Vol. 55. Paris: Lefèvre; Firmin Didot frères, 1831.

Bewell, Alan. *Romanticism and Colonial Disease*. Baltimore, MD: Johns Hopkins University Press, 1999.

Black, Jeremy, and Roy Porter. *A Dictionary of Eighteenth-Century World History*. Oxford: Blackwell, 1994.

Black, William. *An Arithmetical and Medical Analysis of the Diseases and Mortality of the Human Species*. London, 1789.

Blair, William. *Essays on the Venereal Disease and Its Concomitant Affections, Illustrated by a Variety of Cases*. London: J. Johnson, 1798.

Bloch, Marc. *The Royal Touch: Sacred Monarchy and Scrofula in England and France*, translated by J. E. Anderson. London: Routledge & Kegan Paul, 1973.

Blunt, John [Samuel William Fores]. *Man-Midwifery Dissected; or, the Obstetric Family-Instructor . . . in Fourteen Letters. Addressed to A. Hamilton . . . Occasioned by Certain Doctrines Contained in His Letters to Dr. W. Osborn*. London: Published for the author, 1793.

Boerhaave, Hermann. *Boerhaave's Aphorisms: Concerning the Knowledge and Cure of Diseases,* translated by J. Delacoste. London: B. Cowse & W. Innys, 1715.

Boerhaave, Hermann. *A New Method of Chemistry*. London, 1727.

Boerhaave, Hermann. *A Treatise on the Venereal Disease*, translated by John Barker. London: T. Cox et al., 1729.

Boesnier, Pierre de La Touche. *Préservatif contre l'irréligion, ou démonstrations des véritez fondamentales de la religion chrétienne*. The Hague: 1707.

Bond, Donald F., ed. *The Spectator*. Vol. 1. Oxford: Oxford University Press, 1965.

Bond, Donald F., ed. *The Tatler*. Vol. 1. Oxford: Oxford University Press, 1987.

Bonet, Theophile. 1679–1700. *Sepulchretum*. 3 vols. Geneva: Cramer.

Bonnet, Charles. *Considérations sur les corps organises*. Amsterdam: M. M. Rey, [1762] 1768.

Bonnet, Charles. *Contemplation de la nature*. In *Collection complete des oeuvres*. Vol. 4. Neuchatel, France: S. Fauche, [1764] 1781.

Bonnet, Charles. *Essai de psychologie*. In *Collection complete des oeuvres*. Vol. 8. Neuchatel, France: S. Fauche, [1754] 1783.

Bonnet, Charles. *La palingénésie philosophique*. Part 9. Geneva: Philibert et Chirol, 1769.

Bonnet, Charles. *Principes philosophiques sur la cause première et sur son effet,* in *Essai de psychologie . . . Auxquelles on a ajouté des Principles philosophiques*. London: 1755.

Borsay, Anne. *Disability and Social Policy in Britain since 1750: A History of Exclusion*. Basingstoke, UK: Palgrave Macmillan, 2004.

Borsay, Anne. "An Example of Political Arithmetic: The Evaluation of Spa Therapy at the Georgian Bath Infirmary, 1742–1830." *Medical History* 45 (2000): 149–72.

Borsay, Anne. "Returning Patients to the Community: Disability, Medicine and Economic Rationality before The Industrial Revolution." *Disability and Society* 13 (1998): 645–63.

Bossuet, Jacques Bénigne. *De la connoissance de Dieu et de soi-même*. In *Oeuvres completes*. Vol. 23. Paris: L. Vives, [1704] 1864.

Botelho, Lynn. "Old Age and Menopause in Rural Women of Early Modern Suffolk." In *Women and Ageing in British Society Since 1500,* edited by Lynn Botelho and Pat Thane. Harlow, UK: Pearson, 2001.

Boucé, Paul-Gabriel. "Imagination, Pregnant Women and Monsters in Eighteenth-Century England and France." In *Sexual Underworlds of the Enlightenment,* edited by George Rousseau and Roy Porter. Manchester, UK: Manchester University Press, 1987.

Boulton, Jeremy. "Going on the Parish: The Parish Pension and Its Meaning in the London Suburbs, 1640–1724." In *Chronicling Poverty: The Voices and Strategies of the English Poor, 1640–1840,* edited by Tim Hitchcock, Peter King, and Pamela Sharpe. Basingstoke, UK: Macmillan, 1997.

Brewer, John, and Roy Porter eds. *Consumption and the World of Goods*. London: Routledge, 1993.

Brewster, G. "The Incest Theme in Folksong." *F F Communication* 80 (1972): 3–36.

Brisbane, John. *Select Cases in the Practice of Medicine*. London: G. Scott for T. Cadell, 1772.

Broca, Paul. *Sur le Transformisme*. Paris: Reinwald, 1871.

Bröer, Ralf. *Salomon Reisel: (1625–1701); barocke Naturforschung eines Leibarztes im Banne der mechanistischen Philosophie*. Leipzig, Germany: Barth, 1996.

Brown, Mary Ellen. *William Motherwell's Cultural Politics*. Lexington: University of Kentucky Press, 2001.

Brown, Samuel. *An Inaugural Dissertation on the Bilious Malignant Fever: Read at a Public Examination, Held by the Medical Professors, before the Rev. Joseph Willard, S.T.D. President, and the Governors in the University at Cambridge, for the Degree of Bachelor in Medicine, July 10, 1797*. Boston: Manning & Loring, 1797.

Brown, Theodore M. *The Mechanical Philosophy and the "Animal Oeconomy": A Study of the Development of English Physiology during the Seventeenth and Eighteenth Centuries*. New York: Arno Press, 1981.

Brown, Theodore M. "Medicine in the Shadow of the *Principia*." *Journal of the History of Ideas* 48 (1987): 629–48.

Brown, Tom. *Amusements Serious and Comical, Calculated to the Meridian of London*. Paris: Les Editions D'Aujourd'hui, 1700.

Browne, Joseph. *The London Belles: Or, a Description of the most Celebrated Beauties in the City of London*. London: Printed for the author, 1707.

Browne, Thomas. *Hydriotaphia, Urn Burial.* London: 1669.

Brownlow Street Lying-in Hospital, First Minute Book, 1750–1756 (MS H10/CLM/A1/1/1). London Metropolitan Archives, London.

Bryson, Anna. *From Courtesy to Civility: Changing Codes of Conduct in Early Modern England.* Oxford: Clarendon Press, 1998.

Buchan, Peter. *Ancient Ballads and Songs of North of Scotland.* 2 vols. Edinburgh: W. & D. Laing and J. Stevenson, 1828.

Buchan, William. *Domestic Medicine; or, the Family Physician; Being an Attempt to Render the Medical Art More Generally Useful, by Shewing People What Is in Their Own Power Both with Respect to the Prevention and Cure of Diseases. Chiefly Calculated to Recommend a Proper Attention to Regimen and Simple Medicines.* Edinburgh: Balfour, Auld and Smellie, 1769.

Buchan, William. *Observations Concerning the Prevention and Cure of the Venereal Disease. Intended to Guard the Ignorant and Unwary against the Baneful Effects of That Insidious Malady.* London: T. Chapman et al., 1796.

Buffon, Georges. *Histoire naturelle, générale et particulière.* Vols. 1 & 2. Paris: Imprimerie royale, 1749.

B[ulwer], J[ohn]. *Anthropometamorphosis: Man Transform'd: Or, the Artificiall Changling.* London: William Hunt, 1653.

Bunworth, Richard. *A New Discovery of the French Disease, and Running of the Reins: Their Signs, with Plain and Easy Directions of Perfect Curing the Same.* 2nd ed. London: Henry Marsh, 1662.

Burdon, Henry. *The Fountain of Health.* London: The author, 1734.

Burn, James Dawson. *The Autobiography of a Beggar Boy*, edited by David Vincent. London: Europa, [1855] 1978.

Burnard, Trevor, and Kenneth Morgan. "The Dynamics of the Slave Market and Slave Purchasing Patterns in Jamaica, 1655–1788." *Morgan William and Mary Quarterly* 58 (2001): 205–28.

Bussey, Gertrude Carman, ed. *La Mettrie, Man a Machine.* La Salle, IL: Open Court, 1912.

Bynum, William F. *Science and the Practice of Medicine in the Nineteenth Century.* Cambridge: Cambridge University Press, 1994.

Bynum, William F. "Treating the Wages of Sin: Venereal Disease and Specialism in Eighteenth Century Britain." In *Medical Fringe and Medical Orthodoxy, 1750–1850,* edited by Porter, Roy. London: Croom Helm, 1987.

Bynum, William F., and Roy Porter, eds. *William Hunter and the Eighteenth-Century Medical World.* Cambridge: Cambridge University Press, 1985.

Camporesi, Piero. *Juice of Life: The Symbolic and Magic Significance of Blood*, translated by Robert R. Barr. New York: Continuum, 1995.

Camus, Antoine Le. *Abdeker: Or, the Art of Preserving Beauty. Translated from an Arabic Manuscript.* London: A. Millar, 1754.

Cantor, David, ed. *Reinventing Hippocrates.* Aldershot, UK: Ashgate, 2002.

Carlyle, Thomas. *History of Friedrich II of Prussia, Called Frederick the Great.* Vol. 4, book 16: *The Ten Years of Peace. 1746–1756.* London: Chapman and Hall, 1864.

Carmichael, Ann. *Plague and the Poor in Renaissance Florence.* Cambridge: Cambridge University Press, 1986.

The Case of the Inhabitants of the Parish of St Giles in the Fields, as to Their Poor, and a Work-House Designed to Be Built for Employing Them. London: 1730.

Caulfield, James, and R.S. Kirby. *Collection of Four Hundred Portraits of Remarkable, Eccentric and Notorious Personages Printed from the Original Copper Plates of Caulfield's Remarkable Characters, Grainger, and Kirby's Wonderful Museum.* London: Reeves and Turner, 1880.

Chamberlain, Mary, and Ruth Richardson. "Life and Death." *Oral History Journal* 11 (1983): 31–43.

Chaplin, S. "John Hunter and the Anatomy of a Museum." *History Today* 55 (2005): 19–25.

Charlton, Rice. *Cases of Patients Admitted into the Hospital at Bath.* Bath, UK: 1776.

Cheselden, William. *Osteographia.* London: Printed for the author, 1733.

Cheyne, George. *Rules and Observations for the Enjoyment of Health and Long Life.* Leeds, UK: G. Wright & Sons, 1770.

Child, Francis James. *The English and Scottish Popular Ballads.* Vol. 1. Northfield, MN: Loomis House Press, [1882–1898] 2001.

Childs, Fenela. "Prescriptions for Manners in English Courtesy Literature 1690–1760, and their Social Implications." DPhil thesis, Oxford University, 1984.

Chisholm, Colin. *An Essay on the Malignant Pestilential Fever Introduced into the West Indian Islands from Boullam on the Coast of Guinea, as It Appeared in 1793 and 1794.* London: C. Dilly, 1795.

Cipolla, Carlo M. *Miasmas and Diseases: Public Health and the Environment in the Pre-Industrial Age*, translated by Elizabeth Porter. New Haven, CT: Yale University Press, 1992.

Clark, John. *Observations on the Diseases in Long Voyages to Hot Countries, and Particularly on Those Which Prevail in the East Indies.* London: D. Wilson & G. Nicol, 1773.

Cleland, John. *Fanny Hill or Memoirs of a Woman of Pleasure.* Harmondsworth, UK: Penguin, [1748–1749] 1998.

Cleland, John. *Institutes of Health.* London: T. Becket & T. Davies, 1761.

Clericuzio, Antonio. *Elements, Principles and Corpuscles: A Study of Atomism and Chemistry in the Seventeenth Century.* New York: Springer, 1999.

Cody, Lisa Forman. *Birthing the Nation: Sex, Science and the Conception of Eighteenth-Century Britons.* Oxford: Oxford University Press, 2005.

Cody, Lisa. "'The Doctor's in Labour: Or a New Whim Wham from Guildford.'" *Gender and History* 4 (1992): 175–96.

Collier, John. "The Pluralist and Old Soldier." In *The New Oxford Book of Eighteenth-Century Verse,* edited by Roger Lonsdale. Oxford: Oxford University Press, [1763] 1984.

Condillac, Etienne Bonnot de. *Essai sur l'origine des connaissances humaines.* Amsterdam: P. Mortier, 1746.

Cook, Harold. *The Decline of the Old Medical Regime in Stuart London.* Ithaca, NY: Cornell University Press, 1986.

Cook, Harold. *Trials of an Ordinary Doctor: Joannes Groenevelt in Seventeenth-Century London.* Baltimore, MD: Johns Hopkins University Press, 1994.

Corea, Gena. *Man-Made Women: How New Reproductive Technologies Affect Women.* London: Hutchinson, 1985.

Corea, Gena. *The Mother Machine: Reproductive Technologies from Artificial Insemination to Artificial Wombs.* London: Women's Press, 1987.

Coroner's Inquest: John Webb, July 6, 1762. The Library and Muniment Room, Westminster Abbey, London.

Corsi, Pietro. "Before Darwin: Transformist Concepts in European Natural History." *Journal of the History of Biology* 38 (2005): 67–83.

Court of Governors, Bridewell and Bethlam, Minutes: August 23, 1695 to May 2, 1701 (MS 33011/17), Ann Wheatley: 249. Guildhall Library, London.

Court of Governors, Bridewell and Bethlam, Minutes: October 16, 1722 to December 15, 1737 (MS 3011.20), Ann Fretwell alias Davis: 81. Guildhall Library, London.

Court of Governors, Bridewell and Bethlam, Minutes: January 12, 1737–1738 to April 4, 1751 (MS 33011/21), Isaac Fairbrother: 143, Henry Goodwin: 143, William Harding and Ann Middleton: 321. Guildhall Library, London.

Court of Governors, Bridewell and Bethlam, Minutes: May 22, 1751 to December 1, 1761 (MS 3011/22), Julian Harrington: 349. Guildhall Library, London.

Court of the President and Governors for the Poor of London, 1702–1705 (MS New 377C/1/22): Thomas Newby, fol. 200. Corporation of London Records Office, London. London Metropolitan Archives.

Crawford, Patricia. "Attitudes to Menstruation in Seventeenth-Century England." *Past and Present* 91 (1981): 47–73.

Crawford, Patricia. "Printed Medical Advertisements for Women Medical Practitioners in London, 1670–1710." *Bulletin of the Society for the Social History of Medicine* 35 (1994): 66–70.

Cressy, David. *Birth, Marriage, and Death: Ritual, Religion, and the Life-Cycle in Tudor and Stuart England.* Oxford: Oxford University Press, 1997.

Crimmins, James E., ed. *Bentham's Auto-Icon and Related Writings.* Bristol, UK: Thoemmes, 2002.

Critchley, Simon. *On Humour: Thinking in Action.* London: Routledge, 2002.

Cudworth, Ralph. *The True Intellectual System of the Universe: The First Part; wherein All the Reason and Philosophy of Atheism Is Confuted. And its Impossibility Demonstrated.* Vol. 1. London: 1743.

Cunningham, Andrew. "Auto-Icon: Jeremy Bentham's Three Bodies, the Moral Laws of Nature, and the Ideology of Industrial Capitalism." In *The Faces of Nature in Enlightenment Europe,* edited by Lorraine Daston and Gianna Pomata. Berlin: BWV, 2003.

Dacome, Lucia. "Useless and Pernicious Matter: Corpulence in Eighteenth-Century England." In *Cultures of the Abdomen: Diet, Digestion and Fat in the Modern World,* edited by Christopher E. Forth and Ana Carden-Coyne. Basingstoke, UK: Palgrave Macmillan, 2005.

D'Agoty, Jacques Gautier. *Exposition anatomique des maux vénériens, sur les parties de l'homme et de femme, et les remèdes les plus usités dans ces sortes de maladies.* Paris: J. B. Brunet & Demonville, 1773.

Darmon, Pierre. *Damning the Innocent: A History of the Persecution of the Impotent in Pre-Revolutionary France.* New York: Viking, 1986.

Darmon, Pierre. *Trial by Impotence: Virility and Marriage in Pre-Revolutionary France.* London: Chatto & Windus, 1985.

Darwin, Erasmus. "The Loves of Plants." Part II of *The Botanic Garden, A Poem in Two Parts.* Lichfield, UK: 1789.

Darwin, Erasmus. *Zoonomia; or, the Laws of Organic Life.* Dublin: 1794–1796.

Daston, Lorraine, and Katherine Park. *Wonders and the Order of Nature, 1150–1750.* New York: Zone Books, 1998.

Davenport, Randi. "Thomas Malthus and Maternal Bodies Politic: Gender, Race, and Empire." *Women's History Review* 4 (1995): 415–40.

Davidoff, Leonore, and Catherine Hall. *Family Fortunes: Men and Women of the English Middle Class, 1780–1850*. Chicago: University of Chicago Press, 1987.

Debus, Allen George. *The French Paracelsians: The Chemical Challenge to Medical and Scientific Tradition in Early Modern France*. Cambridge: Cambridge University Press, 2002.

Defoe, Daniel. *Journal of the Plague Year*. London: Nutt, 1722.

Deutsch, Helen. "The Body's Moment: Visible Disability, the Essay and the Limits of Sympathy." *Prose Studies* 27 (2005): 11–26.

De Valangin, Francis. *A Treatise on Diet, or the Management of Human Life; by Physicians Called the Six Non-naturals*. London: The author, 1768.

Dickens, Charles. *Martin Chuzzlewit*. Harmondsworth, UK: Penguin, [1844] 1999.

Dickie, Simon. "Hilarity and Pitilessness in the Mid-Eighteenth Century: English Jestbook Humor." *Eighteenth-Century Studies* 37 (2003): 1–22.

Diderot, Denis. *Eléments de physiologie*. Paris: Didier, [1784] 1964.

Diderot, Denis. *Entretien entre d'Alembert et Diderot*. In *Oeuvres Philosophiques*, edited by P. Verniere. Paris: Garnier, [1769] 1961.

Diderot, Denis. *Le rêve de d'Alembert*. In *Oeuvres Philosophiques*, edited by P. Verniere. Paris: Garnier, [1769] 1961.

Diderot, Denis. *Suite de l'entretien*. In *Oeuvres Philosophiques*, edited by P. Verniere. Paris: Garnier, [1784] 1961.

Diderot, Denis, and Jean d'Alembert, eds. "Syphon." In *Encyclopédie, ou dictionnaire raisonné des sciences, des arts et des métiers*. Paris: 1765.

Dixon, Thomas. "Patients and Passions: Languages of Medicine and Emotion, 1789–1850." In *Medicine, Emotion and Disease, 1700–1950,* edited by Fay Bound Alberti. Basingstoke, UK: Palgrave Macmillan, 2006.

Djerassi, Carl. *An Immaculate Misconception: Sex in an Age of Mechanical Reproduction*. London: Imperial College Press, 2000.

Dobson, Jessie. *John Hunter*. Edinburgh: E. & S. Livingstone, 1969.

Dodsley, Robert. *The Blind Beggar of Bethnal Green*. London: 1741.

Doig, Andrew, J. P. S. Ferguson, I. A. Milne, and R. Passmore, eds. *William Cullen and the Eighteenth-Century Medical World: A Bicentenary Exhibition and Symposium Arranged by the Royal College of Physicians of Edinburgh*. Edinburgh: Edinburgh University Press, 1990.

Donne, John. *Devotions upon Emergent Occasions*, edited with commentary by Anthony Raspa. Montreal, Canada: McGill-Queen's University Press, 1975.

Douglas, Mary. "The Construction of the Physician: A Cultural Approach to Medical Fashions." In *The Healing Bond. The Patient-Practitioner Relationship and Therapeutic Responsibility,* edited by Susan Budd and Ursula Sharma. London: Routledge, 1994.

Doyon, André, and Lucien Liaigre. *Jacques Vaucanson, mécanicien de genie*. Paris: PUF, 1966.

Duchesneau, François. *La Physiologie des Lumières: empirisme, modèles, theories*. The Hague, The Netherlands: Nijhoff, 1982.

Duden, Barbara. *The Woman Beneath the Skin: A Doctor's Patients in Eighteenth-Century Germany*. Cambridge, MA: Harvard University Press, 1991.

Elias, Norbert. *The Civilizing Process*. Vol. 1, "The History of Manners." New York: Urizen Books, 1978.

Emch-Dériaz, Antoinette. "The Non-Naturals Made Easy." In *The Popularization of Medicine, 1650–1850,* edited by Roy Porter. London: Routledge, 1992.

Endelman, Todd. *The Jews of Georgian England, 1714–1830: Tradition and Change in a Liberal Society*. 2nd ed. Ann Arbor: University of Michigan Press, 1999.

Evenden, Doreen. *The Midwives of Seventeenth-Century London*. Cambridge: Cambridge University Press, 2000.

Findlen, Paula. "Jokes of Nature and Jokes of Knowledge: The Playfulness of Scientific Discourse in Early Modern Europe." *Renaissance Quarterly* 43 (1990): 292–329.

Finucci, Valeria, and Kevin Brownlee. *Generation and Degeneration: Tropes of Reproduction in Literature and History from Antiquity through Early Modern Europe*. Durham, NC: Duke University Press, 2001.

Fissell, Mary. "The Disappearance of the Patient's Narrative and the Invention of Hospital Medicine." In *British Medicine in an Age of Reform*, edited by Roger French and Andrew Wear. London: Routledge, 1991.

Fissell, Mary. "Gender and Generation: Representing Reproduction in Early Modern England." *Gender & History*, 7 (1995): 433–56.

Fissell, Mary *Vernacular Bodies: The Politics of Reproduction in Early Modern England*. Oxford: Oxford University Press, 2004.

Flegel, Kenneth M. "Changing Concepts of the Nosology of Gonorrhea and Syphilis." *Bulletin of the History of Medicine* 48 (1974): 571–88.

Fletcher, Anthony. *Gender, Sex, and Subordination in England, 1500–1800*. New Haven: Yale University Press, 1996.

Foot, Jesse. *The Life of John Hunter*. London: T. Becket, 1794.

Forster, Johann Reinhold. *The Resolution Journal of Johann Reinhold Forster 1772–1775*, edited by Michael E. Hoare. Vol. 3. London: Hakluyt Society, [1772–1775] 1982.

Foucault, Michel. *The Birth of the Clinic: An Archaeology of Medical Perception*, translated by Alan Sheridan. New York: Vintage, 1973.

Foucault, Michel. *Discipline and Punish: The Birth of the Prison*. New York: Pantheon, 1977.

Foucault, Michel. *The History of Sexuality: Volume I: An Introduction*. New York: Vintage, 1977.

Foucault, Michel. *The Order of Things: An Archaeology of the Human Sciences*. New York: Vintage, 1970.

Fox, Celina. *Londoners*. London: Thames & Hudson, 1987.

Frank, Judith. *Common Ground: Eighteenth-Century Satiric Fiction and the Poor*. Stanford, CA: Stanford University Press, 1997.

French, Roger. "Sickness and the Soul: Stahl, Hoffman and Sauvages on Pathology." In *The Medical Enlightenment of the Eighteenth Century*, edited by Andrew Cunningham and Roger French. Cambridge: Cambridge University Press, 1990.

Freud, Sigmund. *Jokes and Their Relation to the Unconscious*. London: Routledge & Kegan Paul, 1960.

Friedman, Albert B. *The Ballad Revival*. Chicago: University of Chicago Press, 1961.

Fulford, Tim. *Landscape, Liberty, and Authority: Poetry, Criticism, and Politics from Thomson to Wordsworth*. Cambridge: Cambridge University Press, 1996.

Fuller, Isaac. *Iconologia: Or, Moral Emblems*. London. B. Motte, 1709.

Furdell, Elizabeth Lane. "Harris, Walter (1647–1732)." In *Oxford Dictionary of National Biography*. Oxford: Oxford University Press, 2004.

Gallagher, Catherine, and Thomas Laqueur. *The Making of the Modern Body: Sexuality and Society in the Nineteenth Century*. Berkeley: University of California Press, 1987.

Galloway, George. "To the Memory of Gavin Wilson (Boot, Leg and Arm Maker)." In *The New Oxford Book of Eighteenth-Century Verse,* edited by Roger Lonsdale. Oxford: Oxford University Press, [1795] 1984.

Garnett, Thomas. *A Lecture on the Preservation of Health.* London: T. Cadell jnr & W. Davis, 1800.

Gatrell, Vic. *The Hanging Tree.* Oxford: Oxford University Press, 1994.

Gauden, John. *A Discourse of Auxiliary Beauty or Artificiall Hansomeness. In Point of Conscience between Two Ladies.* London: R. Royston, 1656.

Gelbart, Nina. *The King's Midwife: A History and Mystery of Madame du Coudray.* Berkeley: University of California Press, 1998.

Gélis, Jacques. *History of Childbirth: Fertility, Pregnancy, and Birth in Early Modern Europe*, translated by Rosemary Morris. Boston: Northeastern University Press, 1991.

Gentleman's Magazine, September 1748, 413–14.

Gentleman's Magazine, October 1748, 448–51.

Gerzina, Gretchen. *Black England: Life before Emancipation.* London: John Murray, 1995.

Gilman, Sander L. *Health and Illness: Images of Difference.* London: Reaktion, 1995.

Gilman, Sander L. *The Jew's Body.* New York: Cornell University Press, 1991.

Gilman, Sander L. *Making the Body Beautiful: A Cultural History of Aesthetic Surgery.* Princeton & Oxford: Princeton University Press, 2001.

Gilman, Sander L. *Sexuality: An Illustrated History.* New York: Wiley, 1989.

Ginzberg, Ruth. "Uncovering Gynocentric Science." In *Feminism and Science,* edited by Nancy Tuana. Bloomington: Indiana University Press, 1989.

Gittings, Clare. "Eccentric or Enlightened? Unusual Burial and Commemoration in England, 1689–1823." *Mortality* 12 (2007): 322–49.

Glass, Bentley, ed. *The Forerunners of Darwin.* Baltimore, MD: Johns Hopkins University Press, 1968.

Goldsmith, Oliver. "Happiness." *The Bee,* October 13, 1759.

Good, John Mason. *Dissertation on the Diseases of Prisons and Poor Houses.* London: Medical Society of London / C. Dilly, 1795.

Gould, Stephen Jay. *The Mismeasure of Man.* Revised edition. New York: Norton, 1996.

Gowing, Laura. *Common Bodies: Women, Touch and Power in Seventeenth Century England.* New Haven, CT: Yale University Press, 2003.

Graham, James. *A Lecture on the Generation, Increase, and Improvement of the Human Species.* London, 1782.

Graunt, John. *Natural and Political Observations Mentioned in a Following Index, and Made upon the Bills of Mortality.* London: Royal Society, 1676.

Graunt, John. *Natural and Political Observations, Mentioned in a Following Index, and Made upon the Bills of Mortality.* 2nd ed. London: John Martin et al., 1662.

Greenblatt, Stephen. "Mutilation and Meaning." In *The Body in Parts,* David Hillman and Carla Mazzio. New York: Routledge, 1997.

Greenblatt, Stephen, ed. *The Norton Anthology of English Literature.* Vol. 1. New York: W. W. Norton, 2006.

Greenhill, Thomas. *Nekrokedeia: Or, the Art of Embalming.* London: The author, 1705.

Grew, Nehemiah. *Cosmologia Sacra: Or a Discourse of the Universe as It is the Creature and Kingdom of God.* London, 1701.

Grose, Francis. *The Grumbler: Containing Sixteen Essays by the Late Francis Grose, Esq. FAS of London and Perth.* London: 1791.

A Guide to Health, Beauty, Riches and Honour. London: S. Hooper, 1785.

Gunn, Fenja. *The Artificial Face: A History of Cosmetics*. Newton Abbot, UK: David and Charles, 1973.

Haidt, Rebecca. *Embodying Enlightenment: Knowing the Body in Eighteenth-Century Spanish Literature and Culture*. New York: St. Martin's Press, 1998.

Halberstam, Judith, and Ira Livingston, eds. *Posthuman Bodies, Unnatural Acts*. Bloomington: Indiana University Press, 1995.

Hales, Charles. *A Letter Addressed to Caesar Hawkins: Containing New Thoughts and Observations, in the Cure of the Venereal Disease . . . With a Few Extraordinary Cases in That Disease*. London: 1750.

Hall, Thomas. *The Loathsomenesse of Long Haire . . . With the Concurrent Judgement of Divines Both Old and New against It. With an Appendix against Painting, Spots, Naked Breasts etc.* London: J. G., 1654.

Hallett, Mark. *The Spectacle of Difference: Graphic Satire in the Age of Hogarth*. London: Yale University Press, 1999.

"Handsome Wife." *The New Spectator; with the Sage Opinions of John Bull*, June 29, 1784, 7–8.

Hankins, Thomas L., and Robert J. Silverman. *Instruments and the Imagination*. Princeton, NJ: Princeton University Press, 1995.

Harding, Vanessa. *The Dead and the Living in Paris and London, 1500–1670*. Cambridge: Cambridge University Press, 2002.

Harris, Jonathan Gil. *Foreign Bodies and the Body Politic: Discourses of Social Pathology in Early Modern England*. New York: Cambridge University Press, 1998.

Harris, Jonathan Gil. *Sick Economies: Drama, Mercantilism, and Disease in Shakespeare's England*. Philadelphia: University of Pennsylvania, 2004.

Harris, Walter. *A Treatise of the Acute Diseases of Infants. To Which Are Added, Medical Observations on Several Grievous Diseases*, translated (from the Latin) by John Martyn. London: T. Astley, 1742.

Harrison, Mark. *Climates and Constitutions: Health, Race, Environment and British Imperialism in India, 1600–1850*. New Delhi: Oxford University Press, 1999.

Harrison, Mark. " 'The Tender Frame of Man': Disease, Climate and Racial Difference in India and the West Indies, 1760–1860." *Bulletin of the History of Medicine* 70 (1996): 68–93.

Harvey, Karen. "The Substance of Sexual Difference: Change and Persistence in Representations of the Body in Eighteenth-Century England." *Gender and History* 14 (2000): 202–23.

Haskell, Thomas L. "Capitalism and the Origins of the Humanitarian Sensibility" (part 1). *American Historical Review* 90 (1985): 339–61.

Hay, Douglas, Peter Linebaugh, John G. Rule, E.P. Thompson, and Cal Winslow. *Albion's Fatal Tree*. London: Allen Lane, 1975.

Hay, William. *Deformity: An Essay*. London. R. and J. Dodsley, 1754.

Helsinger, Elizabeth. *Rural Scenes and National Representation: Britain, 1815–1850*. Princeton, NJ: Princeton University Press, 1997.

Hendy, James. *A Treatise on the Glandular Disease of Barbadoes, Proving It to Be Seated in the Lymphatic System*. London: Printed for C. Dilly, 1784.

Heyl, Christoph. "Deformity's Filthy Fingers: Cosmetics and the Plague in *Artificiall Embellishments, or Arts Best Directions How to Preserve Beauty or Procure It* (Oxford 1665)." In *Didactic Literature in England 1500–1800: Expertise Constructed*, edited by Natasha Glaisyer and Sara Pennell. Aldershot, UK: Ashgate, 2003.

Heysham, John. *Observations on the Bills of Mortality, in Carlisle, for the Year 1779.* Carlisle: J. Milliken, 1780.

Heysham, John. *An Account of the Jail Fever, or Typhus Carcerum.* London: T. Cadell et al., 1782.

Highfill, Philip H. Jr. Kalman A. Burnim, and Edward A. Longhans. *A Biographical Dictionary of Actors, Actresses, Musicians, Dancers, Managers and Other Stage Personnel in London, 1660–1800.* Vol. 10. Carbondale: Southern Illinois University Press, 1984.

Highmore, Nathaniel. *The History of Generation. Examining the Several Opinions of Divers Authors, Especially That of Sir Kenelm Digby, in His Discourse of Bodies . . . To Which Is Joyned a Discourse of the Cure of Wounds by Sympathy . . . Especially by . . . Sir Gilbert Talbots Powder.* London: Printed by R. N. for J. Martin, 1651.

Hill, John. *Essays in Natural History and Philosophy. Containing a Series of Discoveries, by the Assistance of Microscopes.* London: J. Whiston et al., 1752.

Hill, John, and George S. Rousseau. *The Letters and Papers of Sir John Hill, 1714–1775.* New York: AMS Press, 1982.

Hitchcock, Tim. *Down and Out in Eighteenth-Century London.* London: Hambledon, 2004.

Hitchcock, Tim. *English Sexualities, 1700–1800.* Basingstoke, UK: Macmillan, 1997.

Hitchcock, Tim. "'Unlawfully Begotten on her Body.' Illegitimacy and the Parish Poor in St Luke's, Chelsea." In *Chronicling Poverty: The Voices and Strategies of the English Poor, 1640–1840,* edited by Tim Hitchcock, Peter King, and Pamela Sharpe. Basingstoke, UK: Macmillan, 1997.

Hitchcock, Tim, and John Black. 1999. *Chelsea Settlement and Bastardy Examinations, 1733–1766.* Vol. 33. London: London Record Society, 1999.

Hogarth, William. *The Four Stages of Cruelty.* London: William Hogarth, 1751.

Hogarth, William. *The Harlot's Progress.* London: William Hogarth, 1732.

Holloway, John, ed. *The Euing Collection of English Broadside Ballads in the Library of the University of Glasgow.* Glasgow: University of Glasgow Press, 1971.

Home, Everard. "An Account of a Child with a Double Head. In a Letter from Everard Home to John Hunter." *Philosophical Transactions of the Royal Society of London* 80 (1790): 296–305.

Home, Everard. "An Account of the Dissection of an Hermaphrodite Dog. To Which Are Prefixed, Some Observations on Hermaphrodites in General." *Philosophical Transactions of the Royal Society of London* 89 (1799): 157–78.

Home, Everard. "Some Observations on the Mode of Generation of the Kangaroo, with a Particular Description of the Organs Themselves." *Philosophical Transactions of the Royal Society of London* 85 (1795): 221–38.

Houlbrooke, Ralph. *Death, Religion, and the Family in England, 1480–1750.* Oxford: Clarendon Press, 1998.

Houlbrooke, Ralph. *Death Ritual and Bereavement.* London: Routledge, 1989.

Hudson, Geoffrey L. "Disabled Veterans and the State in Early Modern England." In *Disabled Veterans in History,* edited by David A. Gerber. Ann Arbor: University of Michigan Press, 2000.

Hudson, Nicholas. "From 'Nation' to 'Race': The Origin of Racial Classification in 18th-Century Thought." *Eighteenth-Century Studies* 29 (1996): 247–64.

Huet, Marie-Hélène. *Monstrous Imagination.* Cambridge, MA: Harvard University Press, 1993.

Hunt, Lynn. *The Invention of Pornography: Obscenity and the Origins of Modernity, 1500–1800*. New York: Zone, 1993.
Hunt, Lynn. *Politics, Culture, and Class in the French Revolution*. Berkeley: University of California Press, 1984.
Hunter, J. Paul. *Before Novels: The Cultural Contexts of Eighteenth-Century English Fiction*. New York: Norton, 1990.
Hunter, John. *A Treatise on the Venereal Disease*. London: The author, 1786.
Hunter, John, and Everard Home eds. *The Works of John Hunter. With Notes*. London: Longman, 1835.
Hunter, William. *Anatomia uteri humani gravidi tabulis illustrata . . . The Anatomy of the Human Gravid Uterus Exhibited in Figures*. Birmingham, UK: J. Baskerville, & S. Baker & G. Leigh, etc., 1774.
Hurley, Kelly. "Postmodern Bodies: Reading Like an Alien." In *Posthuman Bodies, Unnatural Acts,* edited by Judith Halberstam and Ira Livingstone. Bloomington: Indiana University Press, 1995.
Huxham, John. *An Essay on Fevers*. 3rd ed. London: J. Hinton, 1757.
Ingram, Dale. *An Historical Account of the Several Plagues That Have Appeared in the World Since the Year 1346. With an Enquiry into the Present Prevailing Opinion, That the Plague Is a Contagious Distemper . . . To Which Are Added a Particular Account of the Yellow Fever, Etc*. London: R. Baldwin & J. Clark, 1755.
Irwin, David. "Sentiment and Antiquity: European Tombs 1750–1830." In *Mirrors of Mortality,* edited by Joachin Whaley. London: Europa, 1981.
Israel, Jonathan. *Enlightenment Contested: Philosophy, Modernity and the Emancipation of Man, 1670–1752*. New York: Oxford University Press, 2006.
Israel, Jonathan. *Radical Enlightenment: Philosophy and the Making of Modernity 1650–1750*. Oxford: Oxford University Press, 2001.
Jacob, François. *The Logic of Life: A History of Heredity*, translated by Betty E. Spillman. New York: Vintage, 1989.
James-Cavan, Kathleen, "'[A]ll in Me Is Nature': The values of deformity in William Hay's *Deformity: An Essay*." *Prose Studies* 27 (2005): 27–38.
Janaway, Robert. "An Introductory Guide to Textiles from 18th and 19th Century Burials." In *Grave Concerns,* edited by Margaret Cox. York, UK: Council for British Archaeology, 1998.
Jarcho, Saul, ed. and trans. *Clinical Consultations and Letters by Ippolito Francesco Albertini, Francesco Torti, and Other Physicians: University of Bologna MS 2089-1*. Boston: Francis A. Countway Library of Medicine, 1989.
Jarcho, Saul, ed. and trans. *The Clinical Consultations of Giambattista Morgagni*. Boston: Francis A. Countway Library of Medicine, 1984.
Jeamson, Thomas. *Artificiall Embellishments. Or Arts Best Directions How to Preserve Beauty or Procure It*. Oxford: William Hall, 1665.
Jenner, Mark. "Body, Image, Text in Early Modern Europe." *Social History of Medicine* 12 (1999): 143–54.
Jenner, Mark. "Civilization and Deodorization? Smell in Early Modern English Culture." In *Civil Histories: Essays Presented to Sir Keith Thomas,* edited by Peter Burke, Brian Harrison, and Paul Slack. Oxford: Oxford University Press, 2000.
Jewson, N.D. "The Disappearance of the Sick-Man from Medical Cosmology, 1770–1870." *Sociology* 10 (1976): 225–44.
J. H. *Essays of Love and Marriage*. London: H. Brome, 1673.

Johnson, Abraham [John Hill]. *Lucina sine concubitu. A Letter Humbly Address'd to the Royal Society; in Which Is Proved by Most Incontestible Evidence, Drawn from Reason and Practice, That a Woman May Conceive and Be Brought to Bed Without Any Commerce with Man*. London: Printed and sold by M. Cooper, 1750.

Johnson, Samuel. "Hermann Boerhaave." *The Gentleman's Magazine*, January–April 1739. Reprinted in *The Works of Samuel Johnson*. Vol. 14. Troy, NY: Pafraets Company, 1903.

Jones, Robert W. *Gender and the Formation of Taste in Eighteenth-Century Britain: The Analysis of Beauty*. Cambridge: Cambridge University Press, 1998.

Kerby-Miller, Charles, ed. *Memoirs of the Extraordinary Life, Works, and Discoveries of Martinus Scriblerus*. New Haven, CT: Yale University Press, 1950.

Kilpatrick, James. *An Essay on Inoculation*. London: J. Huggonson, 1743.

Kindersley, Jemima. *Letters from the Island of Teneriffe, Brazil, the Cape of Good Hope, and the East Indies*. London, 1777.

King, Lester S. *The Medical World of the Eighteenth Century*. Chicago: University of Chicago Press, 1958.

King, Lester S. *The Philosophy of Medicine: The Early Eighteenth Century*. Cambridge, MA: Harvard University Press, 1978.

King, Steven. *Poverty and Welfare in England 1700–1850: A Regional Perspective*. Manchester, UK: Manchester University Press, 2000.

King-Hele, Desmond, ed. *The Letters of Erasmus Darwin*. Cambridge: Cambridge University Press, 1981.

Kirkup, John. "Wiseman, Richard (bap. 1620?, d. 1676)." In *Oxford Dictionary of National Biography*. Oxford: Oxford University Press, 2004.

Klein, Herbert S., and Stanley L. Engerman. "Long-Term Trends in African Mortality in the Transatlantic Slave Trade." *Slavery and Abolition* 18 (1997): 36–48.

Klepp, Susan. "Seasoning and Society: Racial Differences in Mortality in Eighteenth-Century Philadelphia." *William and Mary Quarterly* 51 (1994): 473–506.

Knoeff, Rina. *Hermann Boerhaave (1668–1738): Calvinist Chemist and Physician*. Amsterdam: Royal Netherlands Academy of Arts and Sciences, 2002.

Lacan, Jacques. *The Seminar of Jacques Lacan*. Edited by Jacques-Alain Miller, translated with notes by John Forrester and Sylvana Tomaselli. 2 vols. Cambridge: Cambridge University Press, 1988.

The Ladies Mercury, February 28, 1693, question 1.

Lamarck, Jean-Baptiste. *Philosophie zoologique, ou, Exposition des considérations relative à l'histoire naturelle des animaux*. Paris: Dentu, 1809.

La Mettrie, Julien Offray de. *L'Homme-machine*. In *La Mettrie's L'Homme-machine, A Study in the Origins of an Idea,* edited by Aram Vartanian. Princeton, NJ: Princeton University Press, 1960.

La Mettrie, Julien Offray de. *L'Homme-Plante* [1748]. In *Oeuvres philosophiques*. Vol. 1. Paris: Fayard, 1984.

La Mettrie, Julien Offray de. *Système d'Epicure* [1750]. In *Oeuvres philosophiques*. Vol. 1. Paris: Fayard, 1984.

Lanban, Celeste. *Romantic Vagrancy: Wordsworth and the Simulation of Freedom*. Cambridge: Cambridge University Press, 1995.

Laqueur, Thomas. "Bodies, Details, and the Humanitarian Narrative." In *The New Cultural History: Essays,* edited by Lynn Hunt. Berkeley: University of California Press, 1989.

Laqueur, Thomas. *Making Sex: Body and Gender from the Greeks to Freud.* Cambridge, MA: Harvard University Press, 1990.

Laqueur, Thomas. "Orgasm, Generation, and the Politics of Reproductive Biology." In *The Making of the Modern Body: Sexuality and Society in the Nineteenth Century,* edited by Catherine Gallagher and Thomas Laqueur. Berkeley: University of California Press, 1987.

Laqueur, Thomas. "Sex in the Flesh." *Isis* 94 (2003): 300–306.

Lawlor, Clark, and Akhihito Suzuki. "The Diseases of the Self: Representing Consumption, 1700–1830." *Bulletin of the History of Medicine* 74 (2000): 458–94.

Lawrence, Christopher. "Disciplining Disease: Scurvy, the Navy and Imperial Expansion, 1750–1825." In *Visions of Empire: Voyages, Botany, and Representations of Nature,* edited by David Philip Miller and Peter Hanns Reill. Cambridge: Cambridge University Press, 1996.

Lawrence, Christopher. "The Nervous System and Society in the Scottish Enlightenment." In *Natural Order: Historical Studies of Scientific Culture,* edited by Barry Barnes and Steven Shapin. Beverly Hills, CA: Sage, 1979.

Lawrence, Susan C. *Charitable Knowledge: Hospital Pupils and Practitioners in Eighteenth-Century London.* Cambridge: Cambridge University Press, 1996.

Layard, Daniel Peter. *Directions to Prevent the Contagion of the Jail Distemper, Commonly Called the Jail Fever.* London: James Robson, 1772.

Lees, Lynn Hollen. *The Solidarities of Strangers: The English Poor Laws and the People, 1700–1948.* Cambridge: Cambridge University Press, 1998.

LeFanu, William F., ed., *Betsy Sheridan's Journal.* New Brunswick, NJ: Rutgers University Press, 1960.

Leibniz, Gottfried Wilhelm. *Essais de théodicée: sur la bonté de Dieu, la liberté de l'homme et l'origine du mal.* Paris: Aubier, [1710] 1962.

Lemay, J. A. Leo, and George S. Rousseau. *The Renaissance Man in the Eighteenth Century.* William Andrews Clark Memorial Library Seminar Papers. Los Angeles: University of California Los Angeles, 1978.

Lemée, Pierre. *Julien Offray de La Mettrie, médecin, philosophe, polémiste; sa vie, son oeuvre.* Mortain, France: Editions Mortainais, 1954.

"Letter from Victoria." *Gentleman's Magazine,* June 1751, 255–57; continued July 1751, 318–20.

"Letter to the Editor from Martha Plainface." *Ladies Monthly Museum: Or, Polite Repository of Amusement and Instruction,* December 1799, 430–31.

Lettsom, John Coakley. *Medical Memoirs of the General Dispensary in London, for Part of the Years 1773 and 1774.* London: E. & C. Dilly, 1774.

Levine-Clark, Marjorie. "Engendering Relief: Women, Ablebodiedness, and the Poor Law in Early Victorian England." *Journal of Women's History* 11 (2000): 107–30.

Lind, James. *An Essay on Diseases Incidental to Europeans in Hot Climates.* London: T. Becket & P.A. de Hondt, 1768.

Lind, James. *An Essay on the Most Effectual Means of Preserving the Health of Seamen in the Royal Navy . . . Together with an Essay on the Jail Distemper.* London, 1774.

Linebaugh, Peter. 1975. "The Tyburn Riot against the Surgeons." In *Albion's Fatal Tree,* Douglas Hay, Peter Linebaugh, John G. Rule, E. P. Thompson, and Cal Winslow. London: Allen Lane, 1975.

Litten, Julian. "The English Funeral 1700–1850." In *Grave Concerns,* edited by Margaret Cox. York: Council for British Archaeology, 1998.

Livi-Bacci, Massimo. *A Concise History of World Population*, 3rd ed. Malden, MA: Blackwell, 2001.

Llewellyn, Nigel. *The Art of Death*. London: Reaktion, 1991.

Locke, John. *An Essay Concerning Human Understanding*, edited with an introduction by Peter H. Nidditch. Oxford: Clarendon Press, [1690] 1975.

London Guildhall Library, St. Sepulchre (London Division). 1796. "Minutes of the Committee for Rebuilding the Workhouse," MS 3226 (1–3).

London Sessions Papers: Grand Jury Presentment, October Sessions 1749; Thirty-one examinations of supposed vagrants, December Sessions 1782: Thomas Dargaval, November 16, 1782. Corporation of London Records Office, London. London Metropolitan Archives.

Look e're You Leap: Or, a History of the Lives and Intrigues of Lewd Women. 10th ed. London: Edward Midwinter, ca. 1760.

"Love's Conquest; or the Powerful Force of Beautyes Charmes." Pepys Ballads 3.105 (n.d.). Available at:: http://emc.english.ucsb.edu/ballad_project/large_detail.asp?id=21109 (accessed June 15, 2006).

Lowe, N. F. "The Meaning of Venereal Disease in Hogarth's Graphic Art." In *The Secret Malady: Venereal Disease in Eighteenth-Century Britain and France,* Linda E. Merians. Lexington: University Press of Kentucky, 1996.

Low-Life: or One Half of the World, Knows Not How the Other Half Live. 2nd ed. London: 1749.

Lynn, Walter. *An Essay Towards a More Easie and Safe Method of Cure in the Small Pox. Founded upon Experiments, and a Review of Dr Sydenham's Works*. London: R. Knaplock, 1714.

Lysons, Daniel. *Practical Essays upon Continual and Intermitting Fevers*. Bath: 1783.

MacManners, John. *Death and the Enlightenment*. Oxford: Oxford University Press, 1981.

Magnello, Eileen, and Anne Hardy, eds. *The Road to Medical Statistics*. Amsterdam: Rodopi, 2002.

Mallinson, Jonathan. "What's in a Name? Reflections on Voltaire's *Paméla*." *Eighteenth-Century Fiction* 18 (2005–2006.): 157–68.

Malthus, Thomas Robert. *An Essay on the Principle of Population*. Harmondsworth, UK: Penguin, [1798] 1970.

Mansion House Justice Room, Charge Book, 1728–1733 (MS New 392C): Samuel Gold, May 27, 1730. Corporation of London Records Office, London. London Metropolitan Archives.

Markovits, Francine, ed. *La Mettrie: Oeuvres philosophiques*. Vol. 2. Rennes, France: Fayard, 1987.

Martin, Julian. "Sauvages's Nosology: Medical Enlightenment in Montpellier." In *The Medical Enlightenment of the Eighteenth Century,* edited by Andrew Cunningham and Roger French. Cambridge: Cambridge University Press, 1990.

Martin, Matthew. *Letter to the Right Hon. Lord Pelham on the State of Mendicity in the Metropolis*. London, 1803.

Martin, Morag. "Doctoring Beauty: The Medical Control of Women's *Toilettes* in France, 1750–1820." *Medical History* 49 (2005): 351–68.

Marwick, Arthur. *Beauty in History: Society, Politics and Personal Appearance, c.1500 to the Present*. London: Thames & Hudson, 1988.

The Matrimonial Preceptor. London: Thomas Hope, 1759.

Maupertuis, Pierre-Louis Moreau de. *Essai sur la formation des corps organizes*. Berlin: 1754.

May, Edward. *A Most Certain and True Relation of A Strange Monster or Serpent Found in the Left Ventricle of the Heart of John Pennant, Gentleman, of the Age of 21 Years*. London: Published for the author, 1639.

McAllister, Marie E. "Stories of the Origin of Syphilis in Eighteenth-Century England: Science, Myth, and Prejudice." *Eighteenth-Century Life* 24 (2000): 22–44.

McClure, Ruth. *Coram's Children: The London Foundling Hospital in the Eighteenth Century*. New Haven, CT: Yale University Press, 1981.

McLaren, Angus. "Contraception and Its Discontents: Sigmund Freud and Birth Control." *Journal of Social History* 12 (1979): 513–29.

McLaren, Angus. "The Pleasures of Procreation: Traditional and Biomedical Theories of Conception." In *William Hunter and the Eighteenth-Century Medical World*, edited by William F. Bynum and Roy Porter. Cambridge: Cambridge University Press, 1985.

McLaren, Angus. *Reproductive Rituals: The Perception of Fertility in England from the Sixteenth Century to the Nineteenth Century*. London: Methuen, 1984.

McManners, John. *Death and the Enlightenment*. Oxford: Oxford University Press, 1981.

McMaster, Juliet. *Reading the Body in the Eighteenth-Century Novel*. London: Palgrave, 2004.

McTavish, Lianne. *Childbirth and the Display of Obstetrical Authority in Early Modern France*. Burlington, VT: Ashgate Publishing, 2005.

Mead, Richard. *A Mechanical Account of Poisons*. London: R. Smith, 1708.

Meer, Theo van der. "The Persecutions of Sodomites in Eighteenth-Century Amsterdam: Changing Perceptions of Sodomy." In *The Pursuit of Sodomy: Male Homosexuality in Renaissance and Enlightenment Europe*, edited by Kent Gerard and Gert Hekma. New York: Haworth Press, 1989.

The Merry Dutch Miller. London: E. Crowch, 1672.

Milton, John. *L'Allegro*. In *Milton, The Compete Shorter Poems*. Edited by John Carey. London: Longman, 2006.

Minois, George. *History of Suicide: Voluntary Death in Western Culture*. Baltimore, MD: Johns Hopkins University Press, 1999.

"Minutes of the Committee for Rebuilding the Workhouse," MS 3226 (1–3), 1796, 3. London Guildhall Library, St. Sepulchre (London Division).

Misospilus. *A Wonder of Wonders: Or, a Metamorphosis of Fair Faces Voluntarily Transformed into Foul Visages*. London: R. Smith, 1662.

Misson, Henri. *Memoirs and Observations in His Travels over England. Translated by Mr. Ozell*. London: Browne and others, [1698] 1719.

Mitchell, Juliet. *Siblings*. Cambridge, UK: Polity Press, 2003.

Molleson, Theya, and Margaret Cox. *The Spitalfields Project 2: The Anthropology*. York, UK: Council for British Archaeology, 1993.

Montesquieu, Charles-Louis de Secondat. *The Persian Letters*, translated by C. J. Betts. Harmondsworth, UK: Penguin, [1721] 1973.

Moore, Wendy. *The Knife Man*. London: Bantam, 2005.

Morgan, Fidelis. *The Female Tatler*. London: Everyman, 1992.

Morley, John. *Death, Heaven and the Victorians*. London: Studio Vista, 1971.

"Mr Galliard's Harangue in Praise of Beauty." *Gentleman's Magazine*, February 1736, 69.

Munkhoff, Richelle. "Searchers of the Dead: Authority, Marginality, and the Interpretation of Plague in England, 1574–1665." *Gender and History* 11 (1999): 1–29.

Nakayama, Randall S. *The Life and Death of Mrs Mary Frith: Commonly called Moll Cutpurse.* New York: Garland, [1662] 1993.
National Archives, Kew: Northern Assizes Circuit Depositions.
Needham, John Turberville. *Observations upon the Generation, Composition, and Decomposition of Animal and Vegetable Substances.* London: 1749.
Needham, Joseph. *A History of Embryology.* 2nd ed. Cambridge: Cambridge University Press, 1959.
A New Method of Curing the French-Pox / Written by an Eminent French Author [I.E. G. Ucay]. Together with the Practice and Method of Monsieur Blanchard, as Also Dr. Sydenham's Judgement on the Same. To Which Is Added [a Preface] Annotations and Observations by William Salmon. London: John Taylor & Thomas Newborough, 1690.
Newman, Simon P. *Embodied History: The Lives of the Poor in Early Philadelphia.* Philadelphia: University of Pennsylvania Press, 2003.
N. H. *The Ladies Dictionary.* London: John Dunton, 1694.
Niccoli, Ottavia. "Menstruum Quasi Monstruum: Monstrous Births and Menstrual Taboo in the Sixteenth Century." In *Sex and Gender in Historical Perspective,* edited by Edward Muir and Guido Ruggiero. Baltimore, MD: Johns Hopkins University Press, 1990.
Nihell, Elizabeth. *A Treatise on the Art of Midwifery.* London, 1760.
Norton, Rictor. *Mother Clap's Molly House: The Gay Subculture in England, 1700–1830.* London: GMP, 1992.
Nussbaum, Felicity A. *The Limits of the Human: Fictions of Anomaly, Race and Gender in the Long Eighteenth Century.* Cambridge: Cambridge University Press, 2003.
Nutton, Vivian. "The Seeds of Disease: An Explanation of Contagion and Infection from the Greeks to the Renaissance." *Medical History* 27 (1983): 1–34.
Old Bailey Proceedings Online. http://www.oldbaileyonline.org (accessed November 4, 2005).
Onania: Or, the Heinous Sin of Self-Pollution. 16th ed. London: J. Isted, 1737.
Ong, Walter J. *Orality and Literacy: The Technologizing of the Word.* London: Methuen, 1982.
Osborne, Francis. *Advice to a Son.* Oxford: Henry Hall, 1656.
Paine, Thomas. *The Age of Reason: Being an Investigation of True and Fabulous Theology.* Part 1. Paris, 1794.
Park, Katherine, and Lorraine Daston. "Unnatural Conceptions: The Study of Monsters in Sixteenth- and Seventeenth-Century France and England." *Past and Present* 92 (1981): 20–55.
Park, Katherine, and Robert Nye. "Destiny Is Anatomy." *The New Republic,* Feb 18, 1981, 53–57.
Parker, George. *A View of Society and Manners in High and Low Life; Being the Adventures in England, Scotland, Wales, France, etc., of G. Parker.* London: 1781.
Parsons, James. "A Letter to the President Concerning the Hermaphrodite Shewn in London." *Philosophical Transactions* 47 (1751–1752): 142–45.
Parsons, James. *A Mechanical and Critical Enquiry into the Nature of Hermaphrodites.* London: J. Walthoe, 1741.
Parsons, James. *The Microscopical Theatre of Seeds: Being a Short View of the Particular Marks, Characters, Contents, and Natural Dimensions of All the Seeds.* London: M. Cooper, 1745.

Parsons, James. *Philosophical Observations on the Analogy between the Propagation of Animals and That of Vegetables*. London: 1745.
Paster, Gail Kern. *The Body Embarrassed: Drama and the Disciplines of Shame in Early Modern England*. Ithaca, NY: Cornell University Press, 2003.
Paulson, Ronald. *Hogarth's Graphic Works*. 2 vols. New Haven, CT: Yale University Press, 1965.
Paulson, Ronald. *Hogarth's Harlot: Sacred Parody in Enlightenment England*. Baltimore, MD: Johns Hopkins University Press, 2003.
Paulson, William R. *Enlightenment, Romanticism and the Blind in France*. Princeton, NJ: Princeton University Press, 1987.
Payne, Christiana. "'Murillo-like Rags or Clean Pinafores': Artistic and Social Preferences in the Representation of the Dress of the Rural Poor." *Textile History* 33 (2002): 48–62.
Payne, Lynda. *With Words and Knives: Learning Medical Dispassion in Early Modern England*. Aldershot, UK: Ashgate, 2007.
Pelling, Margaret. "Old Age, Poverty, and Disability in Early Modern Norwich: Work, Remarriage, and Other Expedients." In *Life, Death and the Elderly: Historical Perspectives,* edited by Margaret Pelling and R.H. Smith. London: Routledge, 1991.
Pepys, Samuel. *The Diary of Samuel Pepys*, edited by Robert Latham and William Matthews. 11 vols. London: Bell, 1970–1983.
Perry, Ruth. "Colonizing the Breast: Sexuality and Maternity in Eighteenth-Century England." *Journal of the History of Sexuality* 2 (1991): 204–34.
Perry, Ruth. *Novel Relations: The Transformation of Kinship in English Literature and Culture, 1748–1818*. Cambridge: Cambridge University Press, 2004.
Philoponus. *On Aristotle on the Soul, 2.1–6*, translated by William Charlton. London: Duckworth, 2005.
Pilloud, Séverine, and Micheline Louis-Courvoisier. "The Intimate Experience of the Body in the Eighteenth Century: Between Interiority and Exteriority." *Medical History* 47 (2003): 451–72.
Pinto-Correia, Clara. *The Ovary of Eve: Egg and Sperm and Preformation*. Chicago: University of Chicago Press, 1997.
Pitter, Ruth. "The Old Woman." In *Poems 1926–66*. London: Cresset Press, 1968.
Pluche, Noël Antoine. *Le Spectacle de la nature, ou Entretiens sur les particularités de l'histoire naturelle*. Vol. 5. Paris: 1746.
Pomeranz, Kenneth. *The Great Divergence: China, Europe, and the Making of the Modern World Economy*. Princeton, NJ: Princeton University Press, 2000.
Porter, Roy. "Bodies of Thought: Thoughts about the Body in Eighteenth-Century England." In *Interpretation and Cultural History,* edited by Joan H. Pittock and Andrew Wear. Basingstoke, UK: Macmillan, 1991.
Porter, Roy. *Bodies Politic: Disease, Death and Doctors in Britain, 1650–1900*. London: Reaktion Books, 2001.
Porter, Roy. "The Eighteenth Century." In *The Western Medical Tradition: 800 BC to 1800 AD,* Lawrence I. Conrad, Michael Neve, Vivian Nutton, Roy Porter, and Andrew Wear. Cambridge: Cambridge University Press, 1995.
Porter, Roy. *Enlightenment: Britain and the Creation of the Modern World*. London: Penguin Books, 2001.
Porter, Roy. "'Expressing Yourself Ill': The Language of Sickness in Georgian England." In *Language, Self and Society: The Social History of Language,* edited by Peter Burke and Roy Porter. Cambridge, UK: Polity Press, 1991.

Porter, Roy. *Flesh in the Age of Reason*. London: Allen Lane, 2003.

Porter, Roy. *The Greatest Benefit to Mankind: A Medical History of Humanity*. New York: Norton, 1997.

Porter, Roy. *Health for Sale: Quackery in England 1650–1850*. Manchester, UK: Manchester University Press, 1989.

Porter, Roy. *In Sickness and in Health: The British Experience 1650–1850*. London: Fourth Estate, 1998.

Porter, Roy. "John Hunter: A Showman in Society." *Transactions of the Hunterian Society* 52 (1993–1994): 19–24.

Porter, Roy. *Mind Forg'd Manacles: Madness and Psychiatry in England from Restoration to Regency*. London: Penguin Books, 1987.

Porter, Roy. *Quacks: Fakers and Charlatans in English Medicine*. Stroud: Tempus, 2003.

Porter, Roy. "A Touch of Danger: The Man-Midwife as Sexual Predator." In *Sexual Underworlds of the Enlightenment,* edited by G. S. Rousseau and Roy Porter. Manchester, UK: Manchester University Press, 1987.

Porter, Roy, and William F. Bynum, eds. *Medical Fringe and Medical Orthodoxy, 1750–1850*. London: Croom Helm, 1987.

Porter, Roy, and Lesley Hall. *The Facts of Life: The Creation of Sexual Knowledge in Britain, 1650–1950*. New Haven, CT: Yale University Press, 1995.

Porter, Roy, and George S. Rousseau. *Gout: The Patrician Malady*. New Haven, CT: Yale University Press, 1998.

Post Man and the Historical Account, August 7, 1701, 2.

Poynter, F. N. L. "Hunter, Spallanzani, and the History of Artificial Insemination." In *Medicine, Science, and Culture: Historical Essays in Honor of Owsei Tempkin*, edited by Lloyd G. Stevenson and Robert P. Multhauf. Baltimore, MD: Johns Hopkins University Press, 1968.

Pringle, John. *Observations on the Diseases of the Army*. London: A. Millar et al., 1765.

Prior, Thomas. *An Authentick Narrative of the Success of Tar-water*. London: W. Innys et al., 1746.

Procreation Refin'd. A New Method for the Begetting of Children with Handsome Faces. London: Richard Griffin, ca. 1710.

Proksch, Johann Karl. *Die Litteratur über de venerischen Krankheiten von den ersten Schriften über Syphilis aus dem Ende des Fünfzehnten Jarhhundrets bis zum jahre 1889*. 3 vols. Bonn, Germany: P. Hanstein, 1889–1890.

Puckle, B. *Funeral Customs*. London: Werner Laurie, 1929.

Pullan, Brian. "Plague and Perceptions of the Poor in Early Modern Italy." In *Epidemics and Ideas: Essays on the Historical Perception of Pestilence,* edited by Terence Ranger and Paul Slack. Cambridge: Cambridge University Press, 1995.

Quétel, Claude. *History of Syphilis*, translated by Judith Braddock and Brian Pike. Baltimore, MD: Johns Hopkins University Press, 1990.

Quillet, Claude. *Callipaedia: Or the Art of Getting Beautiful Children,* translated by N. Rowe. 3rd ed. London, 1733.

Rabin, Dana. "Searching for the Self in Eighteenth-Century English Criminal Trials, 1730–1800." *Eighteenth-Century Life* 27 (2003): 85–106.

Ray, John. *The Wisdom of God Manifested in the Works of the Creation*. 1st ed. London: Sam Smith, 1660.

Réaumur, Antoine. *The Art of Hatching and Bringing up Domestick Fowl of All Kinds, at Any Time of the Year, Either by Means of the Heat of Hot-Beds, or That of Common Fire*. London: C. Davis, A. Millar & J. Nourse, 1750.

Reynolds, Frances. *An Enquiry Concerning the Principles of Taste, and of the Origin of our Ideas of Beauty etc.* Los Angeles, CA: Augustan Reprint Society Publications no. 27, [1785] 1951.

Ribton-Turner, C.J. *A History of Vagrants and Vagrancy and Beggars and Begging.* London: Chapman & Hall, 1887.

Richards, Robert J. *Darwin and the Emergence of Evolutionary Theories of Mind and Behavior.* Chicago: University of Chicago Press, 1987.

Richards, Robert J. "The Emergence of Evolutionary Biology of Behaviour in the Early Nineteenth Century." *British Journal for the History of Science* 15 (1982): 241–280.

Richards, Robert J. "Influence of Sensationalist Tradition on Early Theories of the Evolution of Behavior." *Journal of the History of Ideas* 40 (1979): 85–105.

Richards, Robert J. *The Meaning of Evolution: The Morphological Construction and Ideological Reconstruction of Darwin's Theory.* Chicago: University of Chicago Press, 1993.

Richardson, Ruth. "Commerce, Delicacy and the Decline of Old Mortality." Paper presented at the Death and Dying Conference, Dulwich College, London, 1994.

Richardson, Ruth. *Death, Dissection and the Destitute.* London: Penguin, 1988.

Richardson, Ruth. *Death Dissection and the Destitute.* 2nd ed. Chicago: Chicago University Press, 2000.

Richardson, Ruth. "The English Shroud." Paper presented at The Art of Death Conference, Victoria and Albert Museum, London, 1992.

Richardson, Ruth. "Pennant's Serpent." *The Lancet* 357 (2001): 966.

Richardson, Ruth, and James Stevens Curl. "George Frederick Carden and the Genesis of the General Cemetery Company." In *Kensal Green Cemetery,* edited by J.S. Curl. Chichester, UK: Phillimore, 2001.

Richardson, Ruth, and Brian Hurwitz. "Jeremy Bentham's Self Image: An Exemplary Bequest for Dissection." *British Medical Journal* 295 (1987): 195–98.

Rickards, Maurice, and Michael Twyman. *Encyclopaedia of Ephemera.* London: British Library, 2000.

Riley, James C. *The Eighteenth-Century Campaign to Avoid Disease.* London: MacMillan, 1987.

Riley, James C. *Population Thought in the Age of the Demographic Revolution.* Durham, NC: Carolina Academic Press, 1985.

Riskin, Jessica. "The Defecating Duck." *Critical Inquiry* 29 (2003): 599–633.

Riskin, Jessica. "The Divine Optician." *American Historical Review,* forthcoming.

Riskin, Jessica. "Eighteenth-Century Wetware." *Representations* 83 (2003): 97–125.

Robertson, Robert. *Essay on Fevers.* London: The author, 1790.

Roe, Shirley. "Anatomia Animata: The Newtonian Physiology of Albrecht von Haller." In *Transformation and Tradition in the Sciences: Essays in Honor of I. Bernard Cohen,* edited by Everett Mendelsohn. Cambridge: Cambridge University Press, 1984.

Roe, Shirley. *The Natural Philosophy of Albrecht von Haller.* Manchester, NH: Ayer, 1981.

Roger, Jacques. *The Life Sciences in Eighteenth-Century French Thought.* Edited by Keith R. Benson, translated by Robert Ellrich. Stanford, CA: Stanford University Press, 1997.

Rosenfield, Leonora Cohen. *From Beast-Machine to Man-Machine: Animal Soul in French Letters from Descartes to La Mettrie.* Oxford: Oxford University Press, 1941.

Rousseau, George S. "Curiosity and the *Lusus Naturae*: Proteus Hill." In *Curiosity and Wonder from the Renaissance to the Enlightenment,* edited by R. J. W. Evans and Alex Marr. Aldershot, UK: Ashgate, 2006.

Rousseau, George S. "The London Earthquakes of 1750." *Cahiers d'Histoire Mondiale* 11 (1968): 436–54.

Rousseau, George S. *Nervous Acts: Essays on Literature, Culture and Sensibility.* Basingstoke, UK: Palgrave Macmillan, 2004.

Rousseau, George S. "The Pursuit of Homosexuality in the Eighteenth Century: 'Utterly Confused Category' and/or Rich Repository?" In *Eighteenth-Century Life*, edited by Robert P. Maccubbin. Vol. 9. Williamsburg, VA: College of William and Mary, 1985.

Rousseau, George S., and David Haycock. "The Jew of Crane Court: Emanuel Mendes Da Costa (1717–1791), Natural History and Natural Excess." *History of Science* 38 (2000): 127–70.

Rousseau, George S., and David Haycock. "'Voices Calling for Reform': The Royal Society in the Mid-Eighteenth Century—Martin Folkes, John Hill, and William Stukeley." *History of Science* 37 (1999): 377–406.

Rublack, Ulinka. "Pregnancy, Childbirth and the Female Body in Early Modern Germany." *Past and Present* 150 (1996): 84–110.

Rugg, Julie. "From Reason to Regulation." In *Death in England,* edited by Peter Jupp and Clare Gittings. Manchester, UK: Manchester University Press, 1999.

Runge, Laura L. "Beauty and Gallantry: A Model of Polite Conversation Revisited." *Eighteenth-Century Life* 25 (2001): 43–63.

Rushton, Peter. "Lunatics and Idiots: Mental Disability, the Community, and the Poor Law in North-East England, 1600–1800." *Medical History* 32 (1988): 34–50.

Rusnock, Andrea A. *Vital Accounts: Quantifying Health and Population in Eighteenth-Century England and France.* Cambridge: Cambridge University Press, 2002.

Russell, Keith F. *British Anatomy 1525–1800: A Bibliography.* Winchester, UK: St Paul's, [1963] 1987.

Saakwa-Mante, Norris. "Western Medicine and Racial Constitutions: Surgeon John Atkins' Theory of Polygenism and Sleepy Distemper in the 1730s." In *Race, Science and Medicine, 1700–1960,* edited by Bernard Harris and Waltraud Ernst. London: Routledge, 1999.

Sacombe, Jean-François. *Lucine Francéaise, ou Recueil d'observations Médicales, Chirurgicales, Pharmaceutiques, Historiques, Critiques et Littéraires, Relatives é La Science Des Accouchemens.* Paris: Bidault, 1803.

Sacombe, Jean-François. *La Luciniade: Ou, L'art Des Accouchemens, Poème Didactique.* Paris: Garnéry, 1792.

Sacombe, Jean-François. *La Luciniade, Poème en Dix Chants, Sur L'art Des Accouchemens.* Paris: Courcier, 1799.

"Samuel Badham's Account of Himself." In *The Ordinary of Newgate, His Account of the Behaviour, Confession, and Dying Words, of the Malefactors, Who Were Executed at Tyburn, on Wednesday the 6th of August (1740). Being the Fourth Execution in the Mayoralty of the Rt. Hon. Sir John Salter, Knt. Number IV. For the Said Year.* London: Printed and Sold by John Applebee, in Bolt-Court, near the Leg-Tavern, Fleet-street, 1740.

Sappol, Michael. *A Traffic of Dead Bodies.* Princeton, NJ: Princeton University Press, 2002.

Satan's Harvest Home: Or the Present State of Whorecraft, Adultery, Fornication, Procuring, Pimping, Sodomy, and the Game at Flatts. London: Printed for the editor, 1749.

Sbiroli, Lynn Salkin. *Libertine O Madre Illibate.* Venice, Italy: Marcilio, 1989.
Schäffer, Daniel. "'That Senescence Itself Is an Illness': A Transitional Medical Concept of Age and Ageing in the Eighteenth Century." *Medical History* 46 (2002): 525–48.
Schiebinger, Londa. "The Anatomy of Difference: Race and Sex in Eighteenth-Century Science." *Eighteenth-Century Studies* 23 (1990): 387–405.
Schiebinger, Londa. *The Mind Has No Sex? Women in the Origins of Modern Science.* Cambridge, MA: Harvard University Press, 1991.
Schiebinger, Londa. *Nature's Body: Gender in the Making of Modern Science.* Boston: Beacon Press, 1993.
Schiebinger, Londa. *Nature's Body: Sexual Politics and the Making of Modern Science.* London, Pandora, 1994.
Schofield, Roger, David Sven Reher, and Alain Bideau, eds. *The Decline of Mortality in Europe.* Oxford: Clarendon Press, 1991.
Scholz, Susanne. *Body Narratives: Writing the Nation and Fashioning the Subject in Early Modern England.* London: Macmillan, 2000.
Schor, Esther. *Bearing the Dead.* Princeton, NJ: Princeton University Press, 1994.
Schurig, Martin. *Spermatologia Historico-Medica.* Wittenberg, Germany: 1780.
Scott, Walter. *The Heart of Midlothian.* With an introductory essay and notes by Andrew Lang. Vol. 1. Boston: Estes and Lauriat, 1893.
Secord, Arthur Wellesley, ed. *Defoe's Review.* Vol. 2. New York: Columbia University Press, 1938.
Semple, Janet. *Bentham's Prison: A Study of the Panopticon Penitentiary.* Oxford: Clarendon Press, 1993.
Sharp, James. *Instruments of Darkness. Witchcraft in England, 1550–1750.* London: Hamish Hamilton, 1996.
Sharp, Jane. *The Midwives Book. Or the Whole Art of Midwifery Discovered. Directing Childbearing Women How to Behave Themselves.* London: S. Miller, 1671.
Shattuck, Roger. *Forbidden Knowledge. From Prometheus to Pornography.* New York: St. Martin's Press, 1996.
Shaw, Jane. *Miracles in Enlightenment England.* London: Yale University Press, 2006.
Shesgreen, Sean. *Images of the Outcast: The Urban Poor in the Cries of London.* Manchester, UK: Manchester University Press, 2002.
Shire, Helena Mennie, ed. *Poems from Panmure House.* Cambridge, UK: Ninth of May Press, 1960.
Shoemaker, Robert. *Gender in English Society, 1650–1850: The Emergence of Separate Spheres?* London: Longman, 1998.
Short, Thomas. *New Observations, Natural, Moral, Civil, Political, and Medical, on City, Town, and Country Bills of Mortality.* London: T. Longman & A. Millar, 1750.
Shorter, Edward. "The Management of Normal Deliveries and the Generation of William Hunter." In *William Hunter and the Eighteenth-Century Medical World,* edited by William F. Cambridge: Cambridge University Press, 1985.
Shuttleton, David E. "A Culture of Disfigurement: Imagining Smallpox in the Long Eighteenth Century." In *Framing and Imaging Disease in Cultural History,* edited by George S. Rousseau, Miranda Gill, David Haycock, and Malte Herwig. Basingstoke, UK: Palgrave-Macmillan, 2003.
Sibly, Ebenezer. *The Medical Mirror. Or Treatise on the Impregnation of the Human Female: Shewing the Origin of Diseases, and the Principles of Life and Death.* London: Printed for the author and sold by Champante and Whitrow, 1796.

Siena, Kevin. "Pollution, Promiscuity, and the Pox: English Venereology and the Early Modern Discourse on Social and Sexual Danger." *Journal of the History of Sexuality* 8 (1998): 553–74.

Siena, Kevin. *Venereal Disease, Hospitals, and the Urban Poor: London's "Foul Wards."* Rochester, NY: University of Rochester, 2004.

Simpson, Claude M. *The British Broadside Ballad and Its Music.* New Brunswick, NJ: Rutgers University Press, 1966.

Sintelaer, John. *The Scourge of Venus and Mercury, Represented in a Treatise of the Venereal Disease.* London: G. Harris et al., 1709.

Slack, Paul. *From Reformation to Improvement: Public Welfare in Early Modern England.* Oxford: Clarendon Press, 1999.

Slack, Paul. *The Impact of Plague in Tudor and Stuart England.* London: Routledge, 1985.

Sloan, Phillip R. "The Idea of Racial Degeneracy in Buffon's *Histoire Naturelle.*" In *Racism in the Eighteenth Century,* edited by Harold E. Pagliaro. Cleveland, OH: Case Western Reserve University, 1973.

Smellie, William. *A Treatise on the Theory and Practice of Midwifery.* 3 vols. London: D. Wilson et al., 1752–1764.

Smith, John Thomas. *Vagabondiana: Or, Anecdotes of Mendicant Wanderers through the Streets of London.* London: Published for the proprietor and sold by J. and A. Arch, 1817.

Smith, Samuel Stanhope. *An Essay on the Causes of the Variety of Complexion and Figure in the Human Species: To Which Are Added, Strictures on Lord Kaim's Discourse on the Original Diversity Of Mankind.* London: Philadelphia printed, London re-printed for John Stockdale, 1789.

Smith, Warren Hunting. *Horace Walpole's Correspondence: Complete Index / Compiled by Warren Hunting Smith with the Assistance of Edwine M. Martz, Ruth K. McClure and William T. La Moy.* New Haven, CT: Yale University Press, 1983.

Smollett, Tobias. *The Adventures of Roderick Random,* edited by Paul-Gabriel Boucé. Oxford: Oxford University Press, [1748] 1979.

Sontag, Susan. *Against Interpretation: And Other Essays.* New York: Octagon Books, 1986.

Spary, Emma C. "The Performance of Surgery in Enlightenment France." *Endeavour* 23 (1999): 180–3.

Stagg, Kevin. "Representing Physical Difference: The Materiality of the Monstrous." In *Social Histories of Disability and Deformity,* edited by David M. Turner and Kevin Stagg. London: Routledge, 2006.

Stallybrass, Peter, and Allon White. *The Politics and Poetics of Transgression.* Ithaca, NY: Methuen, 1986.

Stannard, David E. *The Puritan Way of Death.* Oxford: Oxford University Press, 1979.

Steinke, Hubert. *Irritating Experiments: Haller's Concept and the European Controversy on Irritability and Sensibility, 1750–90.* Amsterdam: Rodopi, 2005.

Stepan, Nancy Leys. "Race, Gender, Science and Citizenship." *Gender and History* 10 (1998): 26–52.

Sterne, Laurence. *Adventures of Tristram Shandy.* New York: Odyssey, [1759–1767] 1950.

Stimson, Dorothy. *Scientists and Amateurs: A History of the Royal Society.* London: Sigma Books, 1949.

Stock, Gwynne. "Quaker burial." In *Grave Concerns,* edited by Margaret Cox. York, UK: Council for British Archaeology, 1998.

Stolberg, Michael. "A Woman Down to Her Bones: The Anatomy of Sexual Difference in the Sixteenth and Early Seventeenth Centuries." *Isis* 94 (2003): 274–99.

Stone, Lawrence. *The Family, Sex and Marriage in England, 1500–1800.* London: Weidenfeld & Nicolson, 1977.

Stonecastle, Henry. *The Universal Spectator.* Vol. 2. London: J. Pemberton et al., 1736.

Summers, Montague. *A Gothic Bibliography.* London: Fortune Press, 1941.

Swift, Jonathan. *Gulliver's Travels*, edited by Harold Williams. Oxford: Blackwell, [1726] 1965.

Sydenham, Thomas. *The Whole Works of That Excellent Practical Physician, Dr Thomas Sydenham*, translated by John Pechey. 9th ed. London: J. Darby et al., 1729.

Synnott, Anthony. *The Body Social: Symbolism, Self and Society*. London: Routledge, 1993.

Tadmor, Naomi. *Family and Friends in Eighteenth-Century England.* Cambridge: Cambridge University Press, 2001.

Terrall, Mary. *The Man Who Flattened the Earth: Maupertuis and the Sciences in the Enlightenment.* Chicago: University of Chicago Press, 2002.

Terrall, Mary. "Salon, Academy, and Boudoir: Generation and Desire in Maupertuis' Science of Life." *Isis* 87 (1996.): 217–29.

Thibault, Dieudonné. *Mes souvenirs de vingt ans to séjour à Berlin: ou Frédéric le Grand, sa famille, sa cour, son gouvernement, son académie, ses écoles, et ses amis littérateurs et philosophes.* 2nd ed. Vol. 5. Paris: F. Buisson, 1805.

Thomas, Chris. *London's Archaeological Secrets: A World City Revealed.* New Haven, CT: Yale University Press, 2003.

Thomas, Keith. *Religion and the Decline of Magic: Studies in Popular Beliefs in Sixteenth and Seventeenth-Century England.* London: Weidenfeld & Nicolson, [1971] 1997.

Thornton, John L. *Jan Van Rymsdyk*. Cambridge: Oleander, 1982.

Tilley, Morris Palmer. *A Dictionary of the Proverbs in England in the Sixteenth and Seventeenth Centuries.* Ann Arbor: University of Michigan Press, 1950.

Todd, Dennis. *Imagining Monsters: Miscreations of the Self in Eighteenth-Century England.* Chicago: University of Chicago Press, 1995.

Toland, John. *Letters to Serena.* London: Bernard Lintot, 1704.

Trial of Ann Milbourne, Newcastle 1753, ASSI 45 18/5/60, 62. National Archives, London.

A Trip from St James's to the Royal Exchange: The Manners, Customs and Amusements of the Inhabitants of London and Westminster. London: Edward Withers, 1744.

A Trip through Town. Containing Observations on the Humours and Manners of the Age. 4th ed. London: 1735.

Trotter, Thomas. *Observations on the Scurvy.* Edinburgh: C. Elliott, 1786.

Trumbach, Randulph. *Sex and the Gender Revolution: Heterosexuality and the Third Gender in Enlightenment London.* Chicago: University of Chicago Press, 1998.

Tuke, Thomas. *A Treatise against Painting and Tincturing of Men and Women.* London: Edward Marchant, 1616.

Turner, Daniel. *Siphylis: A Practical Dissertation on the Venereal Disease.* 4th ed. London: J. Walthoe et al., 1732.

Turner, David, and Kevin Stagg, eds. *Social Histories of Disability and Deformity*. London: Routledge, 2006.

Tylston, John. *The Country Physician*. Edinburgh: 1701.

Vagrant Books (4), Together with Five Loose Pages and a Bill of Mortality, 1738–1742 (Misc. MS 322.5): Examinations of Henry Buxton, June 12, 1738; Henry Goodwin, March 20, 1741. Corporation of London Records Office, London. London Metropolitan Archives.

Van Elke, Martine. "The Counterfeit Vagrant: The Dynamic of Deviance in the Bridewell Court Records and the Literature of Roguery." In *Rogues and Early Modern English Culture,* edited by Craig Dionne and Steve Mentz. Ann Arbor: University of Michigan Press, 2004.

Vartanian, Aram, ed. *La Mettrie's L'Homme-machine, A Study in the Origins of an Idea*. Princeton, NJ: Princeton University Press, 1960.

Vartanian, Aram. "Trembley's Polyp, La Mettrie and Eighteenth-Century French Materialism." *Journal of the History of Ideas* 11 (1950): 259–86.

Vesalius, Andreas. *De Humanis Corporis Fabrica*. Basle, Switzerland: Printed for the author, 1543.

Vidal, Fernando. "Extraordinary Bodies and the Physicotheological Imagination." In *The Faces of Nature in Enlightenment Europe,* edited by Lorraine Daston and Gianna Pomata. Berlin: BWV, 2003.

A View of London and Westminster: Or the Town Spy. Part 1. London: T. Warner, 1728.

Vila, Ann C. *Enlightenment and Pathology: Sensibility in the Literature and Medicine of Eighteenth-Century France*. Baltimore, MD: Johns Hopkins University Press, 1998.

Voltaire, François. *The Philosophical Dictionary for the Pocket*. London: S. Bladon, 1765.

Wahrman, Dror. *The Making of the Modern Self: Identity and Culture in Eighteenth-Century England*. New Haven, CT: Yale University Press, 2004.

Wainewright, Jeremiah. *A Mechanical Account of the Non-Naturals*. London: J. Clarke, 1737.

Walzer, A. "Logic and Rhetoric in Malthus's *Essay on the Principle of Population,* 1798." *Quarterly Journal of Speech* 73 (1987): 1–17.

Ward, Edward. *The History of the London Clubs; or the Citizens Pastime, Particularly the Lying Club, the Beggars Club, the Yorkshire Club; with a Sermon Preach'd to a Gang of Highway-men*. Part 1. London: J. Bagnall, 1709.

Ward, Ned. *The London Spy,* edited by Paul Hyland. East Lansing, MI: Colleagues Press, [1709] 1993.

Ward, Samuel, and Samuel Clark. *A Warning-Piece to All Drunkards and Health-Drinkers* . . . London: Printed for the author, and sold by Langley Curtis, 1682

Wear, Andrew. "Fear, Anxiety and the Plague in Early Modern England: 'Religious and Medical Responses.'" In *Religion, Health and Suffering,* edited by John R. Hinnells and Roy Porter. London: Kegan Paul International, 1999.

Wear, Andrew. "The History of Personal Hygiene." In *Companion Encyclopedia of the History of Medicine,* vol. 2, edited by W. F. Bynum and Roy Porter. London: Routledge, 1993.

Wear, Andrew. *Knowledge and Practice in English Medicine, 1550–1680*. Cambridge: Cambridge University Press, 2000.

Wear, Andrew. "Making Sense of Health and the Environment in Early Modern England." In *Medicine in Society: Historical Essays,* edited by Andrew Wear. Cambridge: Cambridge University Press, 1992.

Webster, Noah, ed. *A Collection of Papers on the Subject of Bilious Fevers, Prevalent in the United States for a Few Years Past.* New York: 1796.

Wecker, Johann Jacob. *Arts Master-Piece: Or, the Beautifying Part of Physick.* London: Nathaniel Brook, 1660.

Weinstone, Ann. *Avatar Bodies.* Minneapolis: University of Minnesota Press, 2004.

Werner, Alex, ed. *London Bodies: The Changing Shape of Londoners from Prehistoric Times to the Present Day.* London: Museum of London, 1998.

Wheeler, Roxam. *The Complexion of Race: Categories of Difference in Eighteenth-Century British Culture.* Philadelphia: University of Pennsylvania Press, 2000.

Wilkinson, C.H. ed. *The King of the Beggars: Bampfylde-Moor Carew.* Oxford: Clarendon Press, 1931.

Williams, Elizabeth. *A Cultural History of Medical Vitalism in Enlightenment Montpellier.* Aldershot, UK: Ashgate, 2003.

Williams, Elizabeth. "Hysteria and the Court Physician in Enlightenment France." *Eighteenth-Century Studies* 35 (2002): 247–55.

Williams, Elizabeth. *The Physical and the Moral: Anthropology, Physiology and Philosophical Medicine in France, 1750–1850.* Cambridge: Cambridge University Press, 1994.

Wilson, Adrian. *The Making of Man-Midwifery.* Cambridge, MA: Harvard University Press, 1995.

Wilson, Dudley Butler. *Signs and Portents: Monstrous Births from the Middle Ages to the Enlightenment.* London: Routledge, 1993.

Wilson, Kathleen. *The Island Race: Englishness, Empire and Gender in the Eighteenth Century.* London: Routledge, 2002.

Wilson, Philip K. "Exposing the Secret Disease: Recognizing and Treating Syphilis in Daniel Turner's London." In *The Secret Malady: Venereal Disease in Eighteenth-Century Britain and France,* edited by Linda E. Merians. Lexington: University Press of Kentucky, 1996.

Wilson, Philip K. " 'Out of Sight, Out of Mind': The Turner-James Blondel Dispute over the Power of the Maternal Imagination." *Annals of Science* 49 (1992): 63–85.

Wintringham, Clifton. *An Essay on Contagious Diseases.* York, UK: Francis Hildyard, 1721.

Wisdom, Anthony. *Color Atlas of Venereology.* Chicago: Year Book Medical Publications, 1973.

Wiseman, Richard. *Eight Chirurgical Treatises.* Vol. 2. London: B. Tooke, 1719.

Wolf, Naomi. *The Beauty Myth: How Images of Beauty Are Used against Women.* London: Vintage, 1991.

Wollaston, William. *The Religion of Nature Delineated.* Dublin: 1726.

Wolley, Hannah. *The Gentlewoman's Companion.* London: A. Maxwell, 1675.

Woodbridge, Linda. *Vagrancy, Homelessness and English Renaissance Literature.* Urbana: University of Illinois Press, 2001.

Wright, George. *The Lady's Miscellany; or Pleasing Essays, Poems, Stories and Examples for the Instruction and Entertainment of the Female Sex in General, in Every Station of Life.* London: Chapman & Co., 1793.

Wrigley, Edward A., and Roger Schofield. *The Population History of England 1541–1871.* Cambridge: Cambridge University Press, 1989.

Zielonka, Jason S. "'A Man-Midwife'. Etching, Hand Coloured, by S.W. Fores, London, 1793. New Haven, Yale Medical Library, Clements C. Fry Collection." *Journal of the History of Medicine* 30 (1975): 259.

Zimmermann, Johann Georg. *A Treatise on the Dysentery*, translated by C. R. Hopson. London: J. & F. Rivington, 1771.

Zirkle, Conway. "Animals Impregnated by the Wind." *Isis* 25 (1936): 95–130.

CONTRIBUTORS

Lisa Forman Cody is an associate professor of history and associate dean of the faculty at Claremont McKenna College in Los Angeles. She is the author of *Birthing the Nation: Sex, Science, and the Conception of Eighteenth-Century Britons* (2005), which was awarded, among other prizes, the Berkshire Best First Book Prize. She has published several articles on the history of reproduction, midwifery, medicine, cultural politics, and visual culture. She is currently working on a book project tentatively titled *Divided We Stand: Female Independence in the Age of the American Revolution*, a study of divorce, sexual scandal, and politics in the British Isles and Atlantic world.

Laura Gowing is reader in early modern British History, King's College London. She works on women, sex, and the body; her most recent book is *Common Bodies: Women, Touch and Power in Seventeenth Century England* (2003).

Tim Hitchcock is professor of eighteenth-century history, University of Hertfordshire. He has written or edited eleven books on the histories of eighteenth-century poverty, street life, sexuality, and masculinity. His most recent books include *Down and Out in Eighteenth-Century London* (2004) and, with Robert Shoemaker, *Tales from the Hanging Court* (2007). He is also the codirector, with professors Robert Shoemaker and Clive Emsley, of the Old Bailey Online (www.oldbaileyonline.org), a fully searchable edition of all printed trial accounts of the Old Bailey court from 1674 to 1913.

Ruth Perry is professor of literature at Massachusetts Institute of Technology, founding director of the MIT Women's Studies Program, and past president of the American Society for Eighteenth-Century Studies. Her most recent books are *Novel Relations: The Transformation of Kinship in English Literature and Culture 1748–1818* (2004), an edition of Charlotte Lennox's *Henrietta*, edited with Susan Carlile (2008), and a special double issue of *The Eighteenth Century: Theory and Interpretation* titled "Ballads and Songs in the Eighteenth Century." Her current project is a biography of Anna Gordon Brown, an eighteenth-century Scotswoman whose repertoire of ballads was the first collected from a living person.

Carole Reeves is the outreach historian at the Wellcome Trust Centre for the History of Medicine, University College London. She coauthored (with John L. Thornton) *Medical Book Illustration: A Short History* (1983). Her recent works include *Egyptian Medicine* (2001) and an illustrated history of medicine from ancient Greece to the Genome Project for HistoryWorld (www.history world.net). She was a contributing editor to the *Dictionary of Medical Biography* (2007) and is coauthor of *The Children of Craig-y-nos: Life in a Welsh Tuberculosis Sanatorium, 1922–1959* (2009), based on the oral testimonies of people who spent their childhoods in a sanatorium.

Ruth Richardson is senior visiting fellow in history, University of Hertfordshire, and affiliated scholar in the history and philosophy of science, University of Cambridge. She is the author of *Death, Dissection and the Destitute* (1988) and *The Making of Mr Gray's "Anatomy": Bodies, Books, Fortune and Fame* (2008).

Jessica Riskin teaches history in the History Department at Stanford University. She is the author of *Science in the Age of Sensibility: The Sentimental Empiricists of the French Enlightenment* (2002) and the editor of *Genesis Redux: Essays in the History and Philosophy of Artificial Life* (2007). She is currently writing a book about the human–machine model and associated technology, physiology, philosophy, culture, and politics from Descartes to Darwin. The book is under contract with Basic Books; its working title is *Mind Out of Matter.*

George Rousseau is a member of the faculty of modern history at Oxford University. He has published widely in the areas of eighteenth-century literature, science, medicine, and the history of sexuality. Recent books include *Framing and Imagining Disease in Cultural History* (2003), *Nervous Acts: Essays on Literature, Culture and Sensibility* (2004), and *Children and Sexuality: The Greeks to the Great War* (2007). He has edited the letters and papers of Sir John Hill, and his interpretative biography of Hill is forthcoming in 2011.

Kevin Siena is associate professor of history, Trent University in Peterborough, Canada. The author of *Venereal Disease, Hospitals and the Urban Poor,*

London's "Foul Wards," 1600–1800 (2004) and editor of *Sins of the Flesh: Responding to Sexual Disease in Early Modern Europe* (2005), he is currently researching the ways in which the eighteenth-century poor prevented and dealt with illness and also the associations between class and contagion forged during the Enlightenment.

Susan Staves is professor emerita of English and American literature at Brandeis University. Her scholarly interests center on Enlightenment literature and history in the Restoration and eighteenth century, particularly on how cultural ideologies are variously created and represented in texts ranging from comedies to judicial opinions. She is the author of *Players' Scepters: Fictions of Authority in the Restoration* (1979), *Married Women's Separate Property in England, 1660–1833* (1990), and *A Literary History of Women's Writing in Britain, 1660–1789* (2006). With John Brewer, she edited and contributed to *Early Modern Conceptions of Property* (1995).

David M. Turner is senior lecturer in gender history at Swansea University. He has published widely on the history of sex and marriage in early modern Britain, including the monograph *Fashioning Adultery: Gender, Sex and Civility in England 1660–1740* (2002). His current research is focused on the social and cultural history of disability in the seventeenth and eighteenth centuries, and he is coeditor (with Kevin Stagg) of *Social Histories of Disability and Deformity* (2006).

INDEX

www.ingramcontent.com/pod-product-compliance
Ingram Content Group UK Ltd.
Pitfield, Milton Keynes, MK11 3LW, UK
UKHW050944290926
15470UKWH00005B/1622

9 781847 887917